D1482925

Ray C. Mullin

This latest edition of Ray C. Mullin's popular handbook has been expanded and updated to reflect all innovations in electrical regulations, equipment, and wiring practices embodied in the 1981 National Electrical Code (NEC). The NEC provides basic standards for the layout and construction of circuitry and components in residential installations. Mr. Mullin's new edition will assure readers of abiding by the Code's most recent regulations while helping them to improve their own wiring skills.

Two new units have been added to this Seventh Edition to provide even more comprehensive coverage than that of previous editions. An extensive section explains the wiring and installation requirements for heat and smoke detectors, and another section covers in detail the many exacting regulations and safety precautions for the wiring and installation of swimming pools.

Code changes in the following areas are clearly explained:

- Aluminum conductors, connections, and terminals on wiring devices
- Kilowatt-hour consumption calculations
- Submersible pumps
- Wiring through cold air returns
- Installations in closets
- Ground-fault circuit interruption in garages
- Insulation around recessed fixtures
- Open neutrals and conduit bodies
- Grounding ranges and dryers
- Microwave ovens and other modern appliances

A plan-size diagram summarizing the Code's pool-wiring requirements accompanies residential plans in a tear-out section at the back of the book. In conformance with the 1981 NEC, metric (SI) measurements, in addition to the traditional English measurements, are included where applicable. Many new drawings and updated illustrations are also included throughout.

ELECTRICAL WIRING RESIDENTIAL

CODE • THEORY • PLANS
SPECIFICATIONS • INSTALLATION METHODS

Based on 1981 National Electrical Code
Seventh Edition

RAY C. MULLIN

VNR VAN NOSTRAND REINHOLD COMPANY
NEW YORK CINCINNATI TORONTO LONDON MELBOURNE

Library of Congress Catalog Card Number 80-39483

ISBN 0-442-26311-2

Printed in the United States of America

Published by Van Nostrand Reinhold Company
135 West 50th Street, New York, NY 10020, U.S.A.

Van Nostrand Reinhold Limited
1410 Birchmount Road
Scarborough, Ontario MIP 2E7, Canada

Van Nostrand Reinhold Australia Pty. Ltd.
17 Queen Street
Mitcham, Victoria 3132, Australia

Van Nostrand Reinhold Company Limited
Molly Millars Lane
Wokingham, Berkshire, England

16 15 14 13 12 11 10 9 8 7 6 5 4 3 2

Library of Congress Cataloging in Publication Data

Mullin, Ray C
Electrical wiring, residential.

Includes index.
1. Electric wiring, Interior. I. Title.
TK3285.M84 1981 621.319'24 80-39483
ISBN 0-442-26311-2

Preface

The seventh edition of *Electrical Wiring – Residential* is based on the 1981 National Electrical Code (NEC). The many changes in the new Code relating to residential wiring are thoroughly explained in this text. Two new units have been added to this edition. Unit 26 discusses the wiring and installation requirements for heat and smoke detectors. Unit 30 covers in detail the many exacting regulations and safety precautions for the wiring and installation of swimming pools, spas, and hot tubs. A plan-size diagram summarizing the Code's pool wiring requirements (figure 30-6) accompanies the residential plans in the tear-out section at the back of the text.

Many new drawings and updated illustrations and related information make *Electrical Wiring – Residential* the most up-to-date guide to household wiring.

The most recent edition of the National Electrical Code is used as the basic standard for the layout and construction of electrical systems. To gain the greatest benefit from this book, the reader must use and refer to the National Electrical Code on a continuing basis. Certain modifications to the Code rules may be necessary because of the requirements of state and local electrical codes. The reader is encouraged to obtain any variations from the Code as they affect this residential installation.

The 1981 edition of the National Electrical Code represents a major revision over the previous edition. Code changes affecting the following topics are fully explained in the text:

- swimming pools
- spas
- hot tubs
- aluminum conductors
- aluminum connections and terminals on wiring devices
- kilowatt-hour consumption calculations
- smoke and heat detectors
- submersible pumps
- wiring through cold air returns
- installations in closets
- ground-fault circuit interruption in garages
- insulation around recessed fixtures
- open neutrals
- conduit bodies
- grounding ranges and dryers
- microwave ovens and other modern appliances

The 1981 edition of the National Electrical Code introduces metric (SI) measurements in addition to the traditional English measurements. Accordingly, metric measurements are included in this book where applicable. Metric conversions are now shown for the dimensions on the residential plans which accompany the text. Such conversions are considered to be the responsibility of the designer.

This book was prepared by Ray C. Mullin, former electrical circuit instructor for the Electrical Trades, Wisconsin Schools of Vocational, Technical and Adult Education. Mr. Mullin, a former member of the International Brotherhood of Electrical Workers, is presently a member of the International Association of Electrical Inspectors, the Institute of Electrical and Electronic Engineers, Inc., and the National Fire Protection Association, Electrical Section. Mr. Mullin completed his apprenticeship training and has worked as a journeyman and supervisor. He has taught both day and night electrical apprentice and journeyman trade extension courses and has conducted engineering seminars. He is knowledgeable in the what-when-where-why-and-how of electrical installations. Mr. Mullin presents his accumulated knowledge and experience in this book to assist the reader in learning residential wiring in an orderly step-by-step manner. He is currently the Vice President for a large electrical components manufacturer.

Acknowledgments

Sponsoring Editor: William Sprague
Senior Editor: Marjorie A. Bruce
Associate Editor: Frances Larson
Production: L.E.P.I. Graphics

Contributors:

Appreciation is expressed to the following companies for their contributions of data and illustrations:

American Home Lighting Institute
Anchor Electric Division, Sola Basic Industries
Appleton Electric Co.
Arrow-Hart, Inc.
Bussmann Division, a McGraw-Edison Company
Edwin L. Wiegand Division, Emerson Electric Co.
Electri-Flex Co.
General Electric Co.
Honeywell, Inc.
International Association of Electrical Inspectors
Moe Light Division, Thomas Industries
NuTone Division, Scovill Manufacturing Co.
Pass & Seymour, Inc.
Sierra Electric Division, Sola Basic Industries
Square D Co.
Superior Electric Co.
Wiremold Co.

Special thanks to Wolberg Electrical Supply Co., Inc., Albany, New York, for providing numerous photographs of electrical equipment which appear throughout the text.

Applicable tables from the National Electric Code are reproduced with the permission of the National Fire Protection Association, 470 Atlantic Avenue, Boston, Massachusetts 02210.

Contents

unit 1

General Information for Electrical Installations

OBJECTIVES

After studying this unit, the student will be able to

- explain how electrical wiring information is conveyed to the electrician at the construction or installation site.
- demonstrate how the specifications are used in estimating costs and in making electrical installations.
- explain why symbols and notations are used on electrical drawings.
- list the agencies that are responsible for establishing electrical standards and insuring that materials meet the standards.

THE WORKING DRAWINGS

The architect uses a set of working drawings or plans to make the necessary instructions available to the skilled crafts which are to build the structure shown in the plans. The sizes, quantities, and locations of the materials required and the construction features of the structural members are shown at a glance. These details of construction must be studied and interpreted by each skilled construction craft – masons, carpenters, electricians, and others – before the actual work is started.

The electrician must be able to: (1) convert the two-dimensional plans into an actual electrical installation, and (2) visualize the many different views of the plans and coordinate them into a three-dimensional picture, as shown in figure 1-1.

The ability to visualize an accurate three-dimensional picture requires a thorough knowledge of blueprint reading. Since all of the skilled trades use a common set of plans, the electrician must be able to interpret the lines and symbols which refer to the electrical installation and also those which are

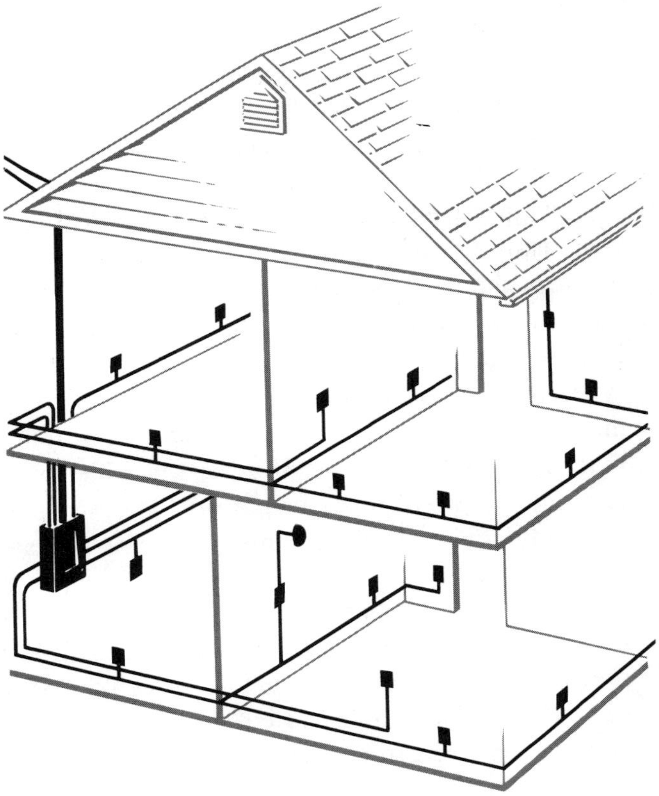

Fig. 1-1 Three-dimensional view of house wiring.

used by the other construction trades. The electrician must know the structural makeup of the building and the construction materials to be used.

SPECIFICATIONS

Working drawings are usually complex because of the amount of information which must be included. To prevent confusing detail, it is standard practice to include with each set of plans a set of detailed written specifications prepared by the architect.

These specifications provide general information to be used by all trades involved in the construction. In addition, specialized information is given for the individual trades. The specifications include information on the sizes, the type, and the desired quality of the standard parts which are to be used in the structure.

Typical specifications include a section on "General Clauses and Conditions" which is applicable to all trades involved in the construction. This section is followed by detailed requirements for the various trades – excavating, masonry, carpentry, plumbing, heating, electrical work, painting, and others.

The plan drawings for the residence used as an example for this text are included in the back of the text. The specifications for the electrical work indicated on the plans are given in the Appendix.

In the electrical specifications, the listing of standard electrical parts and supplies frequently includes the manufacturers' names and the catalog numbers of the specified items. Such information insures that these items will be of the correct size, type, and electrical rating, and that the quality will meet a certain standard. To allow for the possibility that the contractor will not always be able to obtain the specified item, the phrase "or equivalent" is usually added after the manufacturer's name and catalog number.

The specifications are also useful to the electrical contractor in that all of the items needed for a specific job are grouped together and the type or size of each item is indicated. This information allows the contractor to prepare an accurate cost estimate without having to find all of the data in the plans.

SYMBOLS AND NOTATIONS

The architect uses symbols and notations to simplify the drawing and the presentation of information concerning electrical devices and arrangements. One such symbol and notation is shown in the following example.

SYMBOL	NOTATION
 WP	WEATHERPROOF OUTLET

Symbols are described in detail in unit 2. Most of these symbols and notations have a standard interpretation throughout the country. The notation, which is placed on the drawing next to a specific symbol, provides information on the type, size, and quantity of the device required. The notation may be an abbreviation. It may also refer to a specific table for details.

The electrician must be able to interpret these symbols and notations so that the various components can be grouped into the proper circuits. The electrician must be able to visualize the circuits, the distribution centers, the service entrance, and the metering facilities as a complete, three-dimensional installation. Then, this picture must be converted into a completed installation which will meet the approval of the owner, the architect, and the local inspecting authority.

NATIONAL ELECTRICAL CODE (NEC)

Because of the ever-present danger of fire through some failure of the electrical system, the electrician and the electrical contractor must use approved materials and must perform all work in accordance with recognized standards. The National Electrical Code is the basic standard which governs electrical work. The Code contains provisions considered necessary for safety. It states that the installation must be essentially hazard-free, but that such an installation is not necessarily efficient, convenient, or adequate, *Section 90-1(b)*. It is the electrician's responsibility to insure that the installation meets these criteria. In addition to the National Electrical Code, the electrician must also consider local and state codes. The purpose and scope of the National Electrical Code are discussed in *Article 90* of the Code book and should be studied by the student at this time. The Code is revised and updated every three years; the articles and section numbers usually remain the same, as does the subject covered, but the wording may be deleted, changed, or expanded.

Code Terms

The following terms are used throughout the Code. It is important to understand the meanings of these terms.

Identified: (As applied to equipment.) Recognizable as suitable for a specific purpose, function, use, environment, or application, where described in a particular Code requirement. Suitability of use, marked on or provided with the equipment, may include labeling or listing.

Labeled: Equipment or materials to which has been attached a label, symbol, or other identifying mark of an organization acceptable to the authority having jurisdiction and concerned with product evaluation, that maintains periodic inspection of production of labeled equipment or materials and by whose labeling the manufacturer indicates compliance with appropriate standards or performance in a specified manner.

Listed: Equipment or materials included in a list published by an organization acceptable to the authority having jurisdiction and concerned with product evaluation, that maintains periodic inspection of production of listed equipment or materials and whose listing states either that the equipment or material meets appropriate standards or has been tested and found suitable for use in a specified manner.

Approved: Acceptable to the authority having jurisdiction.

Authority Having Jurisdiction: An organization, office, or individual responsible for "approving" equipment, an installation, or a procedure.

Shall: Indicates a mandatory requirement.

Should: Indicates a recommendation or that which is advised but not required.

One of the most far reaching NEC rules is *Section 110-3(b)*. This section states that the use and installation of listed or labeled equipment must conform to any instructions included in the listing or labeling. This means that an electrical system and its associated electrical equipment must be installed and used in accordance with both the National Electrical Code and the Underwriters' Laboratories standards.

In the past, programs such as *Adequate Wiring, House Power, Live Better Electrically, Bronze Medallion,* and *Gold Medallion*, were instituted to supplement established Code standards. The purpose of these programs was to promote the installation of efficient, convenient, and useful home wiring systems. After a period of time, as the recommendations of these programs were gradually written into the National Electrical Code, the programs were phased out.

The NEC tells what is permitted and what is not permitted by using certain key words, as follows:

shall be	compulsory; mandatory; a requirement; must be.
shall have.	the same as "shall be."
shall not	not allowed; not permitted to be done; must not be; against the Code.
shall be permitted	is allowed; may be done; not against the Code.

The National Electrical Code refers to a residence as a *dwelling unit.*

CODE USE OF METRIC (SI) MEASUREMENTS

For the first time, the 1981 National Electrical Code includes both English and metric measurements. The metric system is known as the *International System of Units* (SI).

Metric measurements appear in the Code as follows:

- in the Code paragraphs, the approximate metric measurement appears in parentheses following the English measurement.
- in the Code tables, a footnote shows the SI conversion factors.

A metric measurement is not shown for conduit size, box size, wire size, horsepower designation for motors, and other "trade sizes" that do not reflect actual measurements.

The National Electrical Code is following this conversion schedule:

mega	1 000 000	(one million)
kilo	1 000	(one thousand)
hecto	100	(one hundred)
deka	10	(ten)
the unit	1	(one)
deci	0.1	(one-tenth) (1/10)
centi	0.01	(one-hundredth) (1/100)
milli	0.001	(one-thousandth) (1/1 000)
micro	0.000 001	(one-millionth) (1/1 000 000)
nano	0.000 000 001	(one-billionth) (1/1 000 000 000)

Fig. 1-2 Metric prefixes and their values.

1981 edition – English measurement first, with metric (SI) measurement in parentheses ().

1984 edition – metric measurement first, with English measurement in parentheses ().

Guide to Metric Usage

In the metric system, the units increase or decrease in multiples of 10, 100, 1 000, and so on. For instance, one megawatt (1 000 000 watts) is 1 000 times greater than one kilowatt (1 000 watts).

By assigning a name to a measurement, such as a *watt*, the name becomes the unit. Adding a prefix to the unit, such as *kilo*, forms the new name *kilowatt*, meaning 1 000 watts. Refer to figure 1-2 for prefixes used in the metric system.

The prefixes used most commonly are: *centi, kilo,* and *milli.* Consider that the basic unit is a meter (one). Therefore, a centimeter is 0.01 meter, a kilometer is 1 000 meters, and a millimeter is 0.001 meter.

Some common measurements of length in the English system are shown with their metric equivalents in figure 1-3.

Electricians will find it useful to refer to the conversion factors and their abbreviations shown in figure 1-4.

one inch	=	2.54	centimeters
	=	25.4	millimeters
	=	0.025 4	meter
one foot	=	12	inches
	=	0.304 8	meter
	=	30.48	centimeters
	=	304.8	millimeters
one yard	=	3	feet
	=	36	inches
	=	0.914 4	meter
	=	914.4	millimeters
one meter	=	100	centimeters
	=	1 000	millimeters
	=	1.093	yards
	=	3.281	feet
	=	39.370	inches

Fig. 1-3 Some common measurements of length and their metric equivalents.

inches (in) X 0.025 4	= meters (m)
inches (in) X 0.254	= decimeters (dm)
inches (in) X 2.54	= centimeters (cm)
centimeters (cm) X 0.393 7	= inches (in)
inches (in) X 25.4	= millimeters (mm)
millimeters (mm) X 0.039 37	= inches (in)
feet (ft) X 0.304 8	= meters (m)
meters (m) X 3.280 8	= feet (ft)
square inches (in²) X 6.452	= square centimeters (cm²)
square centimeters (cm²) X 0.155	= square inches (in²)
square feet (ft²) X 0.093	= square meters (m²)
square meters (m²) X 10.764	= square feet (ft²)
square yards (yd²) X 0.836 1	= square meters (m²)
square meters (m²) X 1.196	= square yards (yd²)
kilometers (km) X 1 000	= meters (m)
kilometers (km) X 0.621	= miles (mi)
miles (mi) X 1.609	= kilometers (km)

Fig. 1-4 Useful conversions (English/SI - SI/English) and their abbreviations.

UNDERWRITERS' LABORATORIES, INC. (UL)

Underwriters' Laboratories (UL) is a highly qualified, nationally recognized testing laboratory. Most reputable manufacturers of electrical equipment submit their products to the Underwriters' Laboratories where the equipment is subjected to numerous tests. These tests determine if the products can perform safely under normal and abnormal conditions to meet published standards. After UL determines that a product complies with the specific standard, a manufacturing firm is then permitted to *label* its product with the UL logo. The products are then *listed* in a UL publication.

It must be noted that UL does not *approve* any product. Rather, UL *lists* those products that conform to its safety standards.

Two very useful UL publications are the *Electrical Appliance and Utilization Equipment Directory* and the *Electrical Construction Materials Directory*. If the answer to a question cannot be found readily in the National Electrical Code, then it generally can be found in these two UL publications.

REVIEW

Note: Refer to the National Electrical Code or the plans where necessary.

1. What is the purpose of specifications? __________

2. In what additional way are the specifications particularly useful to the electrical contractor? __________

3. To prevent a plan from becoming confusing because of too much detail, what is done? __________

4. Name four requirements contained in the specifications regarding material.
   a. __________ c. __________
   b. __________ d. __________

5. Name three cautions regarding workmanship found in the specifications. __________

6. What phrase is used when a substitution is permitted for a specific item? __________

7. What is the purpose of an electrical symbol? __________

8. What is a notation? __________

9. Where are notations found? __________

10. List at least twelve electrical notations found on the plans. __________

11. What three parties must be satisfied with the completed electrical installation?
    a. __________ b. __________ c. __________

12. What code sets standards for electrical installation work? __________

13. What authority enforces the standards set by the Code? __________

14. Does the Code provide minimum or maximum standards? __________

15. What do the letters *UL* signify? __________

16. What section of the Code states that all listed or labeled equipment shall be used or installed in accordance with any instructions included in the listing or labeling? __________

17. When the words "shall be" appear in a Code reference, they mean that it (must)(may) be done. (Underline the correct word.)

unit 2

Electrical Symbols and Outlets

OBJECTIVES

After studying this unit, the student will be able to

- identify and explain the electrical outlet symbols used in the plans of the single-family dwelling.
- discuss the types of outlets, boxes, fixtures, and switches used in the residence.
- explain the methods of mounting the various electrical devices used in the residence.

ELECTRICAL SYMBOLS

Electrical symbols used on an architectural plan show the location and type of electrical device required. A typical electrical installation as taken from a plan is shown in figure 2-1.

The National Electrical Code describes an outlet as a point on a wiring system which supplies current to utilization equipment. The term outlet is used broadly by electricians to include noncurrent-consuming switches and similar control devices in a wiring system when estimating the cost of the installation. Each type of outlet is represented on the plans as a symbol. In figure 2-1, the outlets are shown by the symbols ⊖ and -○-. The standard electrical symbols are shown in figure 2-2.

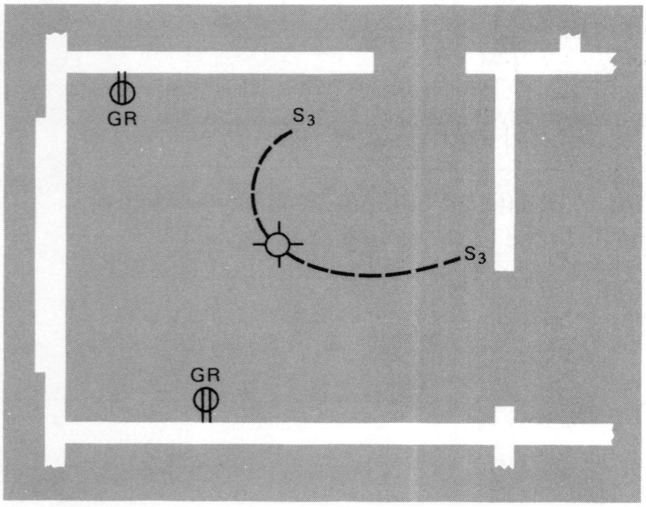

Fig. 2-1 Use of electrical symbols and notations on a floor plan.

The dash lines in figure 2-1 run from the outlet to the switch or switches which control the outlet. These dash lines are usually curved so they cannot be mistaken for invisible edge lines. Outlets shown on the plan without curved dash lines are independent outlets and have no switch control.

A study of the plans for the single-family dwelling shows that many different electrical symbols are used to represent the electrical devices and equipment used in the building.

In drawing electrical plans, most architects, designers, and electrical engineers use symbols approved by the American National Standards Institute (ANSI) wherever possible. However, plans may contain symbols that are not found in these standards. When such unlisted (nonstandard) symbols are used, the electrician must refer to a legend which interprets these symbols. The legend may be included on the plans or in the specifications. In many instances, a notation on the plan will clarify the meaning of the symbol.

Figure 2-2 lists the standard, approved electrical symbols and their meanings. Many of these

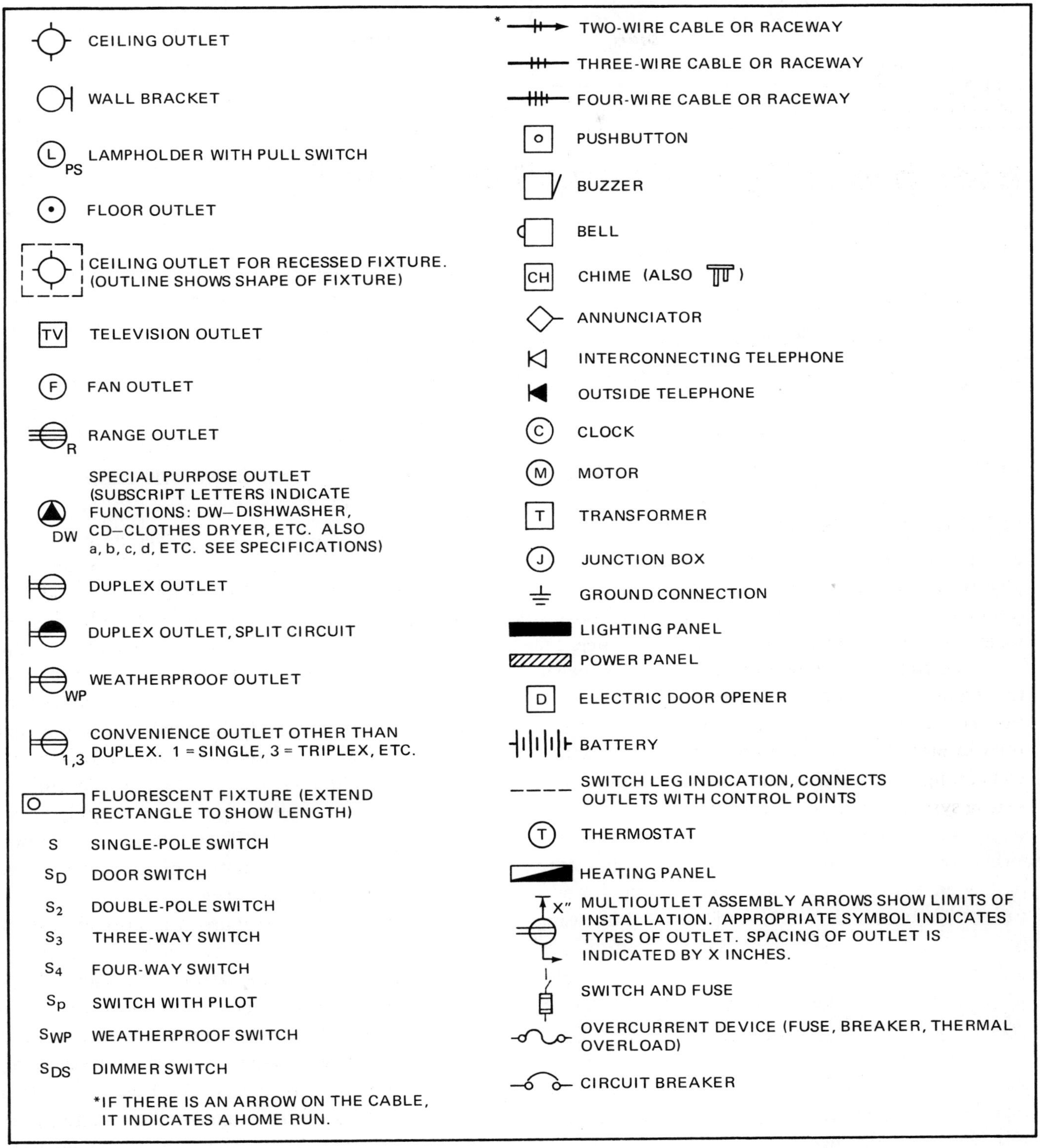

Fig. 2-2 Electrical wiring symbols.

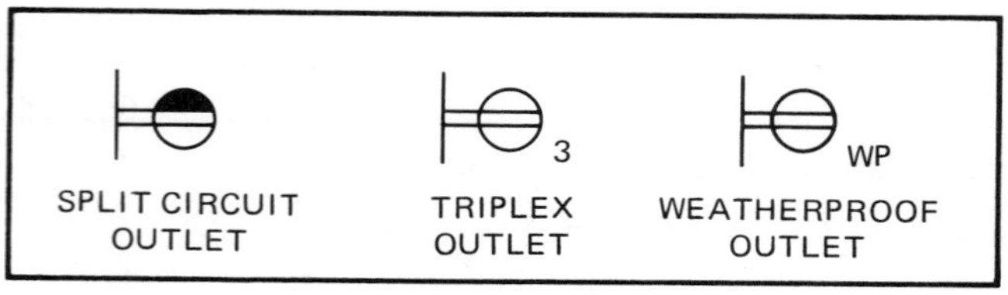

Fig. 2-3 Variations in significance of outlet symbols.

symbols can be found on the accompanying plans of the residence. Note in figure 2-2 that several symbols have the same shape. However, differences in the interior presentation indicate that the meanings of the symbols are different. For example, different meanings are shown in figure 2-3 for the outlet symbol. A good practice to follow in studying symbols is to learn the basic forms first and then add the supplemental information to obtain different meanings.

FIXTURES AND OUTLETS

Architects often include in the specifications a certain amount of money for the purchase of electrical fixtures. The electrical contractor includes this amount in the bid and the choice of fixtures is then left to the homeowner. If the owner selects fixtures whose total cost exceeds the fixture allowance, the owner is expected to pay the difference between the actual cost and the specification allowance. If the fixtures are not selected before the roughing-in stage of wiring the house, the electrician usually installs outlet boxes having standard fixture mounting studs. Most modern lighting fixtures can be fastened to a fixture stud. Other fixtures can be mounted either to an outlet box or plaster cover, using a strap or bar and No. 8-32 screws. In addition, fixtures can be mounted on a standard switch box using No. 6-32 screws. A box must be installed at each outlet or switch location, *Section 300-15*. (Note: All National Electrical Code section references are printed in italics.)

If the owner selects fixtures prior to construction, the architect can specify these fixtures in the plans and/or specifications. Thus, the electrician is provided with advance information on any special framing, recessing, or mounting requirements for the fixtures. This information must be provided in the case of recessed fixtures which require a specific wall or ceiling opening.

Many types of lighting fixtures are presently available. Figure 2-4 shows several typical lighting fixtures that may be found in a dwelling unit. Also shown are the electrical symbols used on plans to designate these fixtures and the type of outlet boxes or switch boxes on which the lighting fixtures can be mounted. A standard convenience outlet is shown, as well. The switch boxes shown here are made of steel. Switch boxes may also be made of plastic, as shown later in the text in figure 22-1. Other types of outlets are covered in later units.

FLUSH SWITCHES

Some of the standard symbols for various types of switches are shown in figure 2-5. Typical connection diagrams are also given. Any sectional switch box or 4-inch square box with a side mounting bracket and raised switch cover can be used to install these switches.

JUNCTION BOXES AND SWITCH BOXES (*ARTICLE 370*)

Junction boxes are sometimes placed in a circuit for convenience in joining two or more cables or conduits. All conductors entering a junction box are joined to other conductors entering the same box to form the proper hookups so that the circuit will operate in the manner intended.

All electrical installations must conform to the National Electrical Code standards requiring that junction boxes be installed in such a manner that the wiring contained in them shall be accessible without removing any part of the building. In house wiring, this requirement limits the use of junction boxes to cellars and open attic spaces because flush blank covers exposed to view detract from the appearance of a room. Of course, an outlet box, such as the one used in the bedroom ceiling fixture, is really a junction box because it contains splices. Removing the fixture makes the box accessible, thereby meeting the Code requirements.

The house wiring system usually is formed of a number of specific circuits. Each circuit consists of a continuous run of cable from outlet to outlet, or from box to box. The residence plans show many branch circuits for general lighting, appliances, electric heating baseboard panels, and other requirements. The specific Code rules for each of these circuits are covered in a later unit.

SYMBOL	TYPE OF FIXTURE OR OUTLET	OUTLET BOXES THAT MAY BE USED
 OR  CEILING OUTLET	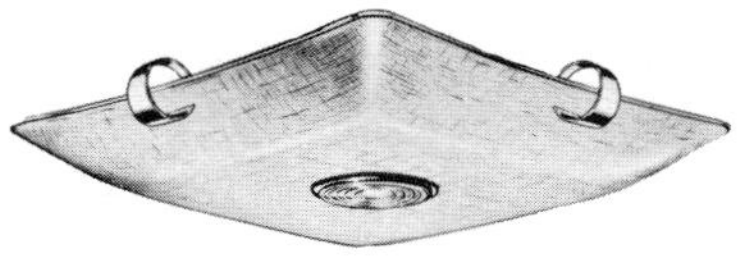 SURFACE MOUNTED	 OCTAGON BOX AND OFFSET BAR HANGER. BOXES ARE AVAILABLE WITH EITHER INTERNAL OR EXTERNAL CLAMPS.
 CEILING OUTLET	 FLUSHED RECESSED TYPE	CABLE TERMINATES IN BOX THAT IS FURNISHED WITH FIXTURE. SPECIAL CODE RULES GOVERNING THE INSTALLATION ARE FOUND IN *SECTIONS 410-64* THROUGH *410-72* OF THE NATIONAL ELECTRICAL CODE.
 WALL OUTLET	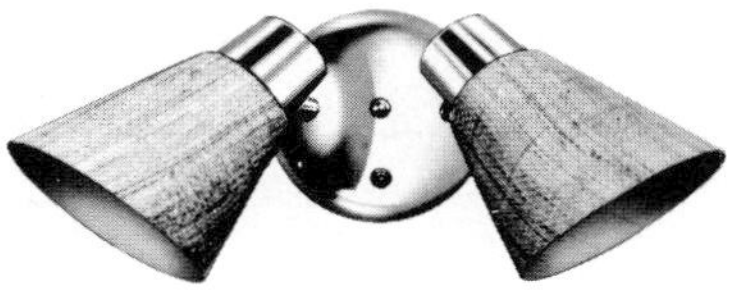	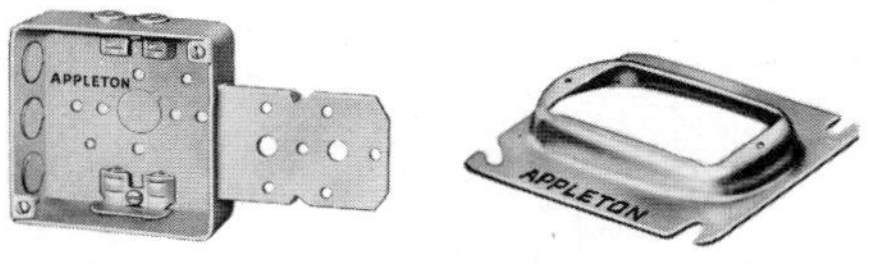 4" SQUARE BOX WITH SIDE BRACKET AND PROPER RAISED PLASTER COVER. 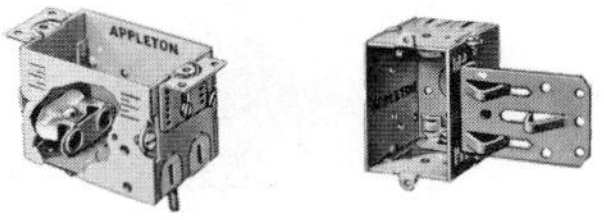 SECTIONAL SWITCH BOXES
 DUPLEX RECEPTACLE OUTLET	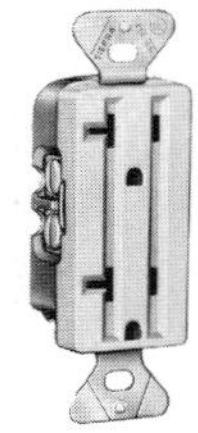 	SAME AS ABOVE

Fig. 2-4 Types of fixtures and outlets.

Ganging Switch Boxes

A flush switch or convenience outlet for residential use fits into a standard 2" x 3" sectional switch box (sometimes called a device box). When two or more switches (or outlets) are located at the same point, the switch boxes are *ganged* or fastened together to provide the required mounts, figure 2-6. Three switch boxes can be ganged together by removing and discarding one side from both the first and third switch boxes and both sides from the second (center) switch box. The boxes are then joined together as shown in figure 2-6. After the switches are installed, the gang is trimmed with a gang plate having the required number of switch handle or convenience outlet openings. These plates are called two-gang wall plates, three-gang wall plates, and so on, depending upon the number of openings.

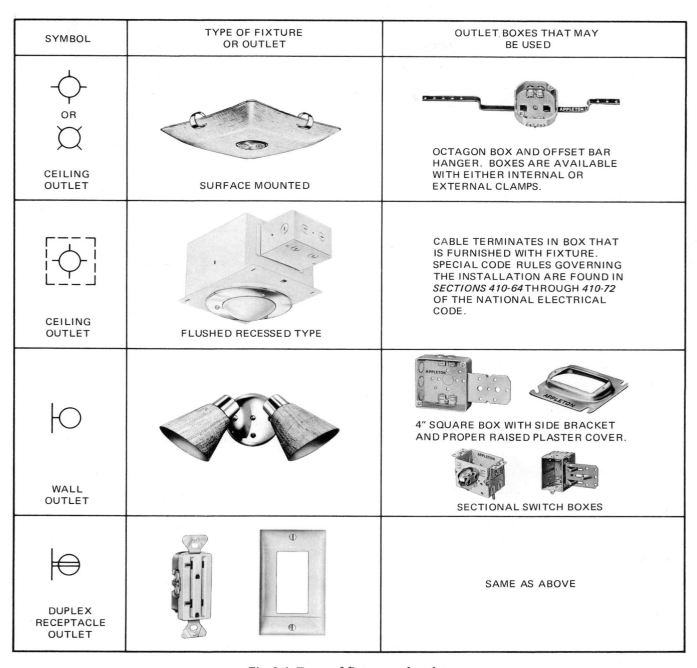

SYMBOL	FLUSH TOGGLE SWITCH	OPERATION	CONNECTIONS
S SINGLE POLE	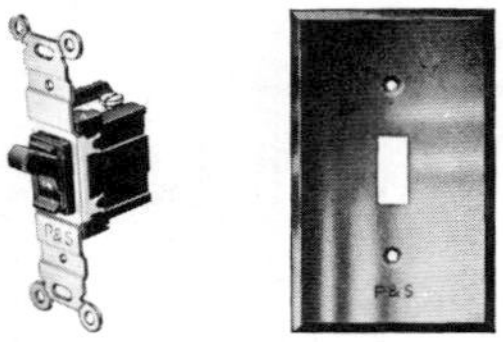	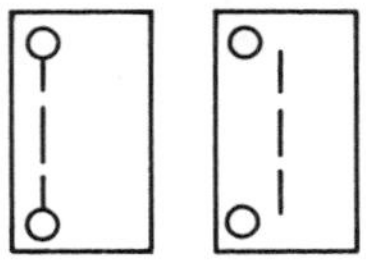 ON OFF	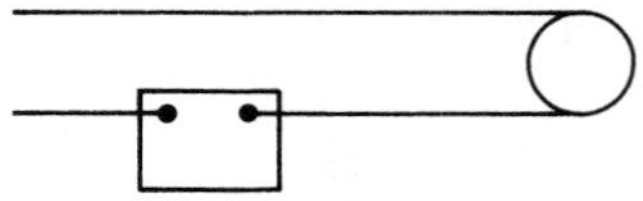
S_2 DOUBLE POLE		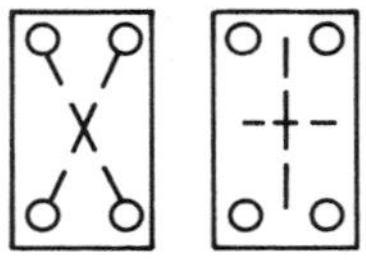 ON OFF	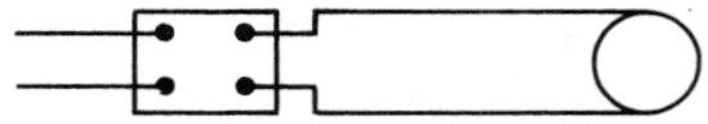
S_3 THREE WAY		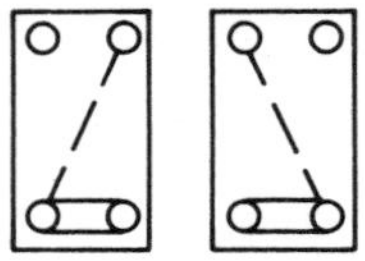 1 2	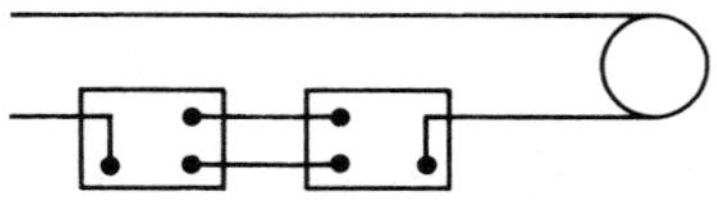
S_4 FOUR WAY		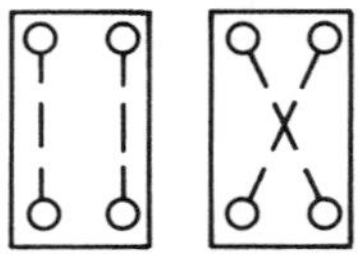 1 2	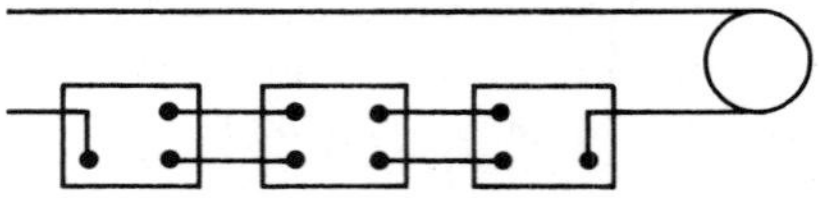
S_p SWITCH AND PILOT LIGHT	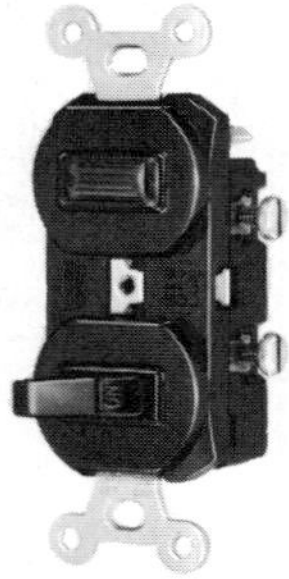	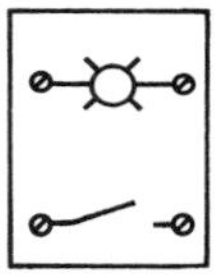 FOR CONTROLLING LIGHTS FROM ONE POINT WITH PILOT LIGHT INDICATION	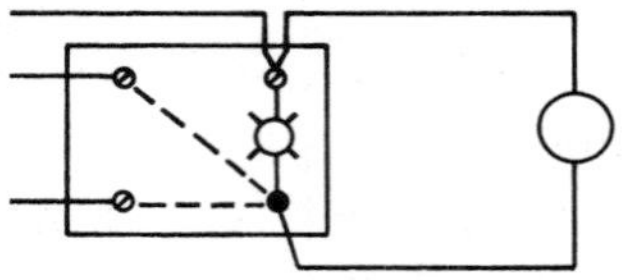 ALSO AVAILABLE IN THREE-WAY TYPE OF CONTROLLING LIGHT FROM TWO POINTS WITH PILOT LIGHT INDICATION

Fig. 2-5 Standard switches and symbols.

Fig. 2-6 Standard flush switches installed in ganged sectional switch boxes.

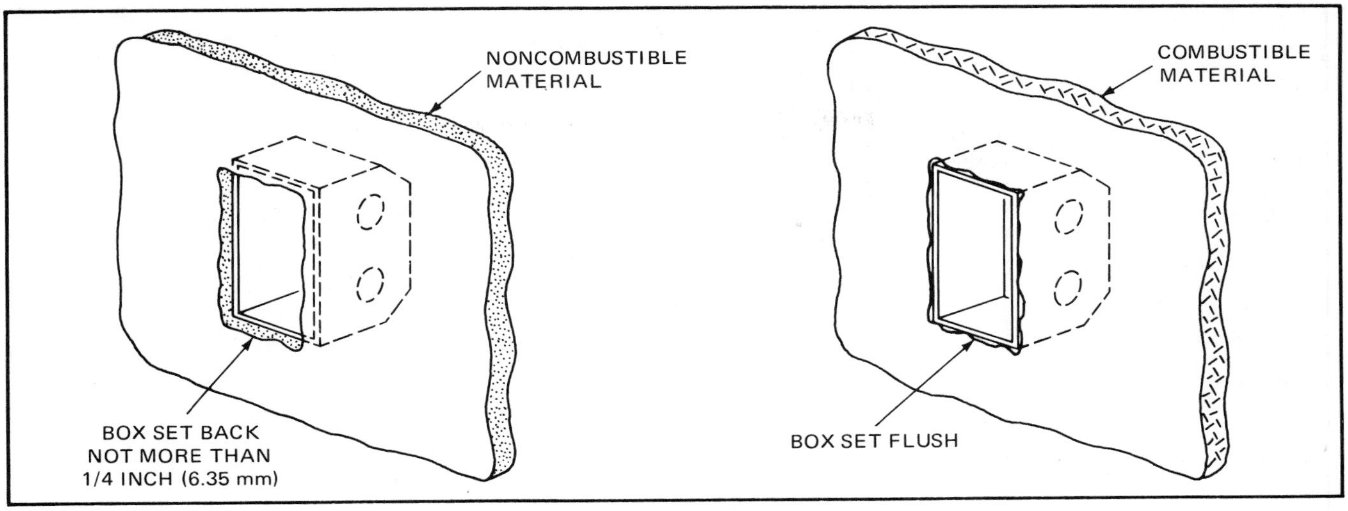

Fig. 2-7 Box position in walls or ceilings, ***Section 370-10.***

The dimensions of a standard sectional switch box (2" x 3") are the dimensions of the opening of the box. The depth of the box may vary from 1 1/2 inches to 3 1/2 inches, depending upon the requirements of the building construction. NEC *Article 370* covers outlet, switch and junction boxes. See figure 2-11 for a complete listing of box dimensions.

Section 370-10 of the Code states that boxes must be mounted so that they will be set back not more than 1/4 inch (6.35 mm) when the boxes are mounted in noncombustible walls or ceilings made of concrete, tile, or similar materials. When the wall or ceiling construction is of combustible material (wood), the box must be set flush with the surface, figure 2-7. These requirements are meant to prevent the spread of fire if a short circuit occurs within the box.

Ganged sectional switch (device) boxes can be installed using a pair of metal mounting strips. These strips are also used to install a switch box between wall studs, figure 2-8. The use of a bracket box in such an installation may result in an offcenter convenience outlet. When an outlet box is to be mounted at a specific location between joists, as for ceiling-mounted fixtures, an offset bar hanger is used (figure 2-4).

The Code states that when a switch box or outlet box is mounted to a stud or ceiling joist by nailing through the box, the nails must be not more than 1/4 inch (6.35 mm) from the back of the box, figure 2-9. This requirement insures that when the nail passes through the box, it does not interfere with the wiring devices in the box.

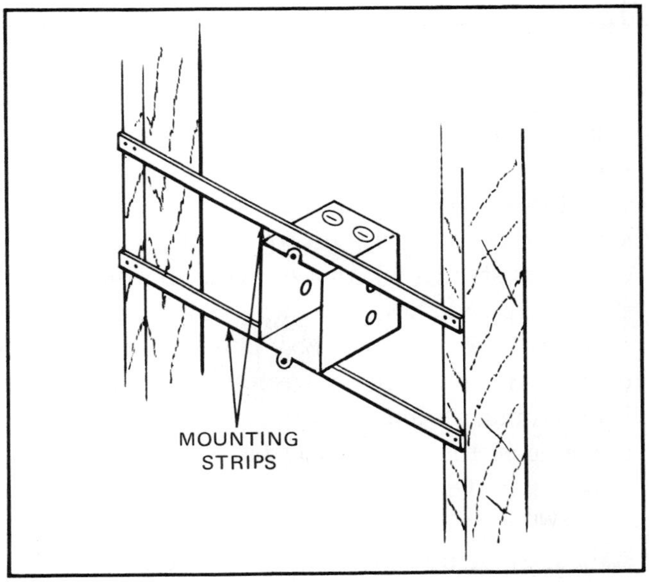

Fig. 2-8 Switch (device) boxes installed between studs using metal mounting strips.

Interchangeable Wiring Devices

In addition to ganged switch boxes, a multiple-switch installation can be made using an interchangeable line of switches. Up to three switches, pilot lights, convenience outlets, or any combination of interchangeable wiring devices can be installed, one above the other, on a single strap in a standard 2" x 3" box opening. A total of six devices can be

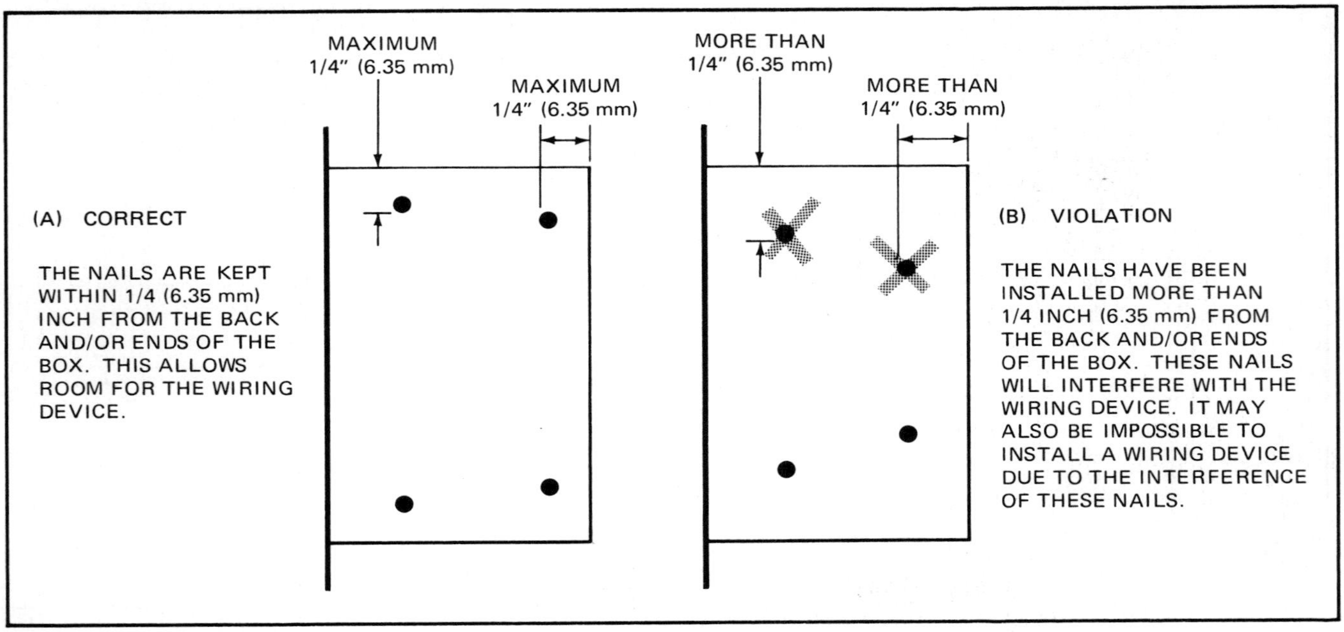
MAXIMUM
1/4" (6.35 mm)
MAXIMUM
1/4" (6.35 mm)
(A) CORRECT
THE NAILS ARE KEPT WITHIN 1/4 (6.35 mm) INCH FROM THE BACK AND/OR ENDS OF THE BOX. THIS ALLOWS ROOM FOR THE WIRING DEVICE.
MORE THAN
1/4" (6.35 mm)
MORE THAN
1/4" (6.35 mm)
(B) VIOLATION
THE NAILS HAVE BEEN INSTALLED MORE THAN 1/4 INCH (6.35 mm) FROM THE BACK AND/OR ENDS OF THE BOX. THESE NAILS WILL INTERFERE WITH THE WIRING DEVICE. IT MAY ALSO BE IMPOSSIBLE TO INSTALL A WIRING DEVICE DUE TO THE INTERFERENCE OF THESE NAILS.

Fig. 2-9 Using nails to install a sectional switch box. *(NEC Section 370-13.)*

installed using two ganged boxes with standard openings.

Interchangeable devices are available in the same types and classifications of switches and outlets as standard wiring devices. In other words, single-pole switches, three-way switches, four-way switches, two-pole switches, pilot lights and other devices are available in an interchangeable style. Both standard and interchangeable devices are available in the silent style that makes little or no noise when the switch is actuated.

When interchangeable devices are to be installed, standard sectional switch boxes generally are not used because of the lack of wiring space in the box. Instead, 4-inch square, 4 11/16-inch square, or 4" x 6" boxes with raised plaster covers are used. (This is not a requirement of the Code, however.) Under certain conditions, 3-inch and 3 1/2-inch deep sectional boxes are large enough for a group of interchangeable devices.

The Code requires that faceplates for switches and receptacles must completely cover the wall opening and must seat against the surface of the wall.

Figure 2-10 shows two interchangeable switches and one pilot light mounted on a single strap in a 4-inch square outlet box. The box has a plaster cover and a bracket. A single-gang, three-hole wall plate is required. This arrangement is useful where limited wall space (such as the space between a door casing and a window casing) prohibits the use of standard two-gang or three-gang wall plates.

SPECIAL-PURPOSE OUTLETS

Special-purpose outlets are usually indicated on the plans. These outlets are described by a notation and are also detailed in the specifications. The plans in the back of this text indicate special-purpose outlets by a triangle inside a circle with subscript letters. In some cases, a subscript number is added to the letter.

When a special-purpose outlet is indicated on the plans or in the specifications, the electrician must check for special requirements. Such a requirement may be a separate circuit, a special 240-volt circuit, a special grounding or polarized receptacle, or other preparation.

NUMBER OF CONDUCTORS

The National Electrical Code specifies the maximum number of conductors allowed in standard outlet boxes and switch boxes, figure 2-11. A conductor running through the box is counted as one conductor. Each conductor originating outside of the box and terminating inside the box is counted as one conductor. Conductors which

Fig. 2-10 Interchangeable devices.

Table 370-6(a). Metal Boxes

Box Dimension, Inches Trade Size or Type	Min. Cu. In. Cap.	Maximum Number of Conductors				
		#14	#12	#10	#8	#6
4 x 1¼ Round or Octagonal	12.5	6	5	5	4	0
4 x 1½ Round or Octagonal	15.5	7	6	6	5	0
4 x 2⅛ Round or Octagonal	21.5	10	9	8	7	0
4 x 1¼ Square	18.0	9	8	7	6	0
4 x 1½ Square	21.0	10	9	8	7	0
4 x 2⅛ Square	30.3	15	13	12	10	6*
4 11/16 x 1¼ Square	25.5	12	11	10	8	0
4 11/16 x 1½ Square	29.5	14	13	11	9	0
4 11/16 x 2⅛ Square	42.0	21	18	16	14	6
3 x 2 x 1½ Device	7.5	3	3	3	2	0
3 x 2 x 2 Device	10.0	5	4	4	3	0
3 x 2 x 2¼ Device	10.5	5	4	4	3	0
3 x 2 x 2½ Device	12.5	6	5	5	4	0
3 x 2 x 2¾ Device	14.0	7	6	5	4	0
3 x 2 x 3½ Device	18.0	9	8	7	6	0
4 x 2⅛ x 1½ Device	10.3	5	4	4	3	0
4 x 2⅛ x 1⅞ Device	13.0	6	5	5	4	0
4 x 2⅛ x 2⅛ Device	14.5	7	6	5	4	0
3¾ x 2 x 2½ Masonry Box/gang	14.0	7	6	5	4	0
3¾ x 2 x 3½ Masonry Box/gang	21.0	10	9	8	7	0
FS — Minimum Internal Depth 1¾ Single Cover/Gang	13.5	6	6	5	4	0
FD — Minimum Internal Depth 2⅜ Single Cover/Gang	18.0	9	8	7	6	3
FS — Minimum Internal Depth 1¾ Multiple Cover/Gang	18.0	9	8	7	6	0
FD — Minimum Internal Depth 2⅜ Multiple Cover/Gang	24.0	12	10	9	8	4

* Not to be used as a pull box. For termination only.

Table 370-6(b). Volume Required Per Conductor

Size of Conductor	Free Space Within Box for Each Conductor
No. 14	2. cubic inches
No. 12	2.25 cubic inches
No. 10	2.5 cubic inches
No. 8	3. cubic inches
No. 6	5. cubic inches

Fig. 2-11 Allowable number of conductors in boxes.

originate and terminate within the box are not counted.

When conductors are the same size, the proper box size can be selected by referring to *Table 370-6(a)*. When conductors are of different sizes, refer to *Table 370-6(b)*.

Tables 370-6(a) and *370-6(b)* do not consider fittings or devices such as fixture studs, cable clamps, hickeys, switches, or receptacles which may be in the box. When the box contains one or more devices, such as fixture studs, cable clamps or hickeys, the number of conductors must be one less than shown in the table for *each* type of device.

Example: A box contains one fixture stud and two cable clamps. The number of conductors permitted in the box shall be two less than shown in the table. (Deduct one conductor for the fixture stud; deduct one conductor for the two cable clamps.) A further deduction of one conductor is made for one or several flush devices mounted on the same strap. For example, if switches or receptacles of an interchangeable line are mounted on the same strap, then one conductor is deducted. A further deduction of one conductor must be made for one or more grounding conductors entering the box.

The Code requires that all boxes *other* than those listed in *Table 370-6(a)* be durably and legibly marked by the manufacturer with their cubic inch capacity, *Section 370-6(b)*. When sectional boxes are ganged together, the volume to be filled is the total cubic inch volume of the assembled boxes. Fittings may be used with the sectional boxes, such as plaster rings, raised covers,

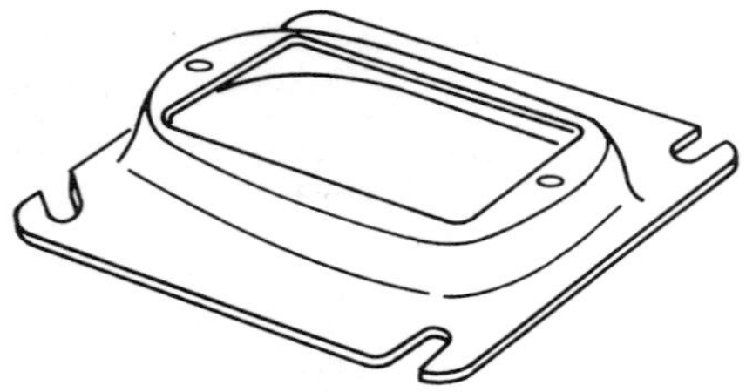

ADDITIONAL WIRING SPACE PROVIDED BY THE RAISED COVER IS: 2" x 3" x 3/4" = 4 1/2 CUBIC INCHES

Fig. 2-12 Raised cover.

and extension rings. When these fittings are marked with their volume in cubic inches, or have dimensions comparable to those boxes shown in *Table 370-6(a)*, then their volume may be considered in determining the total cubic inch volume to be filled [see *Section 370-6(a)*(1)]. Figure 2-12 shows how a 3/4-inch raised cover increases the wiring space above the cubic inch capacity of the box to which it is attached.

The size of equipment grounding conductors is shown in *Table 250-95*. The grounding conductors are the same size as the circuit conductors in cables having No. 14 AWG, No. 12 AWG, and No. 10 AWG circuit conductors. Thus, box sizes can be calculated using *Table 370-6(a)*.

LOCATION OF OUTLETS

There are no hard and fast rules for locating most outlets. A number of conditions determine the proper height for a switch box. For example, the height of the kitchen counter backsplash determines where the switches and convenience outlets are located between the kitchen countertop and the cabinets. The proper height for the convenience outlets located near a baseboard depends upon the height of the baseboard heating units. The residence shown in the plans is heated by electric baseboard units. In most instances, the height of these units from the top of the unit to the finish floor seldom exceeds six inches (152 mm). However, some manufacturers produce baseboard heating units less than six inches (152 mm) high, and others produce units in excess of six inches (152 mm). (See "Location of Electric Baseboard Heaters" in unit 23.

The location of lighting outlets is determined by the amount and type of illumination required to produce the desired lighting effects. It is not the intent of this text to describe how proper and adequate lighting is determined. Rather, the text covers the proper methods of installing the circuits for such lighting. If the student is interested, standards have been developed to guide the design of adequate lighting. The local electric utility company can supply information on these standards.

It is common practice among electricians to consult the plans and specifications to determine the proper heights and clearances for the installation of electrical devices. The electrician then has these dimensions verified by the architect, electrical engineer, designer, or homeowner. This practice avoids unnecessary and costly changes in the locations of outlets and switches as the building progresses.

To insure uniform installation and safety, and in accordance with long-established custom, standard electrical outlets are located as shown in figure 2-13. These dimensions usually are satisfactory. However, the electrician must check the blueprints, specifications, and details for measurements which may affect the location of a

SWITCHES	
	* Inches above floor
Regular.	48-50 (1.22-1.27 m)
Between counter and kitchen cabinets (depends on backsplash)	44-46 (1.12-1.17 m)
CONVENIENCE OUTLETS	
	* Inches above floor
Regular.	12 (305 mm)
Between counter and kitchen cabinets (depends on backsplash)	44-46 (1.12-1.17 m)
In garage	48 (1.22 m)
WALL BRACKETS	
	* Inches above floor
Outside.	66 (1.68 m)
Inside.	60 (1.52 m)
Side of medicine cabinet	60 (1.52 m)

***Note:**
All dimensions given are from the finish floor to the center of the outlet box. Verify all dimensions before roughing in.

Fig. 2-13 Outlet locations.

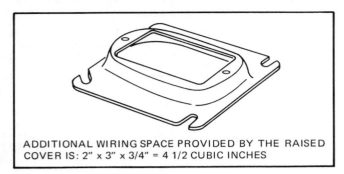

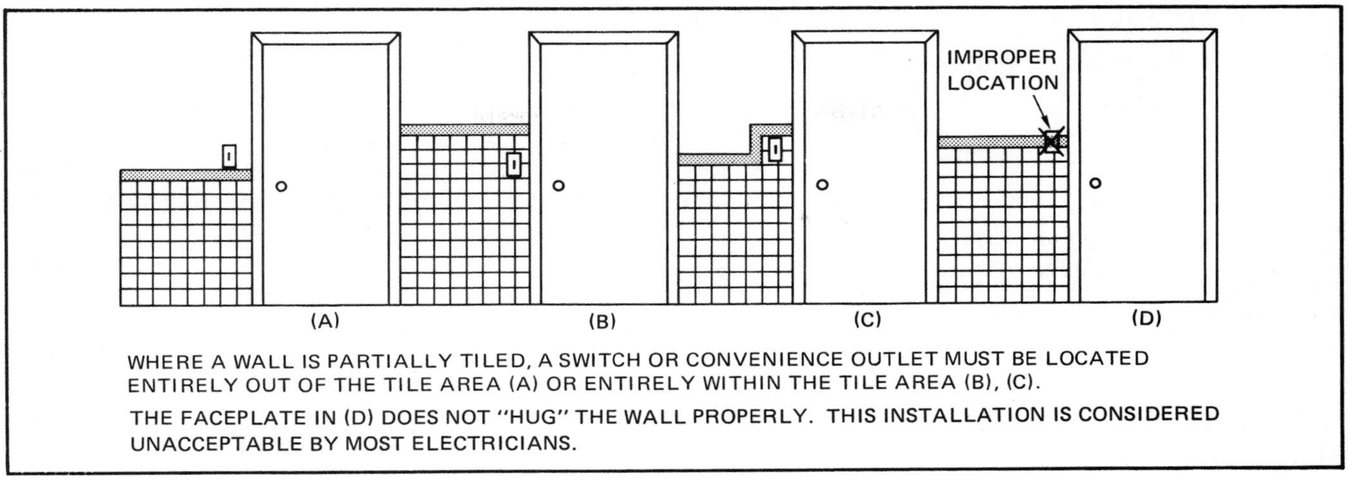

IMPROPER
LOCATION
(A)
(B)
(C)
(D)
WHERE A WALL IS PARTIALLY TILED, A SWITCH OR CONVENIENCE OUTLET MUST BE LOCATED ENTIRELY OUT OF THE TILE AREA (A) OR ENTIRELY WITHIN THE TILE AREA (B), (C).
THE FACEPLATE IN (D) DOES NOT "HUG" THE WALL PROPERLY. THIS INSTALLATION IS CONSIDERED UNACCEPTABLE BY MOST ELECTRICIANS.

Fig. 2-14 Locating an outlet on a tiled wall.

particular outlet or switch. The cabinet spacing, available space between the countertop and the cabinet, and the tile height may influence the location of the outlet or switch. For example, if the top of the wall tile is exactly 48 inches (1.22 m) from the finish floor line, a wall switch should not be mounted 48 inches (1.22 m) to center. This is considered poor workmanship. The switch should be located entirely within the tile area or entirely out of the tile area, figure 2-14. This situation requires the full cooperation of all crafts involved in the construction job.

Faceplates

Faceplates for switches shall be installed so as to completely cover the wall opening and seat against the wall surface.

Faceplates for receptacles shall be installed so as to completely cover the opening and seat against the mounting surface. The mounting surface might be the wall, or it might be the gasket of a weatherproof box.

In either case, the intent of the Code is to prevent access to live parts that could cause electrical shock.

REVIEW

Note: Refer to the National Electrical Code or the plans where necessary.

PART 1 – ELECTRICAL FEATURES

1. What does a plan show about electrical outlets? ______________________________

____

2. What is an outlet? ____

____

3. Match the following switch types with the proper symbol.

a. single pole	S_P
b. three way	S_4
c. four way	S
d. single pole with pilot light	S_3

4. The plans show dash lines running between switches and various outlets. What do these dash lines indicate? ______________________________

____

5. Why are dash lines usually curved? ______

6. a. What are junction boxes used for? ______

   b. Are junction boxes normally used in wiring the first floor? Explain. ______

   ______

   c. Are junction boxes normally used to wire exposed portions of the basement? Explain. ______

   ______

7. What is the usual pattern of wiring for residential jobs? ______

   ______

8. How are standard sectional switch boxes mounted? ______

   ______

9. a. What is an offset bar hanger? ______

   ______

   b. What types of boxes may be used with offset bar hangers? ______

   ______

10. What methods may be used to mount lighting fixtures to an outlet box fastened to an offset bar hanger? ______

    ______

11. What advantage does a 4-inch octagon box have over a 3 1/4-inch octagon box?

    ______

    ______

12. What is the size of the opening of a switch box for a single device? ______

    ______

13. The space between a door casing and a window casing is 3 1/2 inches (88.9 mm). Two switches are to be installed at this location. What type of switches will be used?

    ______

    ______

14. Three switches are mounted in a three-gang switch box. The wall plate for this assembly is called a ______ plate.

15. The use of cable clamps inside a box ______ the number of conductors allowed by one.

16. a. How high above the finish floor are the switches located in the garage?

    ______

    b. In the living room? ______

17. How high above the finish floor are the convenience receptacles in the garage located? ______ In the living room? ______

18. Outdoor convenience outlets are located ______ inches above grade.

19. In the space provided, draw the correct symbol for each of the descriptions listed in (a) through (r).

a. ________	Lighting panel	j. ________	Special purpose outlet
b. ________	Clock outlet	k. ________	Fan outlet
c. ________	Duplex outlet	l. ________	Range outlet
d. ________	Outside telephone	m. ________	Power panel
e. ________	Single-pole switch	n. ________	Three-way switch
f. ________	Motor	o. ________	Push button
g. ________	Duplex outlet, split circuit	p. ________	Thermostat
h. ________	Lampholder with pull switch	q. ________	Electric door opener
i. ________	Weatherproof outlet	r. ________	Multioutlet assembly

20. The edge of a box installed in a combustible wall must be ________________ with the finished surface.

21. List the maximum number of No. 12 AWG conductors permitted in a
    (a) 4″ x 1 1/2″ octagon box. ________________
    (b) 4 11/16″ x 1 1/2″ square box. ________________
    (c) 3″ x 2″ x 3 1/2″ device box. ________________

22. When a switch box is nailed to a stud, and the nail runs through the box, the nail must not interfere with the wiring space. To accomplish this, keep the nail ____________ .
    (a) halfway between the front and rear of the box
    (b) a maximum of 1/4 inch (6.35 mm) from the front edge of the box
    (c) a maximum of 1/4 inch (6.35 mm) from the rear of the box

PART II – STRUCTURAL FEATURES

Note: Refer to the plans.

1. What is the size of the building lot? ________________
2. The front of the building is facing in what compass direction? ________________
3. How far from the curb line is the building located? ________________
4. What is the approximate distance between the garage wall and the side property line? ________________
5. What is the difference in the grade between the high and low points on the property? ________________
6. To what scale is the basement plan drawn? ________________
7. How are the wall footings indicated on the basement plan? ________________

8. What is the size of the footing for the lally column in the garage? ________________

9. Of what material is the basement fireplace constructed? ________________

10. Of what material are the adjoining basement partition walls constructed? ________________

11. To what kind of material will the terrace bracket light be fastened? ________________

12. What type of finish floor is the recreation room to have? ________________

13. Are the main partition walls in the basement to be plastered? ________________

14. Is there a floor drain indicated on the basement plan? If so, where? ________________

____

15. Give the direction of the floor joists in the basement ceiling. ________________

16. Give the size and spacing of the floor joists in the basement ceiling. ________________

17. Give the thickness of the concrete block walls in the basement. ________________

18. What material is indicated for the main foundation walls? ________________

19. Where is the 80-gallon water tank located? ________________

20. Where is the artesian well located? ________________

21. Is an electric light outlet called for in the ceiling of the shower stall in the basement lavatory? ________________

22. To what kind of masonry wall will the convenience outlets in the basement be fastened?

____

23. Is the ceiling in the basement to be plastered? (See Sections A-A and B-B).

____

24. What is the ceiling height in the basement from the underside of the floor joists to the concrete floor? ________________

25. What is the ceiling height under the garage portion of the basement? ________________

26. Is the electric light outlet indicated at the foot of the stairs in the basement fastened to the wood ceiling joist or the concrete ceiling? ________________

27. What type of ceiling does the pump room have? ________________

28. What is the purpose of the I-beam shown in the recreation room? ________________

29. In what room are the laundry trays located? ________________

30. What is the overall length of the building as shown on the first floor plan?

____

31. What is the depth of the garage, front to back? Give the measurement to the inside edge of the studs. ________________

32. How wide is the concrete apron? (NE to SW) ________________

33. What is the size of the first floor ceiling joists in the house and garage? ________________

____

34. Give the size of the girder in the garage and state whether it is wood or steel.

____

35. Is the entire garage area to have a concrete floor? ________________

36. How many lally columns are indicated in the garage? ______

37. What does the plumbing symbol in the garage represent? ______

38. How many plumbing fixtures or units are indicated on the first floor plan, excluding the garage? (Exclude food waste disposer and dishwasher.) ______

39. How many feet of baseboard electric heating units are indicated in the living and dining room areas? ______

40. Give the size of the studding used in the partition between the bathroom and rear bedroom. ______

41. Are all the first floor ceiling joists running in the same direction (front to back)?

______

42. What is the size of the front entrance door? ______

43. What type and size of window over the kitchen sink does the plan call for? ______

44. In what part of the dwelling is access to the attic provided? ______

45. What type of girder is supporting the ceiling joist over the living and dining rooms?

______

46. What is the distance between the top of the bedroom floor joists and the bottom side of the ceiling joists? ______

47. Where does the stone veneer on this building extend from the ground to the cornice?

______

48. How high is the stone work from the finished floor line on the kitchen wall?

______

49. To what material, wood or masonry, are the following lights to be fastened?
a. Front entrance light ______ b. Front garage light ______

unit 3

Determining the Number of Circuits Required

OBJECTIVES

After studying this unit, the student will be able to

- calculate the occupied floor area of a dwelling.
- determine the total load requirements in amperes for general lighting.
- determine the minimum number of lighting branch circuits required.
- calculate the number of small appliance branch circuits required.
- use the National Electrical Code to determine the basic requirements for the various branch circuits in a dwelling.

It is standard practice in the design and planning of dwelling units to permit the electrician to plan the circuits. Thus, the residence plans do not include layouts for the various branch circuits. The electrician may follow the general guidelines established by the architect. However, any wiring systems designed and installed by the electrician must conform to the standards of the National Electrical Code, as well as local and state code requirements.

BRANCH-CIRCUIT COMPUTATIONS

The fundamental rules for determining the number of receptacles and lighting circuits needed for dwelling units are as follows:

1. The maximum load shall not exceed 80% of the branch-circuit rating where the load is expected to continue for three hours or more, *Section 220-2(a)*.
2. The lighting load per square foot of floor area is 3 watts, *Section 220-2(b)*.
3. Divide the load in watts determined in steps 1 and 2 by the ampacity of the circuit(s) to be used. The resulting number is the minimum number of circuits required.

To determine the number of branch circuits required, first find the total load of the branch circuits carrying general lighting outlets and convenience outlets for which a definite wattage is not stated. To this load value, add the total load of those circuits for which the load is determined by the fixed wattage rating of the device served by the circuit. For example, the load of an electric range circuit, an oil burner circuit, or other device circuit is determined by the rating of the range, oil burner, trash compactor or food waste disposer.

CALCULATING FLOOR AREA

First Floor

To estimate the total load in watts per square foot (W/ft^2) for a dwelling, the occupied floor area of the dwelling must be calculated. Note in the residence plans that the first floor area has an irregular shape. In this case, the simplest method of calculating the occupied floor area is to determine the total floor area using the outside dimensions of the dwelling. Then, the areas of the following spaces are subtracted from the total area: open porches, garages, or other unfinished or unused spaces if they are not adaptable for future use, *Section 220-2(b)*.

Many open porches, terraces, patios and similar areas are commonly used as recreation and entertainment areas. Therefore, adequate lighting and receptacle outlets must be provided.

The Code does not clearly define what is to be done with the thickness of walls and partitions when calculating the total floor area. The method used in this unit disregards wall and partition dimensions. Thus, the total area is obtained using the outside dimensions of the building and subtracting those areas (except walls and partitions) which are not to be included.

Figure 3-1 is a portion of the floor plan minus the garage area. The method of calculating the floor area is indicated in the figure. Note that certain numbers are rounded off to simplify the calculations. However, the resulting value is accurate enough to determine the load requirements.

Basement Area

The basement area normally is not included in the calculations. In the residence plans, however, certain sections of the basement are indicated as being occupied, figure 3-2.

The method of calculating the total occupied floor area of the basement is shown in figure 3-2. When the combined areas of the terrace, storage room, and the unexcavated portions of the basement are deducted from the total overall area, the net floor area of the basement is obtained.

The combined occupied area of the dwelling is found by adding the total occupied first floor area, 1428 square feet from figure 3-1, and the total occupied basement floor area, 1154 square feet from figure 3-2.

First floor	1428 ft^2
Basement	1154 ft^2
Total	2582 ft^2

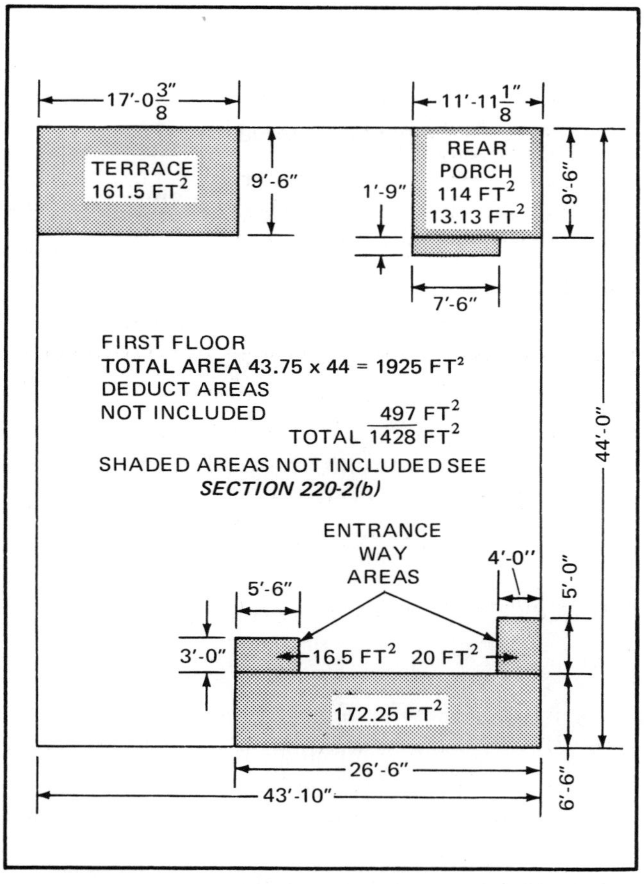

Fig. 3-1 Determining occupied area on the first floor.

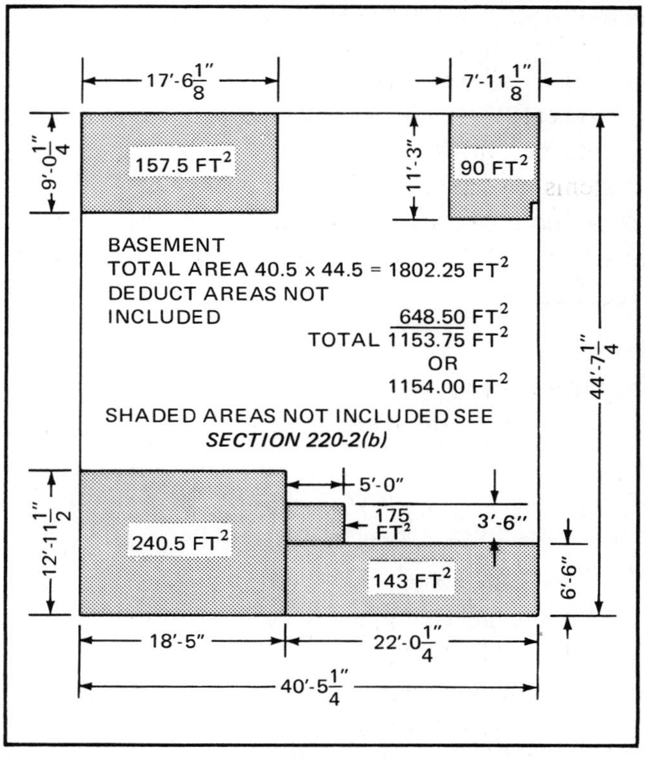

Fig. 3-2 Determining occupied area in the basement.

DETERMINING THE MINIMUM NUMBER OF LIGHTING CIRCUITS

Table 220-2(b) shows that the load permitted for dwelling units is 3 watts per square foot (0.093 m^2) of occupied area for general lighting. The calculated load for the total occupied area, 2582 ft^2 (240 m^2), is:

$$2582\ (ft^2) \times 3\ (W/ft^2) = 7746\ W$$

The total required amperage is determined as follows:

$$\text{Amperes (A)} = \frac{\text{watts (W)}}{\text{volts (V)}} = \frac{7746}{120} = 64.55\ A$$

where 120 V = ac line voltage for lighting circuits.

Table 310-16 shows that the maximum circuit rating is 15 amperes when No. 14 AWG conductors are used. (Unit 4 describes the sizes and ratings of conductors.) The residence specifications state that larger No. 12 AWG wire is to be used for general lighting circuits, rather than No. 14 AWG wire. However, the circuits are to remain as 15-ampere circuits (*Section 210-3*). The minimum number of circuits required is equal to the total amperage divided by the maximum ampere rating of each circuit.

$$\frac{64.55}{15} = 4.3 \text{ or } 5 \text{ circuits minimum}$$

Five circuits is the *minimum* number of circuits allowed for the total load. More circuits may be used if desired. The residence plans show that twelve general lighting circuits are actually used. All of the general lighting circuits in the residence are to be wired with No. 12 AWG conductors and protected by 15-ampere Type SC fuses. Overcurrent protection can also be provided by Type S fuses or circuit breakers, depending on the type of electric panels installed by the electrician. Detailed explanations of all branch circuits are given in the following units. Fuses and circuit breakers are covered in unit 27.

DETERMINING THE NUMBER OF SMALL APPLIANCE BRANCH CIRCUITS

Sections 220-16(a) and *(b)* state that an additional load of not less than 1500 watts shall be included for each two-wire small appliance circuit. According to *Section 220-3(b)(1)*, these appliance circuits shall feed only receptacle outlets in the kitchen, pantry, dining room, and breakfast room of dwelling units. *Section 220-3(b)(1)* also states that for these areas, two or more 20-ampere branch circuits shall be provided and such circuits shall have no other outlets. An exception is a clock outlet which may be connected to either an appliance circuit or a lighting circuit. No. 12 AWG wire is used in these small appliance circuits to minimize the voltage drop in the circuit. The use of this larger wire, rather than No. 14 AWG wire, helps improve appliance performance and lessens the danger of overloading the circuits.

Automatic washers draw a large amount of current during certain portions of their operating cycles. Thus, *Section 220-3(c)* requires that at least one additional 20-ampere branch circuit must be provided to supply the laundry receptacle outlet(s) as required by *Section 210-52(e)*.

To provide for true electrical living, the residence plans show:

- four 20-ampere circuits supplying the kitchen receptacles.
- one 20-ampere circuit supplying the dining room receptacles.
- four 20-ampere circuits supplying the four wall receptacles in the utility room (washer and freezer included).

  Note: A single receptacle installed on an individual branch circuit shall have a rating of not less than the rating of the branch circuit, *Section 210-21(b)(1)*.
- one 20-ampere circuit supplying the receptacles in the workshop.

In summary, the residence will have a total of ten 20-ampere appliance circuits and twelve 15-ampere lighting circuits.

A complete list of the branch circuits for this residence is found in unit 27. The circuit directories for Main Panel A and Load Center B show clearly the number of lighting, appliance, and special-purpose circuits provided.

It is recommended that the student now study the requirements of the following sections of the National Electrical Code.

- *Section 210-7* – methods of connecting grounding-type receptacles in new and existing installations
- *Section 210-8* – ground-fault circuit protection for receptacles installed outdoors when there is direct grade level access to the dwelling, in garages, in bathrooms, and on construction sites
- *Sections 210-50* and *210-52* – spacing of convenience receptacles
- *Section 250-57* – methods of grounding fixed equipment
- *Section 250-74* – connecting the receptacle grounding terminal to the box
- *Section 250-114* – continuity and attachment of grounding conductors to boxes

When an existing nongrounding-type receptacle is replaced with a grounding receptacle, a properly installed grounding conductor must be provided. If it is impractical to reach a source of ground, a nongrounding-type receptacle shall be used, *Section 210-7(d).* Note that this exception applies only to replacements, not to new installations.

OUTLETS FOR BRANCH CIRCUITS

The residence plans show that most of the convenience receptacles are connected to the general lighting branch circuits. Exceptions are the receptacle outlets in the kitchen, dining room, workshop,

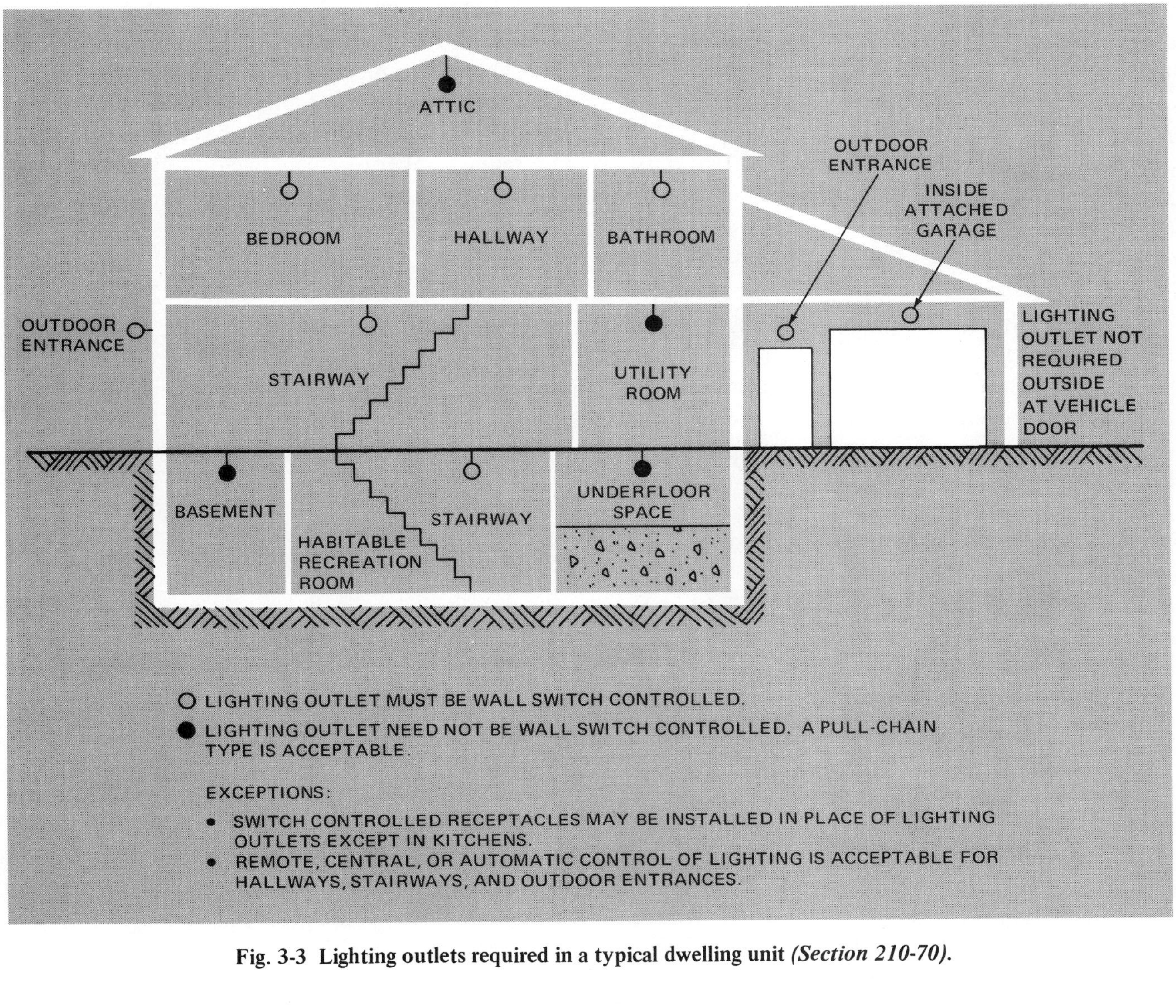
ATTIC
OUTDOOR ENTRANCE
INSIDE ATTACHED GARAGE
BEDROOM
HALLWAY
BATHROOM
OUTDOOR ENTRANCE
STAIRWAY
UTILITY ROOM
LIGHTING OUTLET NOT REQUIRED OUTSIDE AT VEHICLE DOOR
BASEMENT
STAIRWAY
UNDERFLOOR SPACE
HABITABLE RECREATION ROOM
LIGHTING OUTLET MUST BE WALL SWITCH CONTROLLED.
LIGHTING OUTLET NEED NOT BE WALL SWITCH CONTROLLED. A PULL-CHAIN TYPE IS ACCEPTABLE.
EXCEPTIONS:
• SWITCH CONTROLLED RECEPTACLES MAY BE INSTALLED IN PLACE OF LIGHTING OUTLETS EXCEPT IN KITCHENS.
• REMOTE, CENTRAL, OR AUTOMATIC CONTROL OF LIGHTING IS ACCEPTABLE FOR HALLWAYS, STAIRWAYS, AND OUTDOOR ENTRANCES.

Fig. 3-3 Lighting outlets required in a typical dwelling unit ***(Section 210-70).***

and utility room. These outlets are connected to special 20-ampere small appliance circuits. While NEC *Sections 210-50* and *210-52* do give general requirements for the location of these outlets, the owner may indicate preferred locations.

GENERAL REQUIREMENTS FOR RECEPTACLE AND LIGHTING OUTLETS

For dwelling units, *Section 210-52* states that receptacle outlets must be installed:

- so no point along the floor line in any space is more than 6 feet (1.83 m), measured horizontally, from an outlet in that space.
- in any wall space 2 feet (610 mm) or more in width.
- for each counter space wider than 12 inches (305 mm) in kitchens and dining rooms.
- adjacent to the basic location in bathrooms; at least one receptacle outlet; must be GFCI protected (see unit 6).
- outdoors; at least one receptacle outlet; must be GFCI protected (see unit 6).
- in basement; at least one receptacle outlet in addition to the laundry outlet.
- in garage; at least one receptacle outlet; must be GFCI protected (see unit 6).
- in laundry room; at least one outlet for laundry equipment, within 6 feet (1.83 m) of appliance.

Note that receptacle outlets on fixtures or appliances, except as noted in the next paragraph, are in addition to the previous items.

The requirements for lighting outlets are shown in figure 3-3. In addition, factory installed receptacles in permanent electric baseboard heaters *may* be counted in the number of receptacles required for the wall space used by the heater. These receptacles *may not* be connected to the heater circuit. Refer to unit 23 for a more detailed discussion of this topic.

The requirements stated in this unit are minimum requirements. Note that the residence covered in this text exceeds the Code requirements for the minimum number of receptacle outlets and lighting outlets.

REVIEW

Note: In the following questions, the student may refer to the National Electrical Code to obtain the answers. To help the student find these answers, references to Code sections are included after the questions. These references are not given in later reviews where it is assumed that the student is familiar with the Code.

1. What is the meaning of computed load? ______________________________

____

2. How are branch circuits classified? *Section 210-3* ______________________

____

3. a. How is the rating of the branch circuit protective device affected when the conductors used are of a larger size than called for by the Code? ______________

   b. How is the circuit classification affected? *Section 210-3* ______________

4. What dimensions are used when measuring the area of a building? *Section 220-2(b)*

____

____

5. What spaces are not included in the floor area when computing the load in watts per square foot? *Section 220-2(b)* ______________________________

6. What is the unit load per square foot for dwelling units? *Table 220-2(b)* __________

7. According to *Section 210-50*, a laundry equipment outlet must be placed within ________ feet (________ meters) of the intended location of the laundry equipment.

8. How is the total load in watts for lighting purposes determined? *Section 220-2(b)*

____

____

9. How is the total lighting load in amperes determined? ____

____

10. How is the required number of lighting branch circuits determined? ____

____

11. What is the minimum number of 15-ampere lighting circuits required if the dwelling has an occupied area of 4000 square feet? Show all calculations. ____

____

12. How many branch lighting circuits are provided in this dwelling? ________________

13. Why is No. 12 AWG wire used rather than No. 14 AWG wire for branch lighting circuits in this residence? ____

14. What is the minimum load allowance for small appliance circuits for dwellings? *Section 220-16(a)* ____

15. What is the smallest size wire that can be used in a branch circuit rated at 20 amperes?

____

16. How is the load determined for outlets supplying the specific appliances? *Section 220-2(c)* ____

17. What type of circuits must be provided for receptacle outlets in the kitchen, pantry, family room, laundry, dining room, and breakfast room? *Section 220-3(b)*

____

____

18. How is the minimum number of receptacle outlets determined for most occupied rooms? *Section 210-50* ____

19. The electrical plans show that Bedroom No. 1 contains six (6) convenience receptacles. According to *Section 210-50*, what is the minimum number of convenience receptacles permitted in this bedroom? ____

20. In a single-family dwelling, what types of overload protection for circuits are used?

____

____

21. Is a grounded conductor included in this dwelling for:
    a. a 120-volt, 2-wire branch circuit? ________________
    b. a 240-volt, 2-wire branch circuit? ________________
    c. a 240-volt, 3-wire branch circuit? ________________

22. May receptacle outlets in electric baseboard heaters be connected to the heater circuit?

____

23. The Code indicates the rooms in a dwelling that are required to have switched lighting outlets or switched receptacles. Write *yes* (switch required) or *no* (switch not required) for the following areas:
    a. attic ______________________
    b. stairway ______________________
    c. crawl space ______________________
    d. hallway ______________________
    e. bathroom ______________________

unit 4

Conductor Sizes and Types

OBJECTIVES

After studying this unit, the student will be able to

- define the terms used to size and rate conductors.
- discuss the subject of aluminum conductors.
- describe the types of cables used in most dwelling unit installations.
- list the installation requirements for each type of cable.
- describe the uses and installation requirements for electrical conduit systems.
- describe the requirements for grounding the service entrance.
- describe the use and installation requirements for flexible metal conduit.

CONDUCTORS

Throughout this text, all references to conductors are for copper conductors, unless otherwise stated.

Wire Size

The copper wire used in electrical installations is graded for size according to the American Wire Gauge Standard (AWG). The wire diameter in the AWG standard is expressed as a whole number. AWG sizes vary from fine, hairlike wire used in coils and small transformers to very large diameter wire required in industrial wiring to handle heavy loads.

The wire may be a single strand (solid conductor), or it may consist of many strands. Each strand of wire acts as a separate conducting unit. Conductors which are No. 8 AWG and larger generally are stranded. The wire size used for a circuit depends upon the maximum current to be carried. The National Electrical Code states that the minimum conductor size permitted in house wiring is No. 14 AWG. Exceptions to this rule are covered in *Section 210-19(c)* and *Article 725* for the wires used in lighting fixtures, bell wiring, and remote-control low-energy circuits.

Ampacity

Ampacity expresses the current-carrying capacity of a wire in amperes. This value depends on the area of the wire. Since the area of a circle is proportional to the square of the diameter, the ampacity of a round wire varies with the diameter squared. The diameter of a wire is given by a unit called a mil. A *mil* is defined as one-thousandth of an inch (0.001 inch). Mils squared are known as circular mils. The circular mil area of a wire determines the current-carrying capacity of the wire. In other words, the larger the circular mil area of a wire, the greater is its current-carrying capacity. AWG sizes are also expressed in mils and range from No. 40 (10 circular mils) to No. 4/0 (211 600 circular mils). Wire sizes larger than No. 4/0 are expressed in circular mils only. For a further discussion on conductor ampacities, correction factors, and derating factors, see unit 15.

CONDUCTOR APPLICATIONS CHART

Conductor Size	Applications
No. 16 and No. 18 AWG	cords, low-voltage control circuits, bell and chime wiring
No. 14 and No. 12 AWG	normal lighting circuits, and circuits supplying receptacle outlets
No. 10, No. 8, No. 6, and No. 4 AWG	clothes dryers, ovens, ranges, cooktops, water heaters, heat pumps, central air conditioners, furnaces, feeders to sub-panels
No. 3, No. 2, No. 1 (and larger) AWG	main service entrances, feeders to sub-panels

ALUMINUM CONDUCTORS

The conductivity of aluminum is not as great as that of copper for a given wire size. For example, checking NEC *Table 310-16* and *footnotes*, it is found that the allowable load current of No. 12 TW copper wire is 25 amperes, whereas the allowable load current of No. 12 TW aluminum or copper-clad aluminum wire is only 20 amperes. As another example, a No. 8 TW copper wire has an ampacity of 40 amperes, whereas a No. 8 TW aluminum or copper-clad aluminum wire has an ampacity of only 30 amperes. It is important to check the *footnotes* to *Table 310-16* for maximum permitted load current ratings of No. 14, No. 12, and No. 10 AWG conductors.

Resistance is an important consideration when installing aluminum conductors. An aluminum conductor has a higher resistance compared to a copper conductor for a given wire size which, therefore, causes a greater voltage drop.

Voltage Drop (E_d) = Amperes (I) × Resistance (R)

Common Connection Problems

Some common problems associated with aluminum conductors when not properly connected may be summarized as follows:

- a corrosive action is set up when dissimilar wires come in contact with one another when moisture is present.
- the surface of aluminum oxidizes as soon as it is exposed to air. If this oxidized surface is not broken through, a poor connection results. When installing aluminum conductors, particularly in large sizes, an inhibitor is brushed onto the aluminum conductor, then the conductor is scraped with a stiff brush where the connection is to be made. The process of scraping the conductor breaks through the oxidation, and the inhibitor keeps the air from coming into contact with the conductor. Thus, further oxidation is prevented. Aluminum connectors of the compression type usually have an inhibitor paste already factory installed inside of the connector.
- aluminum wire expands and contracts to a greater degree than does copper wire for an equal load. This factor is another possible cause of a poor connection. Crimp connectors for aluminum conductors are usually longer than those for comparable copper conductors, thus resulting in greater contact surface of the conductor in the connector.

PROPER INSTALLATION PROCEDURES

Proper, trouble-free connections for aluminum conductors require terminals, lugs, and/or connectors which are suitable for the type of conductor being installed.

Terminals on receptacles and switches must be suitable for the conductors being attached. The following chart shows how the electrician can identify these terminals.

AMPACITY	MARKING ON TERMINAL OR CONNECTOR	CONDUCTOR PERMITTED
15- or 20-ampere receptacles and switches	CO/ALR	aluminum, copper, copper-clad aluminum
15- and 20-ampere receptacles and switches	NONE	copper, copper-clad aluminum
30-ampere and greater receptacles and switches	AL/CU	aluminum, copper, copper-clad aluminum
30-ampere and greater receptacles and switches	NONE	copper only
Screwless pressure terminal connectors of the push-in type	NONE	copper or copper-clad aluminum
Wire connectors	AL/CU	aluminum, copper, copper-clad aluminum
Wire connectors	NONE	copper only
Wire connectors	AL	aluminum only

Wire Connections

When splicing wires or connecting a wire to a switch, fixture, circuit breaker, panelboard, meter socket or other electrical equipment, the wires may be twisted together, soldered, then taped. Usually, however, some type of wire connector is required.

Wire connectors are known in the trade by such names as *screw terminal, pressure terminal connector, wire connector, wing nut, wire nut, Scotchlok, split-bolt connector, pressure cable connector, solderless lug, soldering lug, solder lug,* and others. Soldering-type lugs are not often used today. Solderless connectors, designed to establish connections by means of mechanical pressure, are quite common. Examples of some types of wire connectors, and their uses, are shown in figure 4-1.

As with the terminals on wiring devices (switches and receptacles) wire connectors must be marked *AL* when they are to be used with aluminum conductors. This marking is found on the connector itself, or it appears on or in the shipping carton.

Connectors marked *AL/CU* are suitable for use with aluminum, copper, or copper-clad aluminum conductors. This marking is found on the connector itself, or it appears on or in the shipping carton.

Connectors not marked *AL* or *AL/CU* are for use with copper conductors only.

Unless specially stated on or in the shipping carton, or on the connector itself, conductors made of copper, aluminum, or copper-clad aluminum may not be used in combination in the same connector. Combinations, when permitted, are usually limited to dry locations only.

CONNECTORS USED TO CONNECT WIRES TOGETHER IN COMBINATIONS OF NO. 18 AWG THROUGH NO. 6 AWG. THEY ARE TWIST-ON, SOLDERLESS, AND TAPELESS.

WIRE CONNECTORS VARIOUSLY KNOWN AS WING NUT, WIRE NUT, AND SCOTCHLOK.

CONNECTORS USED TO CONNECT WIRES TOGETHER IN COMBINATIONS OF NO. 16, NO. 14, AND NO. 12 AWG. THEY ARE CRIMPED ON WITH A SPECIAL TOOL, THEN COVERED WITH A SNAP-ON INSULATING CAP.

CRIMP-TYPE WIRE CONNECTOR AND INSULATING CAP.

SOLDERLESS CONNECTORS ARE AVAILABLE IN SIZES NO. 14 AWG THROUGH 500 MCM. THEY ARE USED FOR ONE SOLID OR ONE STRANDED CONDUCTOR ONLY, UNLESS OTHERWISE NOTED ON THE CONNECTOR OR ON ITS SHIPPING CARTON. THE SCREW MAY BE OF THE STANDARD SCREWDRIVER SLOT-TYPE, OR IT MAY BE FOR USE WITH AN ALLEN WRENCH.

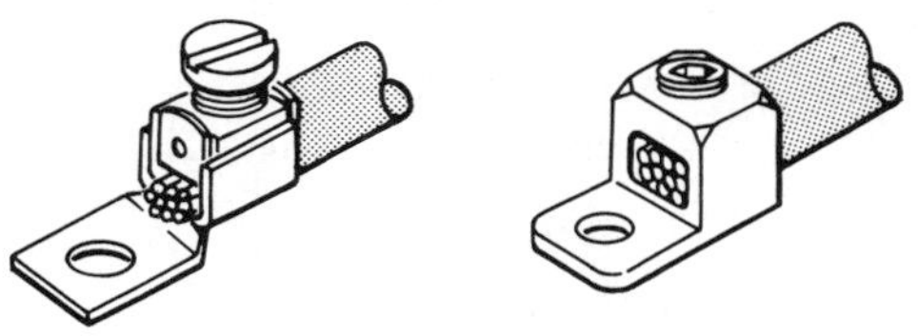

SOLDERLESS CONNECTORS

COMPRESSION CONNECTORS ARE USED FOR NO. 8 AWG THROUGH 1000 MCM. THE WIRE IS INSERTED INTO THE END OF THE CONNECTOR, THEN CRIMPED ON WITH A SPECIAL COMPRESSION TOOL.

COMPRESSION CONNECTOR

SPLIT-BOLT CONNECTORS ARE USED FOR CONNECTING TWO CONDUCTORS TOGETHER, OR FOR TAPPING ONE CONDUCTOR TO ANOTHER. THEY ARE AVAILABLE IN SIZES NO. 10 AWG THROUGH 1000 MCM. THEY ARE USED FOR TWO SOLID AND/OR TWO STRANDED CONDUCTORS ONLY, UNLESS OTHERWISE NOTED ON THE CONNECTOR OR ON ITS SHIPPING CARTON.

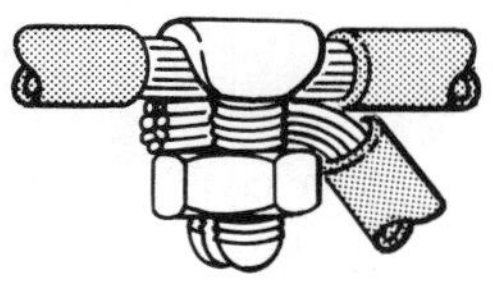

SPLIT-BOLT CONNECTOR

Fig. 4-1 Types of wire connectors.

Insulation of Wires

The Code requires that all wires used in electrical installations be insulated, *Section 310-2*. Exceptions to this rule are clearly indicated in the Code.

The insulation commonly used on wires and cables is either a rubber compound with a metallic outer braid or a thermoplastic material. *Table 310-13* lists the various conductor insulations and applications. The insulation completely surrounds the metal conductor, has a uniform thickness, and runs the entire length of the wire or cable.

The allowable ampacities of copper conductors are given in *Table 310-16* for various types of insulation. A deviation from the ampacity ratings given in *Table 310-16* is permitted for single-phase, three-wire residential services only. For conductors having RH, RHH, RHW, THW, THWN, THHN, and XHHW insulation, the ratings shown in Table 4-1 are permitted.

Table 310-16. Allowable Ampacities of Insulated Conductors Rated 0-2000 Volts, 60° to 90°C

Not More Than Three Conductors in Raceway or Cable or Earth (Directly Buried), Based on Ambient Temperature of 30°C (86°F)

Size	Temperature Rating of Conductor, See Table 310-13								Size
	60°C (140°F)	75°C (167°F)	85°C (185°F)	90°C (194°F)	60°C (140°F)	75°C (167°F)	85°C (185°F)	90°C (194°F)	
AWG MCM	TYPES †RUW, †T, †TW, †UF	TYPES †FEPW, †RH, †RHW, †RUH, †THW, †THWN, †XHHW, †USE, †ZW	TYPES V, MI	TYPES TA, TBS, SA, AVB, SIS, †FEP, †FEPB, †RHH †THHN, †XHHW*	TYPES †RUW, †T, †TW, †UF	TYPES †RH, †RHW, †RUH, †THW †THWN, †XHHW, †USE	TYPES V, MI	TYPES TA, TBS, SA, AVB, SIS, †RHH, †THHN, †XHHW*	AWG MCM
	COPPER				ALUMINUM OR COPPER-CLAD ALUMINUM				
18	....	....	....	14	....	....	....	....	....
16	....	....	18	18	....	....	....	....	....
14	20†	20†	25	25†	....	....	....	....	....
12	25†	25†	30	30†	20†	20†	25	25†	12
10	30†	35†	40	40†	25†	30†	30	35†	10
8	40	50	55	55	30	40	40	45	8
6	55	65	70	75	40	50	55	60	6
4	70	85	95	95	55	65	75	75	4
3	85	100	110	110	65	75	85	85	3
2	95	115	125	130	75	90	100	100	2
1	110	130	145	150	85	100	110	115	1
0	125	150	165	170	100	120	130	135	0
00	145	175	190	195	115	135	145	150	00
000	165	200	215	225	130	155	170	175	000
0000	195	230	250	260	150	180	195	205	0000
250	215	255	275	290	170	205	220	230	250
300	240	285	310	320	190	230	250	255	300
350	260	310	340	350	210	250	270	280	350
400	280	335	365	380	225	270	295	305	400
500	320	380	415	430	260	310	335	350	500
600	355	420	460	475	285	340	370	385	600
700	385	460	500	520	310	375	405	420	700
750	400	475	515	535	320	385	420	435	750
800	410	490	535	555	330	395	430	450	800
900	435	520	565	585	355	425	465	480	900
1000	455	545	590	615	375	445	485	500	1000
1250	495	590	640	665	405	485	525	545	1250
1500	520	625	680	705	435	520	565	585	1500
1750	545	650	705	735	455	545	595	615	1750
2000	560	665	725	750	470	560	610	630	2000
	CORRECTION FACTORS								
Ambient Temp. °C	For ambient temperatures over 30°C, multiply the ampacities shown above by the appropriate correction factor to determine the maximum allowable load current.								Ambient Temp. °F
31-40	.82	.88	.90	.91	.82	.88	.90	.91	86-104
41-45	.71	.82	.85	.87	.71	.82	.85	.87	105-113
46-50	.58	.75	.80	.82	.58	.75	.80	.82	114-122
51-60	....	.58	.67	.71	....	.58	.67	.71	123-141
61-70	....	.35	.52	.58	....	.35	.52	.58	142-158
71-80	....	....	.30	.41	....	....	.30	.41	159-176

† The load current rating and the overcurrent protection for conductor types marked with an obelisk (†) shall not exceed 15 amperes for 14 AWG, 20 amperes for 12 AWG, and 30 amperes for 10 AWG copper; or 15 amperes for 12 AWG and 25 amperes for 10 AWG aluminum and copper-clad aluminum.

* For dry locations only. See 75°C column for wet locations.

COPPER CONDUCTOR (AWG) FOR INSULATION OF RH-RHH-RHW-THW-THWN-THHN-XHHW	ALUMINUM OR COPPER-CLAD ALUMINUM CONDUCTOR (AWG)	SERVICE AMPACITY RATING
4	2	100
3	1	110
2	1/0	125
1	2/0	150
1/0	3/0	175
2/0	4/0	200

Table 4-1.

The insulation covering wires and cables used in house wiring is usually rated at 600 volts or less. Exceptions to this statement are low-voltage bell wiring and fixture wiring. Conductors are also rated as to the temperatures they can withstand. (Refer to *Table 310-16.*) The standard type T or TW insulation is rated at 60° Celsius (C) or 140° Fahrenheit (F) and is suitable for the temperatures normally found in most residential installations. However, there are instances where an insulation rated at a higher temperature may be necessary, such as in the case of certain types of recessed fixtures.

If the temperature is expected to exceed 60°C, it may be necessary to install cable having insulation rated to handle the higher temperatures, such as types A, AF, SF, RH, or other similar insulation. Refer to *Table 310-13.*

It is recommended that the reader study NEC *Article 310, Conductors for General Wiring.*

NONMETALLIC-SHEATHED CABLE (*ARTICLE 336*)

Description

Nonmetallic-sheathed cable is defined as a factory assembly of two or more insulated conductors having an outer sheath of moisture-resistant, flame-retardant, nonmetallic material. This cable is avail-with two or three current-carrying conductors. The conductors range in size from No. 14 through No. 2 for copper conductors, and from No. 12 through No. 2 for aluminum or copper-clad aluminum conductors. Two-wire cable contains a black conductor and a white conductor. Three-wire cable contains black, white, and red conductors.

The current-carrying conductors have type T or type TW insulation. Nonmetallic-sheathed cable is also available with an uninsulated copper conductor which is used for grounding purposes only, figure 4-2. This grounding conductor is not intended for use as a current-carrying circuit wire.

Equipment grounding requirements are specified in *Sections 250-42, 250-43, 250-44,* and *250-45*. It is for the reasons stated in these sections that all boxes and fixtures in the residence are to be grounded.

Table 250-95 of the Code lists the sizes of the grounding conductors used in cable assemblies.

Fig. 4-2 Nonmetallic-sheathed cable with an uninsulated copper conductor.

Table 250-95. Minimum Size Equipment Grounding Conductors for Grounding Raceway and Equipment

Rating or Setting of Automatic Overcurrent Device in Circuit Ahead of Equipment, Conduit, etc., Not Exceeding (Amperes)	Size	
	Copper Wire No.	Aluminum or Copper-Clad Aluminum Wire No.*
15	14	12
20	12	10
30	10	8

* See installation restrictions in Section 250-92(a).

Note that the copper grounding conductor shown in the portion of *Table 250-95* is the same size as the circuit conductors for 15-, 20-, and 30-ampere ratings.

USE OF NONMETALLIC-SHEATHED CABLE (*ARTICLE 336*)

Underwriters' Laboratories, Inc. lists two classifications of nonmetallic-sheathed cable: type NM and type NMC. Both types of cable may be used on circuits of 600 volts or less. In addition, both types of cable have a flame-retardant and moisture-resistant outer covering. Table 4-2 shows the uses permitted for type NM and type NMC cable.

Installation

Nonmetallic-sheathed cable is the least expensive of the various wiring methods. It is relatively light in weight and easy to install. It is widely used for dwelling unit installations on circuits of 600 volts or less. The installation of both types of nonmetallic-sheathed cable must conform to the following requirements of NEC *Article 336*. Refer to figure 4-3.

- The cable must be strapped or stapled not more than 12 inches (305 mm) from a box or fitting.
- The intervals between straps or staples must not exceed 4 1/2 feet (1.37 m).
- The cable must be protected against physical damage where necessary.
- The cable must not be bent to a radius less than five times its diameter.
- The cable must not be used in circuits of more than 600 volts, *Section 310-13.*

Additional requirements for the installation of nonmetallic-sheathed cable are given in *Articles 200, 210, 220, 240, 250, 300,* and *310.*

ARMORED CABLE (*ARTICLE 333*)

Description

The National Electrical Code describes type AC armored cable. Underwriters' Laboratories lists types AC and ACT armored cable (which are the same as the cable described by the Code). The terms are used interchangeably by electricians. Many electricians call this type of cable *BX.*

Section 333-1 describes armored cable as an assembly of insulated conductors in a flexible metallic enclosure, figure 4-4. Armored cable is designated either as type AC or type ACT depending upon the conductor insulation. Type AC cable contains conductors insulated with Code-grade rubber covered with a flame-retardant and moisture-

FOR TYPICAL RESIDENTIAL WIRING TYPE NM AND NMC CABLE	TYPE NM	TYPE NMC
• May be used on circuits of 600 volts or less	Yes	Yes
• Has flame-retardant and moisture-resistant outer covering	Yes	Yes
• Has fungus-resistant and corrosion-resistant outer covering	No	Yes
• May be used to wire one- and two-family dwellings, or multifamily dwellings that do not exceed three floors above grade	Yes	Yes
• May be installed exposed or concealed in damp location	Yes	Yes
• May be embedded in masonry, concrete, etc.	No	No
• May be exposed to corrosive fumes	No	Yes
• May be installed in dry, hollow voids in masonry blocks and similar locations	Yes	Yes
• May be installed in moist, damp, hollow voids in masonry blocks and similar locations	No	Yes
• May be used as service-entrance cable	No	No
• Must be protected against damage	Yes	Yes
• May be run in shallow chase in masonry and covered with plaster	No	Yes
• May be applied at not over 60°C	Yes	Yes

Table 4-2 Uses permitted in typical residential wiring for type NM and type NMC nonmetallic-sheathed cable.

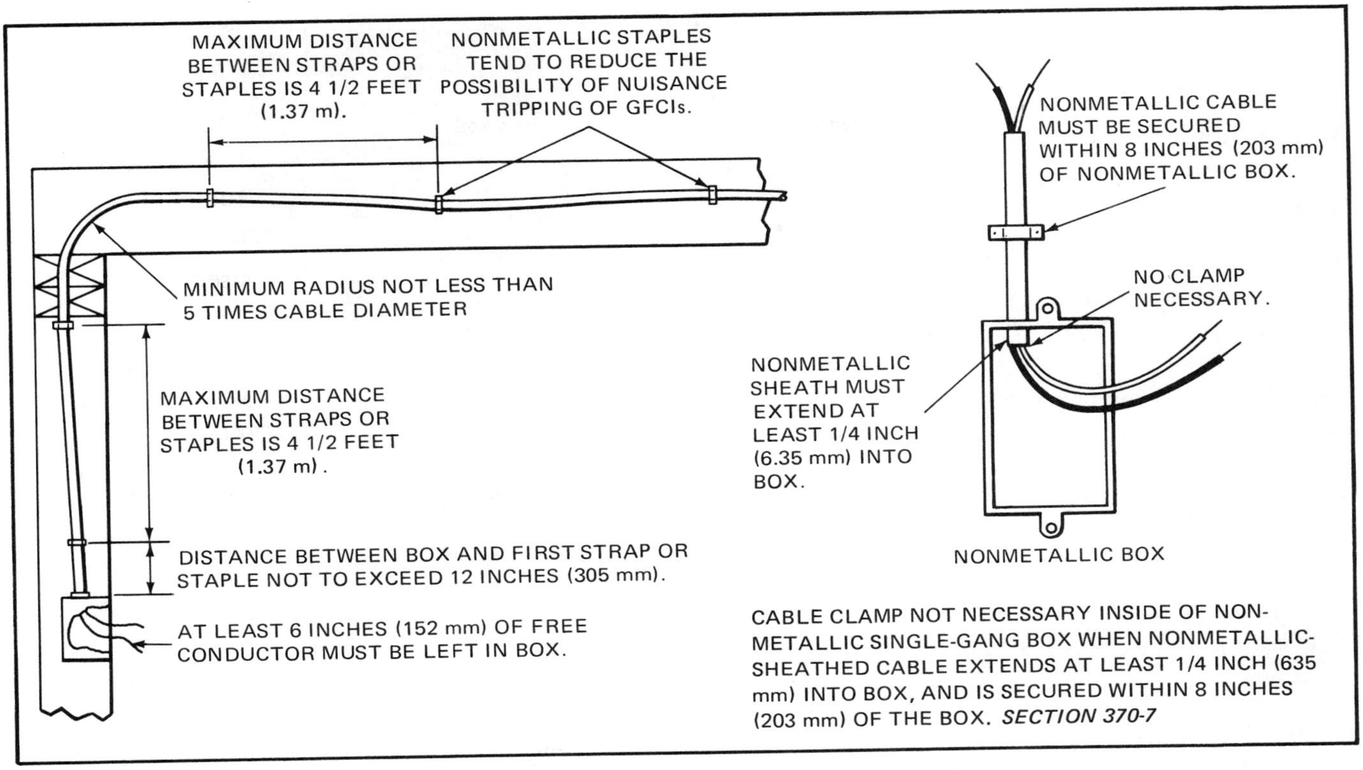
MAXIMUM DISTANCE BETWEEN STRAPS OR STAPLES IS 4 1/2 FEET (1.37 m).
NONMETALLIC STAPLES TEND TO REDUCE THE POSSIBILITY OF NUISANCE TRIPPING OF GFCIs.
MINIMUM RADIUS NOT LESS THAN 5 TIMES CABLE DIAMETER
MAXIMUM DISTANCE BETWEEN STRAPS OR STAPLES IS 4 1/2 FEET (1.37 m).
DISTANCE BETWEEN BOX AND FIRST STRAP OR STAPLE NOT TO EXCEED 12 INCHES (305 mm).
AT LEAST 6 INCHES (152 mm) OF FREE CONDUCTOR MUST BE LEFT IN BOX.
NONMETALLIC CABLE MUST BE SECURED WITHIN 8 INCHES (203 mm) OF NONMETALLIC BOX.
NO CLAMP NECESSARY.
NONMETALLIC SHEATH MUST EXTEND AT LEAST 1/4 INCH (6.35 mm) INTO BOX.
NONMETALLIC BOX
CABLE CLAMP NOT NECESSARY INSIDE OF NON-METALLIC SINGLE-GANG BOX WHEN NONMETALLIC-SHEATHED CABLE EXTENDS AT LEAST 1/4 INCH (635 mm) INTO BOX, AND IS SECURED WITHIN 8 INCHES (203 mm) OF THE BOX. *SECTION 370-7*

Fig. 4-3 Installation of nonmetallic cable and armored cable.

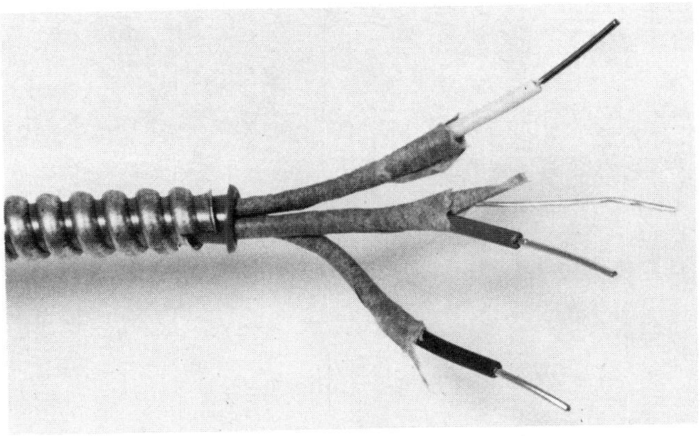

Fig. 4-4 Flexible armored cable.

resistant finish. Most armored cable is manufactured with type T or type TW insulation and is called type ACT cable.

Armored cable is generally available with two or three conductors in sizes from No. 14 AWG to No. 1 AWG inclusive.

- Two-wire armored cable contains one black and one white conductor.
- three-wire armored cable contains one each of black, white, and red conductors.

Armored cable (except ACL) must have an internal bonding strip of copper or aluminum in close contact with the armor for its entire length. This type of cable is used in a grounded system because of the metal bonding strip and the flexible steel armor. The armor also provides mechanical protection to the conductors.

Use and Installation

Armored cable can be used in more applications than nonmetallic-sheathed cable. Certain conditions govern the use of types AC and ACT armored cable in dwelling unit installations. AC and ACT armored cable

- may be used on circuits and feeders for applications of 600 volts or less.
- may be used for open and concealed work in dry locations.
- may be run through walls and partitions.
- may be embedded in the plaster finish on masonry walls or run through the hollow spaces of such walls if these locations are not considered damp or wet (see *Locations, Article 100*).

- must be secured within 12 inches (305 mm) from every outlet box or fitting and at intervals not exceeding 4 1/2 feet (1.37 m).
- must not be bent to a radius of less than five times the diameter of the cable (see figure 4-3).
- must have an approved fiber insulating bushing (antishort) at the cable ends to protect the conductor insulation.

Armored cable is *not* approved for use in the following situations:

- underground installations.
- burying in masonry, concrete, or fill of building during construction.
- installation in any location which is exposed to weather.
- installation in any location exposed to oil, gasoline, or other materials which have a destructive effect on the insulation.

Nonmetallic-sheathed cable and armored cable each has advantages which make it suitable for particular types of installations. However, the type of cable to be used in a specific situation depends largely on the wiring method permitted or required by the local building code.

The specifications at the end of this text require that all boxes and fixtures be grounded. Therefore, if nonmetallic-sheathed cable is to be installed in this residence, it must have a grounding conductor.

The Code, in *Section 370-20(d)*, requires that all metal boxes for use with nonmetallic cable must have provisions for the attachment of the grounding conductor. Figure 4-5 shows a gang-type switch box that is tapped for a screw by which the grounding conductor may be connected underneath. Figure 4-6 shows an outlet box, also with provisions for attaching grounding conductors. Figure 4-7 illustrates the use of a small grounding clip.

Section 410-20 of the Code requires that there be a provision whereby the equipment grounding conductor can be attached to the exposed metal parts of lighting fixtures.

Fig. 4-5 Gang-type switch box.

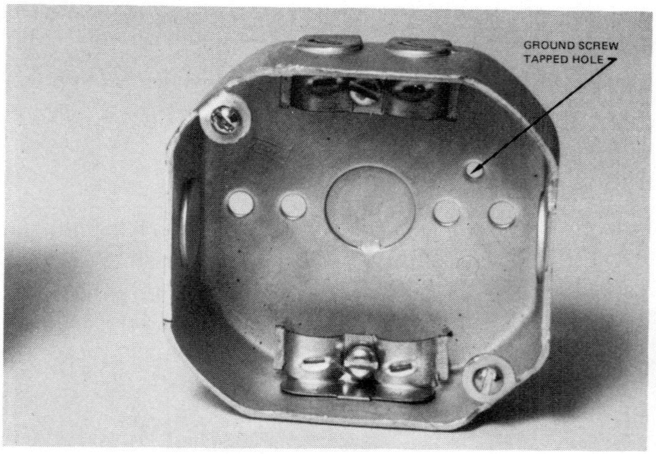

Fig. 4-6 Outlet box.

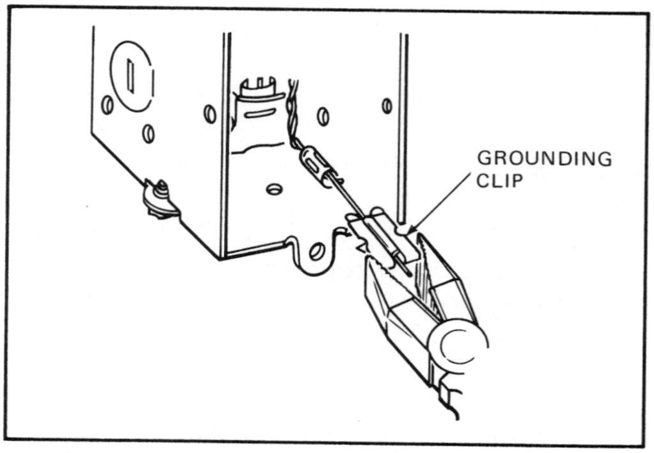

Fig. 4-7 Method of attaching grounding clip to switch box.

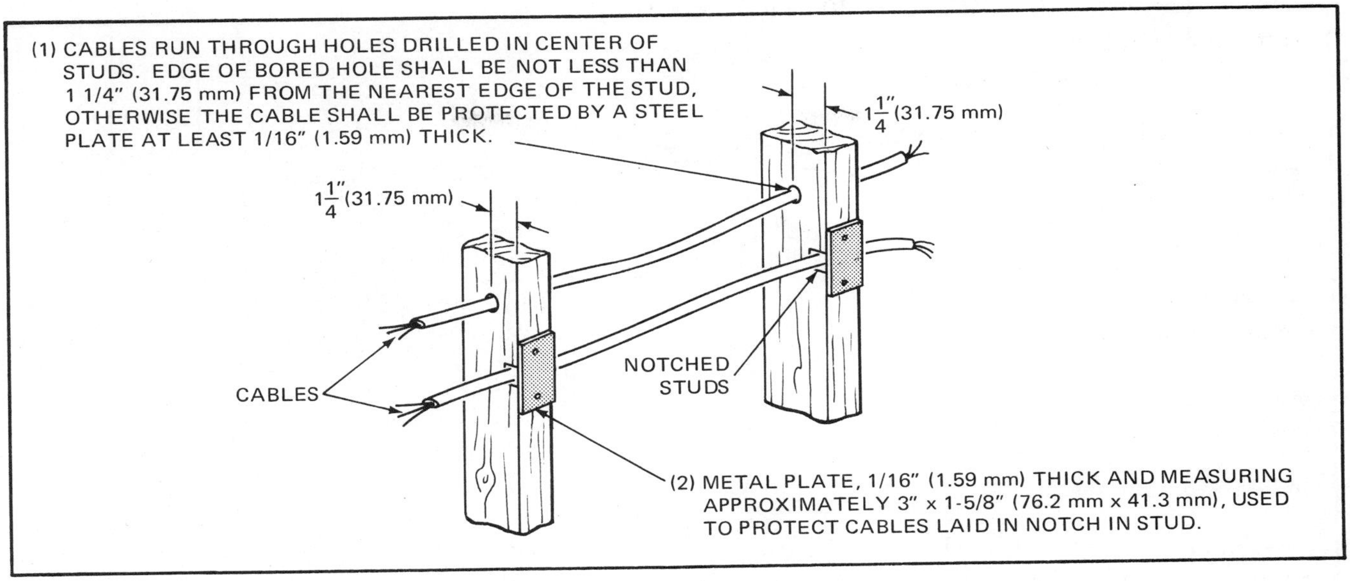

(1) CABLES RUN THROUGH HOLES DRILLED IN CENTER OF STUDS. EDGE OF BORED HOLE SHALL BE NOT LESS THAN 1 1/4" (31.75 mm) FROM THE NEAREST EDGE OF THE STUD, OTHERWISE THE CABLE SHALL BE PROTECTED BY A STEEL PLATE AT LEAST 1/16" (1.59 mm) THICK.
1 1/4" (31.75 mm)
1 1/4" (31.75 mm)
CABLES
NOTCHED STUDS
(2) METAL PLATE, 1/16" (1.59 mm) THICK AND MEASURING APPROXIMATELY 3" x 1-5/8" (76.2 mm x 41.3 mm), USED TO PROTECT CABLES LAID IN NOTCH IN STUD.

Fig. 4-8 Methods of protecting cable, ***Section 300-4.***

INSTALLING CABLES THROUGH BUILDING MEMBERS (*SECTION 300-4*)

To complete the wiring of a residence, cables must be run through studs, joists, and rafters. One method of installing cables in these locations is to run the cables through holes drilled at the approximate centers of wood building members, or at least 1 1/4 inches (31.75 mm) from the nearest edge where practical, figure 4-8. Holes bored through the center of standard 2 x 4s meet the requirements of *Section 300-4*.

Another installation method is to lay the cable in notches cut in the studs or joists. A metal plate at least 1/16 inch (1.59 mm) thick must cover the notch to protect the cable from nails (*Section 300-4*).

See unit 12 for the methods of cable installation and protection in attic areas. *Sections 333-10* and *336-7* of the Code refer to cables run through studs, joists, and rafters. These sections, in turn, refer to *Section 300-4* for more detailed information.

INSTALLATION OF CABLES THROUGH DUCTS

Section 300-22 of the Code is extremely strict as to what types of wiring methods are permitted for installation of cables through ducts or plenum chambers. These stringent rules are for fire safety.

The Code requirements have been relaxed slightly, however, to permit types NM and NMC cable to be installed in joist and stud spaces in dwellings, *Section 300-22, Exception No. 4*. The exception precludes joist or stud spaces in dwelling units when wiring or equipment passes through such spaces perpendicular to the long dimensions of such spaces, as illustrated in figure 4-9.

CONNECTORS FOR INSTALLING NONMETALIC-SHEATHED AND ARMORED CABLE

The connectors shown in figure 4-10 are used to fasten nonmetallic-sheathed cable and armored cable to the boxes and cabinets in which they terminate. These connectors clamp the cable securely to each outlet box. Many boxes have built-in clamps and do not require separate connectors.

ELECTRICAL METALLIC TUBING (*ARTICLE 348*), INTERMEDIATE METAL CONDUIT (*ARTICLE 345*), AND RIGID METAL CONDUIT (*ARTICLE 346*)

Many communities do not permit the installation of cable of any type in residential buildings. These communities require the installation of a raceway system of wiring, such as electrical metallic tubing (thinwall), intermediate metal conduit, or rigid metal conduit.

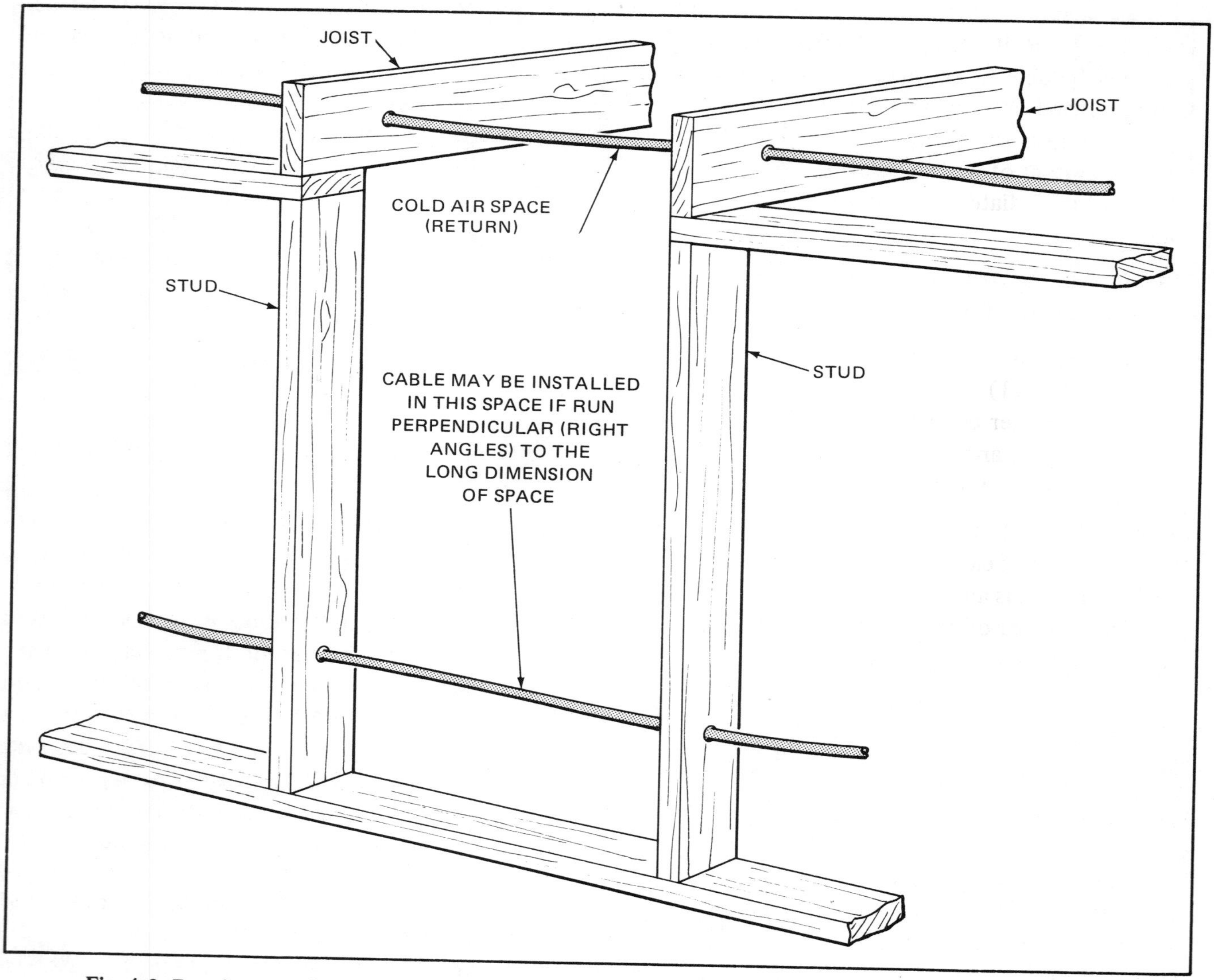

Fig. 4-9 Requirements for cable installation through joist or stud spaces, *Section 300-22, Exception No. 4.*

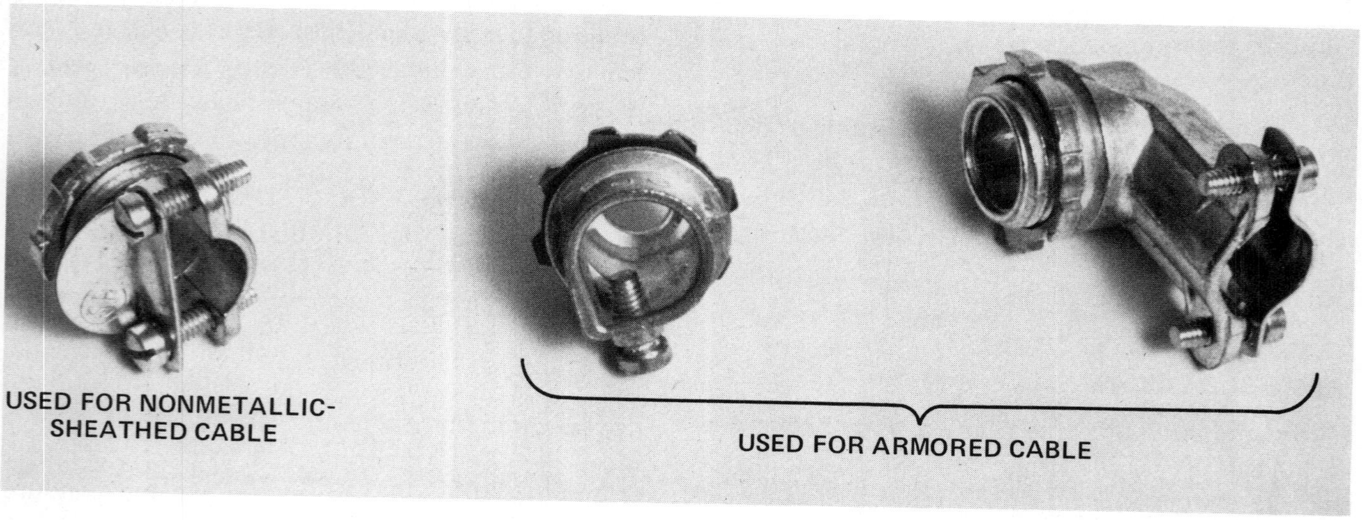

Fig. 4-10 Cable connectors.

Sufficient data is given in this text for both cable and conduit wiring methods. Thus, the student will be able to complete the type of installation permitted or required by local codes.

According to installation requirements for a raceway system of wiring, electrical metallic tubing (EMT), intermediate metal conduit, and rigid metal conduit

- may be buried in concrete or masonry and may be used for open or concealed work.
- may *not* be installed in cinder, concrete, or fill unless (1) protected on all sides by a layer of noncinder concrete at least 2 inches (50.8 mm) thick, and (2) the conduit is at least 18 inches (457 mm) below the fill.

In general, conduit must be supported within 3 feet (914 mm) of each box or fitting and at 10-foot (3.05 m) intervals along runs.

The number of conductors permitted in EMT, intermediate metal conduit, and rigid conduit is given in *Table 1, Chapter 9* of the Code. Heavy-wall rigid conduit provides greater protection against mechanical injury to conductors than does either EMT or intermediate metal conduit which have much thinner walls (*Articles 345, 346,* and *348*).

The residence plans show that rigid conduit is to be used for the service entrance and for 1 1/4-inch conduit run from Main Panel A to Load Center B in the utility room. *Section 373-6(c)* requires that insulating bushings or equivalent be used where No. 4 or larger conductors enter a raceway. In addition to the requirements of *Section 373-6(c)*, a bonding-type bushing must be used where the service-entrance conduit enters Main Panel A. (See *Sections 250-72* and *250-73* for the method of bonding service equipment.)

The plans also indicate that electrical metallic tubing is to be used in the workshop, storage room, and utility room. EMT is also to be used for the terrace convenience outlets and the bracket light, and for the run between the rear garage door and the three-way switch, convenience outlet, and post light. Since the recreation room, basement lavatory, and basement stairwell have finished ceilings and walls, they will be wired with cable.

Many of the National Electrical Code rules are applicable to both cable and conduit installations. Some Code requirements are applicable only to cable installations. For example, there are special requirements for the attachment of the grounding conductors and protection from physical damage must be provided. Many of the Code requirements governing the installation of cables do not apply to a raceway system. Properly installed electrical metallic tubing provides good continuity for equipment grounding, a means of withdrawing or pulling in additional conductors, and reasonable protection against physical damage to the conductors.

FLEXIBLE CONNECTIONS (*ARTICLES 350* AND *351*)

The installation of certain equipment requires flexible connections, both to simplify the installation and to stop the transfer of vibrations. In residential wiring, flexible connections are used to wire attic fans, food waste disposers, dishwashers, air conditioners, recessed fixtures, and similar equipment.

The three types of flexible conduit used for these connections are: flexible metal conduit, figure 4-11(A); flexible liquidtight metal conduit, figure 4-11(B); and flexible liquidtight nonmetallic conduit, figure 4-11(C). Figure 4-12 shows many of the types of connectors used with flexible metal conduit, flexible liquidtight metal conduit, and flexible liquidtight nonmetallic conduit.

Article 350 of the Code covers the use and installation of flexible metal conduit. This type of conduit is similar to armored cable, except that the conductors are installed by the electrician. For armored cable, the cable armor is wrapped around the conductors at the factory to form a complete cable assembly.

Some of the more common installations using flexible metal conduit are shown in figure 4-13. Note that the flexibility required to make the installation is provided by the flexible metal conduit. The figure calls attention to the National Electrical Code and Underwriters' Laboratories restrictions on the use of flexible metal conduit with regard to relying on the metal armor as a grounding means.

The use and installation of liquidtight flexible metal conduit are described in *Article 351* of the

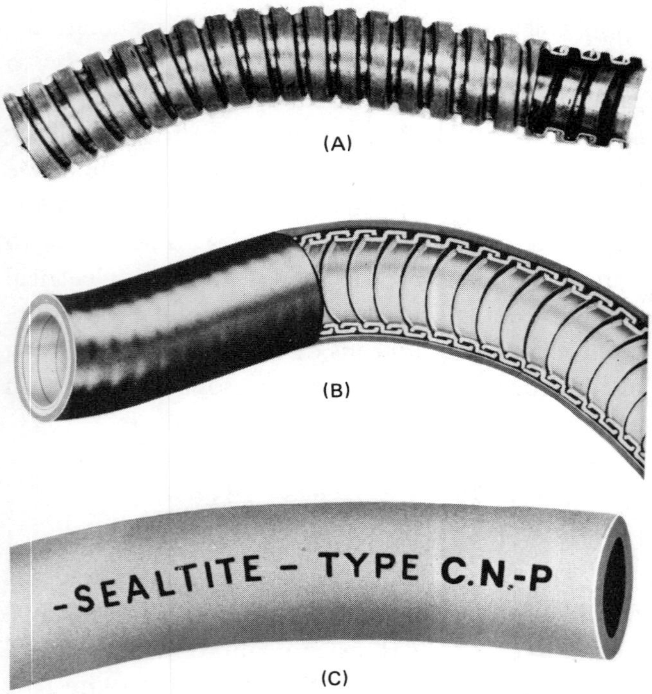

Fig. 4-11 (A) Flexible metal conduit. (B) Flexible liquidtight metal conduit. (C) Flexible liquidtight nonmetallic conduit.

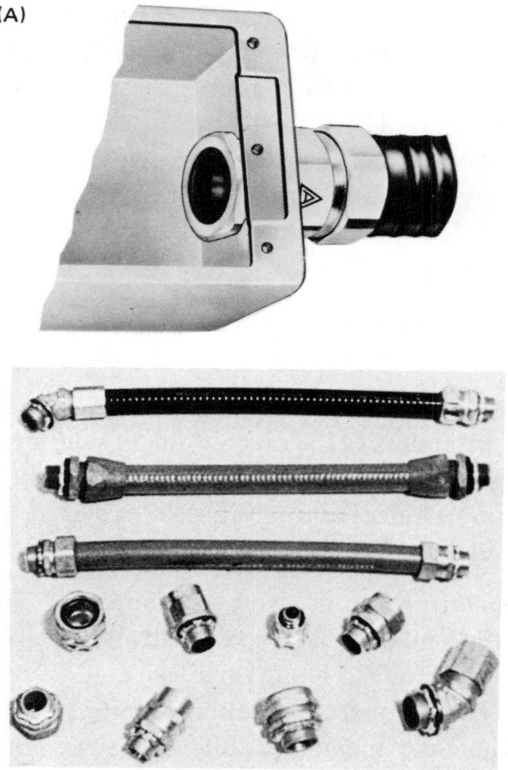

Fig. 4-12 (A) Liquidtight fitting. (B) Various types of connectors.

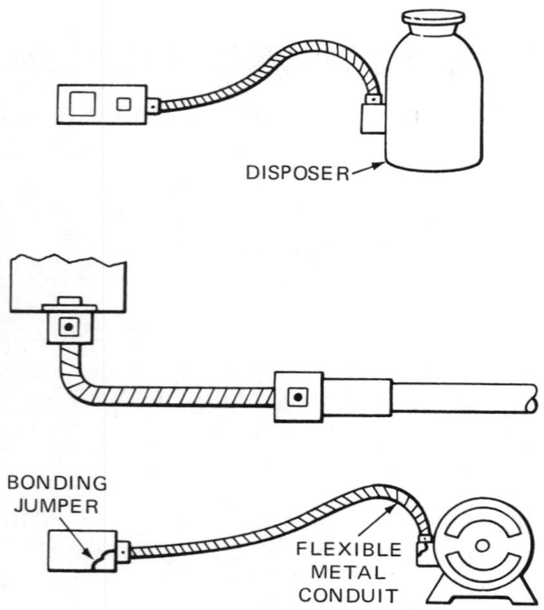

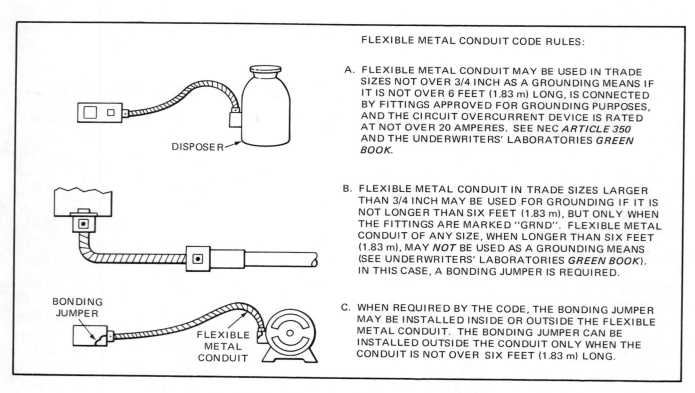

FLEXIBLE METAL CONDUIT CODE RULES:

A. FLEXIBLE METAL CONDUIT MAY BE USED IN TRADE SIZES NOT OVER 3/4 INCH AS A GROUNDING MEANS IF IT IS NOT OVER 6 FEET (1.83 m) LONG, IS CONNECTED BY FITTINGS APPROVED FOR GROUNDING PURPOSES, AND THE CIRCUIT OVERCURRENT DEVICE IS RATED AT NOT OVER 20 AMPERES. SEE NEC *ARTICLE 350* AND THE UNDERWRITERS' LABORATORIES *GREEN BOOK.*

B. FLEXIBLE METAL CONDUIT IN TRADE SIZES LARGER THAN 3/4 INCH MAY BE USED FOR GROUNDING IF IT IS NOT LONGER THAN SIX FEET (1.83 m), BUT ONLY WHEN THE FITTINGS ARE MARKED "GRND". FLEXIBLE METAL CONDUIT OF ANY SIZE, WHEN LONGER THAN SIX FEET (1.83 m), MAY *NOT* BE USED AS A GROUNDING MEANS (SEE UNDERWRITERS' LABORATORIES *GREEN BOOK*). IN THIS CASE, A BONDING JUMPER IS REQUIRED.

C. WHEN REQUIRED BY THE CODE, THE BONDING JUMPER MAY BE INSTALLED INSIDE OR OUTSIDE THE FLEXIBLE METAL CONDUIT. THE BONDING JUMPER CAN BE INSTALLED OUTSIDE THE CONDUIT ONLY WHEN THE CONDUIT IS NOT OVER SIX FEET (1.83 m) LONG.

Fig. 4-13 Installations using flexible metal conduit.

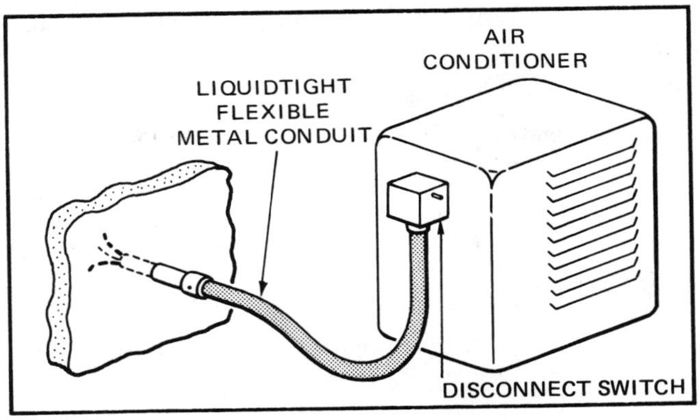

Fig. 4-14 Use of liquidtight flexible metal conduit.

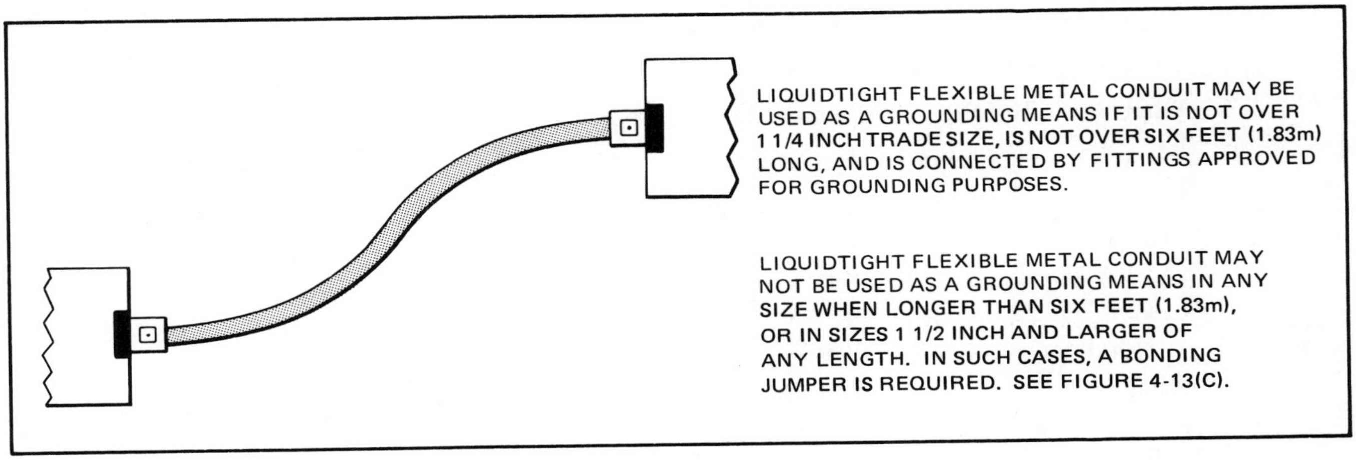

Fig. 4-15 Limitations on the use of liquidtight flexible metal conduit.

Code. Liquidtight flexible metal conduit has a "tighter" fit of its spiral turns as compared to standard flexible metal conduit. Liquidtight conduit also has a thermoplastic outer jacket that is liquidtight. Liquidtight flexible metal conduit is commonly used as the flexible connection to a central air-conditioning unit located outdoors, figure 4-14.

Figure 4-15 shows the limitations placed on the use of liquidtight flexible metal conduit as a grounding means. These limitations are given in the Underwriters' Laboratories Standards.

GROUNDING CONDUCTOR

The grounding conductor specified in the plans is a single uninsulated No. 4 wire encased in a flexible steel sheath, figure 4-16. The Code specifies that a No. 4 or larger grounding conductor may be attached to the surface on which it is carried without the use of knobs, tubes, or insulators. Protection is not required unless the ground wire is expected to be exposed to severe physical damage. If it is to be free from exposure to physical damage, No. 6 grounding conductor may be run along the surface of the building construction. Otherwise, the conductor must be enclosed in rigid conduit, electrical metallic tubing, or cable armor. Grounding conductors smaller than No. 6 must be enclosed in rigid conduit, electrical metallic tubing, rigid nonmetallic conduit, or cable armor, *Section 250-92*.

Fig. 4-16 Armored ground wire.

The plans show that the ground wire runs from the top of Main Panel A, is concealed in the basement lavatory and recreation room ceilings, and then is exposed on the utility room ceiling and pump room. The ground wire is to be connected in the pump room to the water pipe leading to the well. (This cable is out of sight in the utility room because it is to be stapled along the sides of the joists which run in the same direction as the planned cable run.)

Service-entrance grounding is discussed in unit 27.

REVIEW

Note: Refer to the Code or the plans where necessary.

1. What is the largest size of solid building wire cable that is generally used rather than the stranded variety? ____________
2. What is the minimum wire size that may be installed in any dwelling? ____________
3. What exceptions, if any, are there to the answer given for question 2? ____________
4. What determines the ampacity of a wire? ____________
5. What unit of measurement is used for the diameter of wires? ____________
6. What unit of measurement is used for the cross-sectional area of wires? ____________
7. What is the maximum voltage rating of all building wire and cable? ____________
8. Indicate the ampacity of the following Type TW (copper) conductors. (Refer to *Table 310-169*)
   a. 14 AWG ________ amperes
   b. 12 AWG ________ amperes
   c. 10 AWG ________ amperes
   d. 8 AWG ________ amperes
   e. 6 AWG ________ amperes
   f. 4 AWG ________ amperes
9. What is the maximum operating temperature of the following conductors? Give the answer in degrees Fahrenheit and degrees Celsius.
   a. Type A ____________
   b. Type RH ____________
   c. Type RHH ____________
   d. Type TW ____________
10. What are the colors of the conductors in nonmetallic-sheathed cable for
    a. two-wire cable: ____________ , ____________
    b. three-wire cable: ____________ , ____________ , ____________
11. For nonmetallic-sheathed (type NM) cable, can the uninsulated conductor be used for purposes other than grounding? ____________
12. What size ground wire is used with the following sizes of nonmetallic-sheathed cable?
    a. 14 AWG ____________
    b. 12 AWG ____________
    c. 10 AWG ____________
    d. 8 AWG ____________
13. Under what condition may nonmetallic-sheathed cable (type NM) be fished in the hollow voids of masonry block walls? ____________
14. a. What is the maximum distance permitted between straps on a cable installation?
    ____
    b. What is the maximum distance permitted between a box and the first strap?
    ____
15. What is the difference between type AC and type ACT cable? ____________
16. Type ACT cable may be bent to a radius of not less than ________ times the diameter of the cable.

17. When armored cable is used, what protection is provided at the cable ends?

____

18. What protection must be provided when installing a cable in a notched stud or joist?

____

____

19. For installing directly in a concrete slab, (armored cable, nonmetallic-sheathed cable, conduit) may be used. (Underline the correct method of installation.)

20. Describe the Code requirements for the mechanical protection of grounding conductors.

____

____

____

21. The edge of a bored hole in a stud shall be not less than ________ inches from the edge of the stud.

22. A 1 1/4-inch conduit will be run between two electric panels as shown in the residence plans. Where are these panels located?______________________

23. a. Is nonmetallic-sheathed cable permitted in your area for residential occupancies?

____

b. From what source is this information obtained?______________________

24. Is it permitted to use flexible metal conduit over 6 feet (1.83 m) in length as a grounding means? (Yes) (No)

25. Liquidtight flexible metal conduit may serve as a grounding means in sizes up to and including ________ inches where used with approved fittings.

26. The allowable current-carrying capacity (ampacity) of aluminum wire, or the maximum permitted load current in the case of No. 14, No. 12, and No. 10 AWG conductors, is less than that of copper wire for a given size, insulation, and temperature of 86°F. Refer to *Table 310-16* and *footnotes* and complete the following table:

WIRE	COPPER AMPACITY	ALUMINUM AMPACITY
No. 12 TW*		
No. 10 TW*		
No. 3 THW		
0000 THWN		
500 MCM THWN		

*Enter both ampacity and permitted load current values.

27. It is permissible for an electrician to connect aluminum, copper, or copper-clad aluminum conductors together in the same connector. True ________ False ________

28. Terminals of switches and receptacles marked *CO/ALR* are suitable for use with ____________, ____________, and ____________ conductors.

29. Wire connectors marked *AL/CU* are suitable for use with ____________, ____________, and ____________ conductors.

30. A wire connector bearing no marking or reference to *AL, CU,* or *ALR* is suitable for use with (copper) (aluminum) conductors only. Underline the correct answer.

unit 5

Switch Control of Lighting Circuits

OBJECTIVES

After studying this unit, the student will be able to

- identify the grounded and ungrounded conductors in cable or conduit.
- identify the various types of toggle switches for lighting circuit control.
- select a switch with the proper rating for the specific installation conditions.
- describe the operation that each type of toggle switch performs in typical lighting circuit installations.
- demonstrate the correct wiring connections for each type of switch per Code requirements.

The electrician installs and connects various types of lighting switches. To do this, both the operation and the method of connection of each type of switch must be known. In addition, the electrician must understand the meanings of the current and voltage ratings marked on lighting switches, as well as the National Electrical Code requirements for the installation of these switches.

CONDUCTOR IDENTIFICATION

Before making any wiring connections to devices, the electrician must be familiar with the ways in which conductors are identified. For alternating-current circuits, NEC *Articles 200* and *210* require that the grounded (identified) circuit conductor have an outer covering that is either white or a natural gray. In multiwire circuits, the grounded circuit conductor is also called a *neutral* conductor.

The ungrounded (unidentified) conductor of a circuit must be marked in a color other than green, white or gray. This conductor generally is called the *hot* conductor. A shock is felt if this conductor and the grounded conductor are held at the same time, or if this conductor and a grounded surface such as a water pipe are touched at the same time.

Section 210-5 of the Code specifies the color coding for the branch-circuit grounded conductors. The grounded circuit conductor must be identified using a continuous white or natural gray color. The hot ungrounded conductor may be any color except green, white, or gray.

Figure 5-1 shows the color coding used for cable wiring. In conduit wiring, the hot conductors may be any color other than white, gray, or green. Green is reserved for the grounding conductor only.

TOGGLE SWITCH (*ARTICLE 380*)

The most frequently used switch in lighting circuits is the toggle flush switch or snap switch, figure 5-2. When mounted in a flush switch box, the switch is concealed in the wall with only the insulated handle or toggle protruding.

Toggle Switch Ratings

Underwriters' Laboratories classifies toggle switches used for lighting circuits as *General-Use*

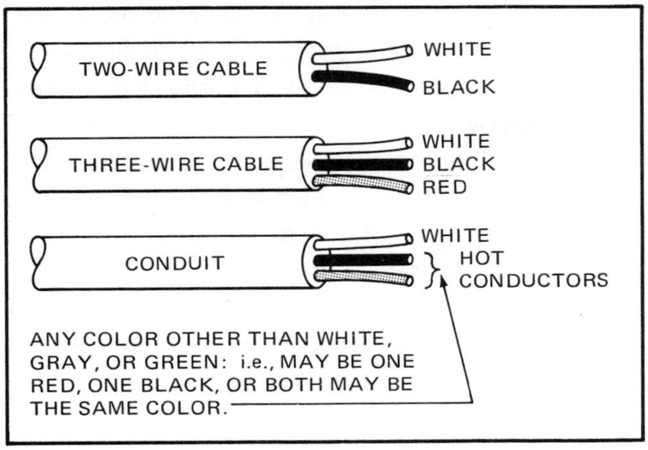

Fig. 5-1.

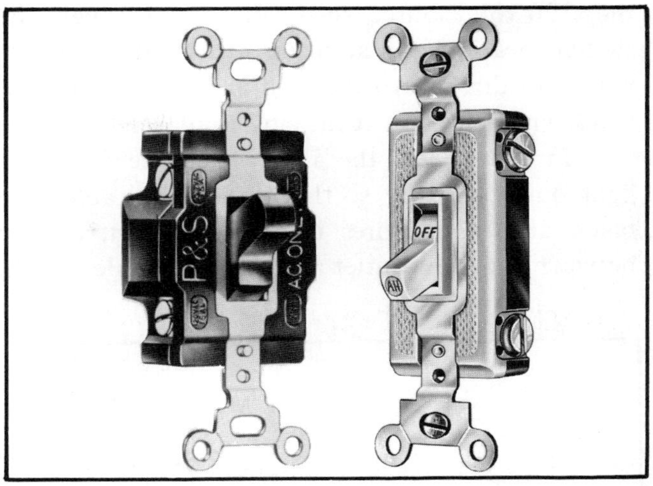

Fig. 5-2 Toggle flush switches.

Snap Switches. These switches are divided into two categories.

Category 1 contains those ac/dc general-use snap switches which are used to control

- alternating-current or direct-current circuits.
- resistive loads not to exceed the ampere rating of the switch at rated voltage.
- inductive loads not to exceed one-half the ampere rating of the switch at rated voltage.
- tungsten filament lamp loads not to exceed the ampere rating of the switch at 125 volts when marked with the letter "T".

(A tungsten filament lamp draws a very high current at the instant the circuit is closed. As a result, the switch is subjected to a severe current surge.)

The ac/dc general-use snap switch normally is not marked ac/dc. However, it is always marked with the current and voltage rating, such as 10A-125V, or 5A-250V-T.

Category 2 contains those ac general-use snap switches which are used to control

- alternating-current circuits only.
- resistive, inductive, and tungsten filament lamp loads not to exceed the ampere rating of the switch at 120 volts.
- motor loads not to exceed 80 percent of the ampere rating of the switch at rated voltage, but not exceeding two horsepower.

Ac general-use snap switches may be marked ac only, or they may also be marked with the current and voltage ratings. A typical switch marking is 15A, 120-277V ac. The 277-volt rating is required on 277/480-volt systems.

Terminals of switches rated at 20 amperes or less, when marked *CO/ALR* are suitable for use with aluminum, copper, and copper-clad aluminum conductors. Switches not marked *CO/ALR* are suitable for use with copper and copper-clad conductors only.

Screwless pressure terminals of the conductor push-in type may be used with copper and copper-clad aluminum conductors only. These push-in type terminals are not suitable for use with ordinary aluminum conductors.

Further information on switch ratings is given in NEC *Section 380-14* and in the underwriter's Laboratories *Electrical Construction Materials List.*

TOGGLE SWITCH TYPES

Toggle switches are available in four types: single-pole switch, three-way switch, four-way switch, and double-pole switch.

Single-pole Switch

A single-pole switch is used when a light or group of lights or other load is to be controlled from one switching point, figure 5-3. The switch is identified by its two terminals and the toggle which is marked ON/OFF. The single-pole switch is connected in series with the ungrounded or hot wire feeding the load.

Figure 5-3 shows a single-pole switch controlling a light from one switching point. Note that the 120-volt source feeds directly through the switch location. In addition, the identified white wire goes directly to the load and the unidentified black wire is broken at the single-pole switch.

In figure 5-4, the 120-volt source feeds the light outlet directly so that a two-wire cable with black and white wires is used as a switch loop between the light outlet and the single-pole switch. The use of a white wire in a single-pole switch loop is covered in *Section 200-7.* The unidentified or black conductor must connect between the switch and the load as in figure 5-4.

Figure 5-5 shows another application of a single-pole switch. The feed is at the switch which controls the light outlet. The convenience outlet is independent of the switch.

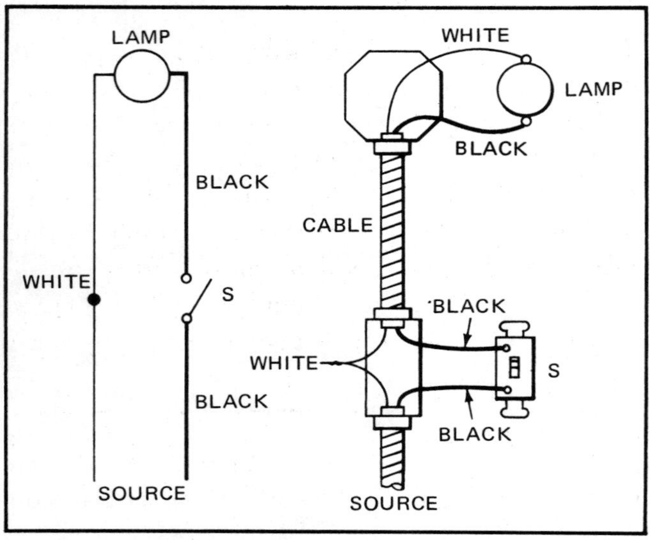

Fig. 5-3 Single-pole switch in circuit with feed at switch.

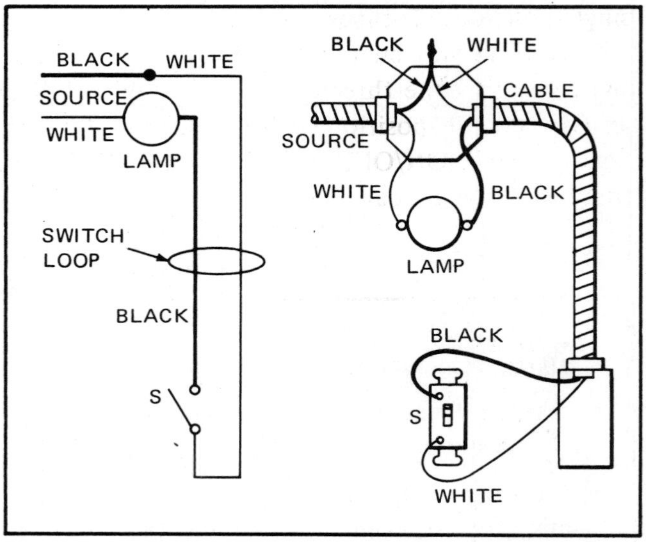

Fig. 5-4 Single-pole switch in circuit with feed at light.

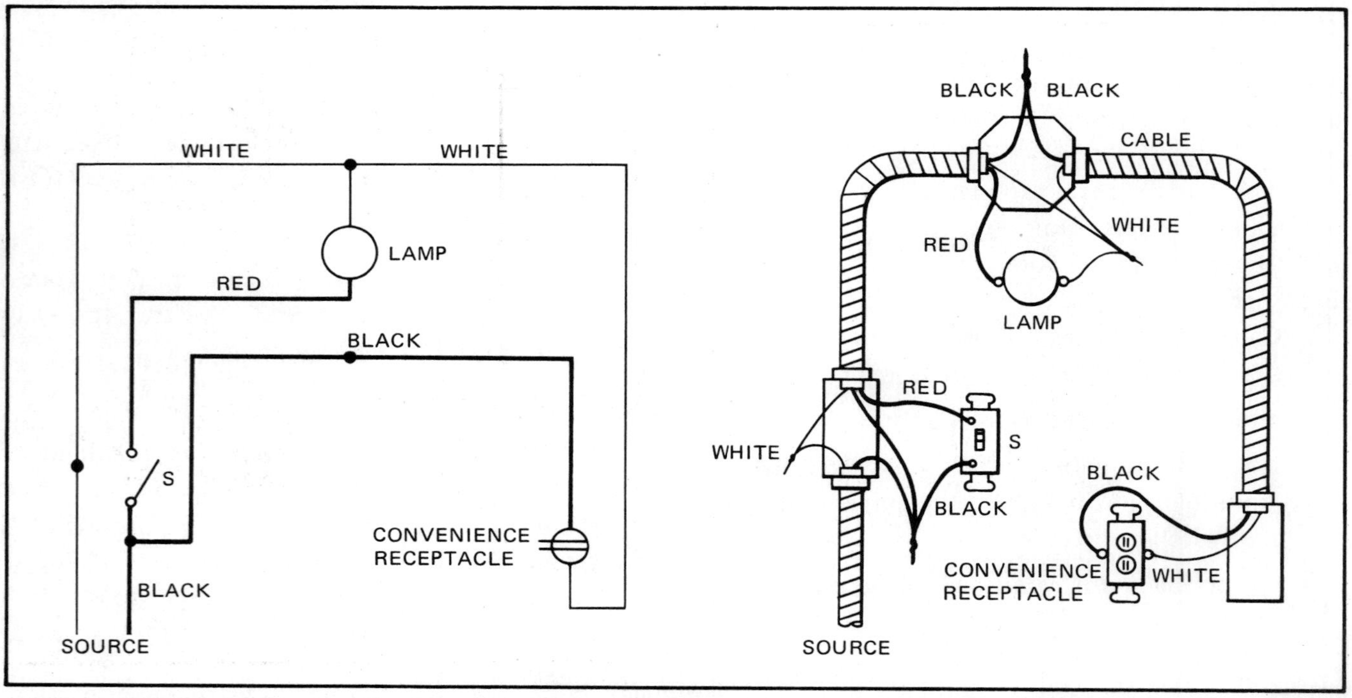

Fig. 5-5 Ceiling outlet controlled by single-pole switch with live convenience receptacle and feed at switch.

Three-way Switch

A three-way switch has one terminal to which the switch blade is always connected. This terminal is called the *common* terminal. The two other terminals are called *traveler wire terminals.* The switch blade can be connected between the common terminal and either one of the traveler terminals. Figure 5-6 shows the two positions of the three-way switch. Note that a three-way switch is actually a single-pole, double-throw switch.

A three-way switch differs from a single-pole switch in that the three-way switch does not have an ON or OFF position. Thus, the switch handle does not have ON/OFF markings, figure 5-7. The three-way switch can be identified further by its three terminals. The common terminal is darker in color than the two traveler wire terminals which are natural brass in color.

Three-way Switch Control with Feed at Switch. Three-way switches are used when a load (or loads) is to be controlled from two different switching points. As shown in figure 5-8, two three-way switches are used. Note that the feed is at the first switch control point.

Three-way Switch Control with Feed at Light. The circuit in figure 5-9 uses three-way switch control with the feed at the light. For this circuit, the white wire in the cable must be used as part of the three-way switch loop. The unidentified or black wire is used as the return wire to the light outlet, in compliance with NEC *Section 200-7, Exception 2.* This exception makes it unnecessary to paint the terminal of the identified conductor at the switch

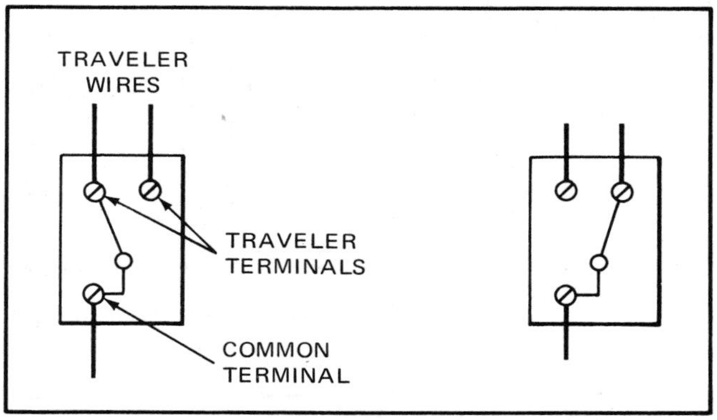

Fig. 5-6 Two positions of a three-way switch.

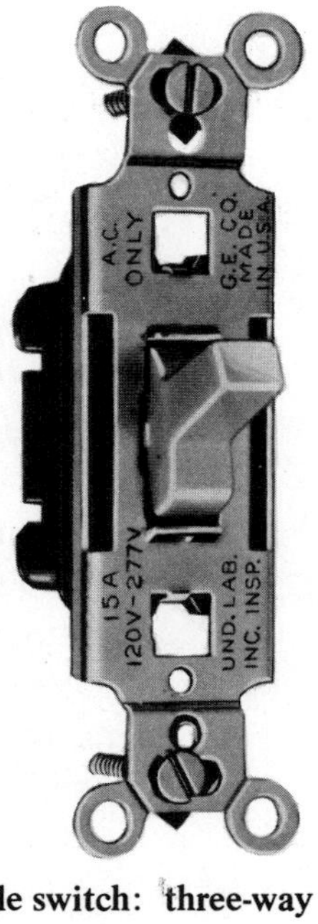

Fig. 5-7 Toggle switch: three-way flush switch.

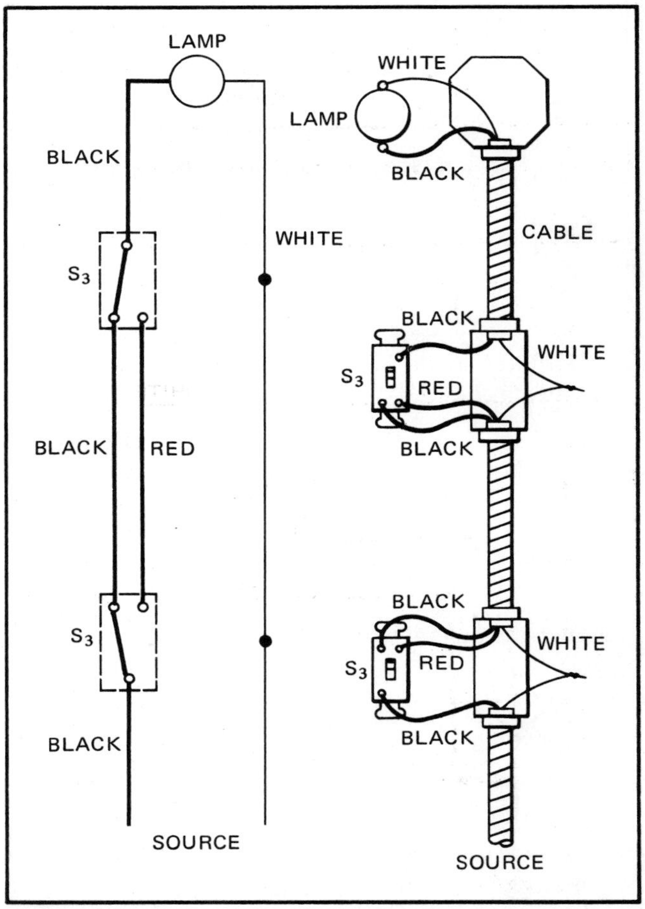

Fig. 5-8 Circuit with three-way switch control and feed at the first switch control point.

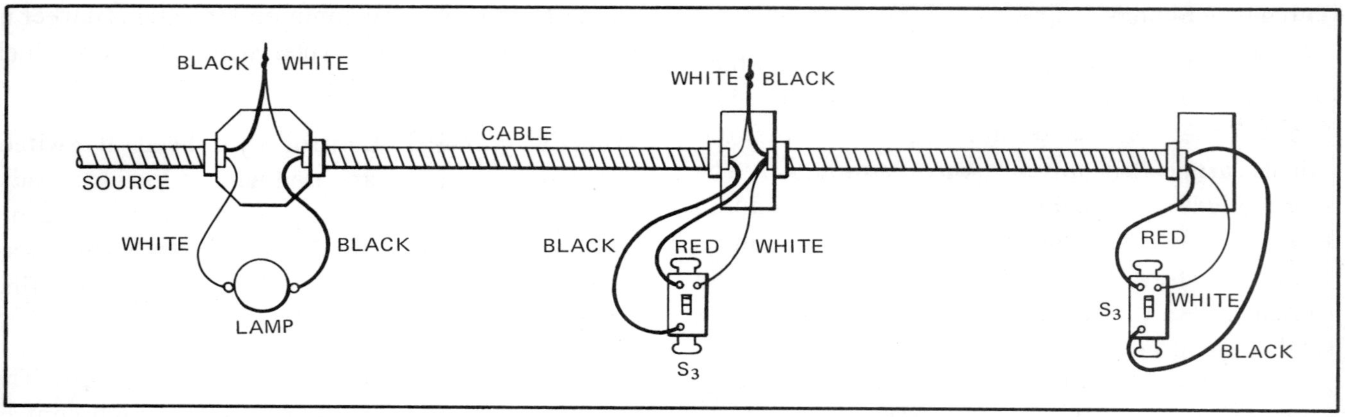

Fig. 5-9 Circuit with three-way switch control and feed at light.

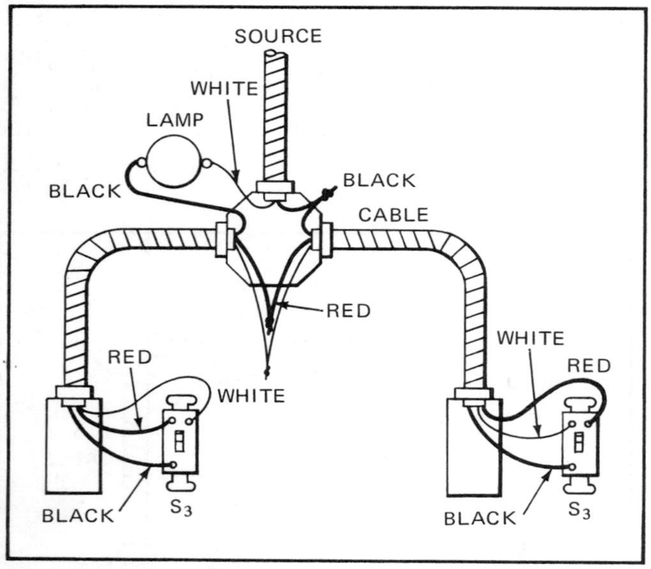

Fig. 5-10 Alternate circuit with three-way switch control and feed at light.

outlet when wiring single-pole, three-way, or four-way switch loops.

Alternate Configuration for Three-way Switch Control with Feed at Light. Figure 5-10 shows another arrangement of components in a three-way switch circuit. The feed is at the light with cable runs from the ceiling outlet to each of the three-way switch control points located on either side of the light outlet.

Examples of the installation of three-way switches at two switching points in residential wiring include the control of a hall light by switches in both upstairs and downstairs locations, and a garage light controlled from both the house and the garage.

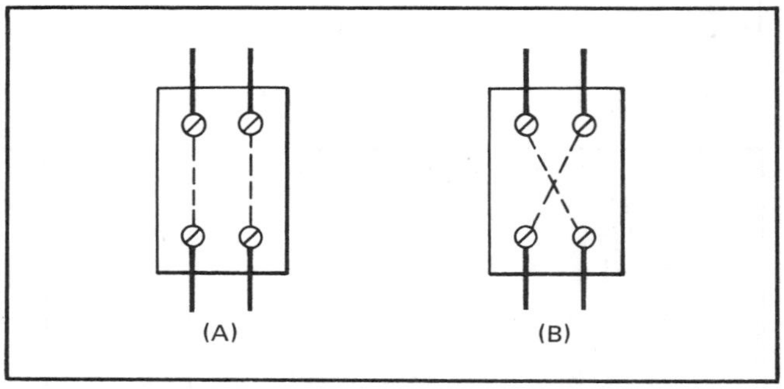

Fig. 5-11 Two positions of a four-way switch.

Four-way Switch

The four-way switch is constructed so that the switching contacts can alternate their positions as shown in figure 5-11. The four-way switch has two positions, but neither position is ON or OFF. The four-way switch can be identified readily by its four terminals and by the fact that the toggle does not have ON or OFF markings.

Four-way switches are used when a load or group of loads must be controlled from more than two switching points. To accomplish this, three-way switches are connected to the source and to the load. The switches at all other control points, however, must be four-way switches.

Figure 5-12 shows how a lamp can be controlled from any one of three switching points. Care must be used in connecting the traveler wires to the proper terminals of the four-way switch: the two traveler wires from one three-way switch are connected to the two terminals on one side of the four-way switch, and the other two traveler wires from the second three-way switch are connected to the two terminals on the other side of the four-way switch.

Double-pole Switch

A double-pole switch may be used when two separate circuits must be controlled with one switch, figure 5-13. This type of switch, however, is rarely used for lighting circuits in residential work.

A double-pole (2-pole) switch may also be used to provide 2-pole disconnecting means for a 240-volt load, figure 5-14.

Double-pole toggle switches are not commonly used in residential work. Double-pole (2-pole) disconnect switches, however, are used quite often in residences for the furnace, water pump motors, and other 240-volt feeders.

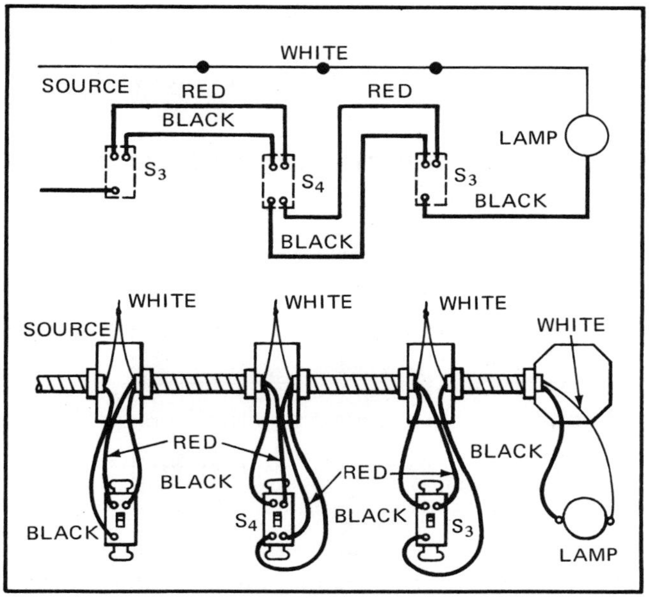

Fig. 5-12 Circuit with switch control at three different locations.

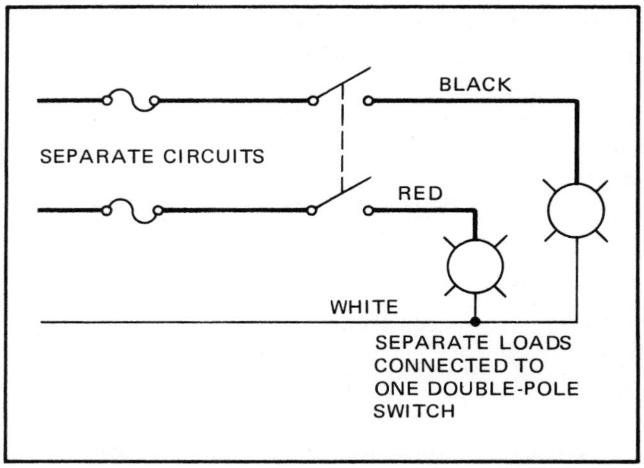

Fig. 5-13 Application of a double-pole switch.

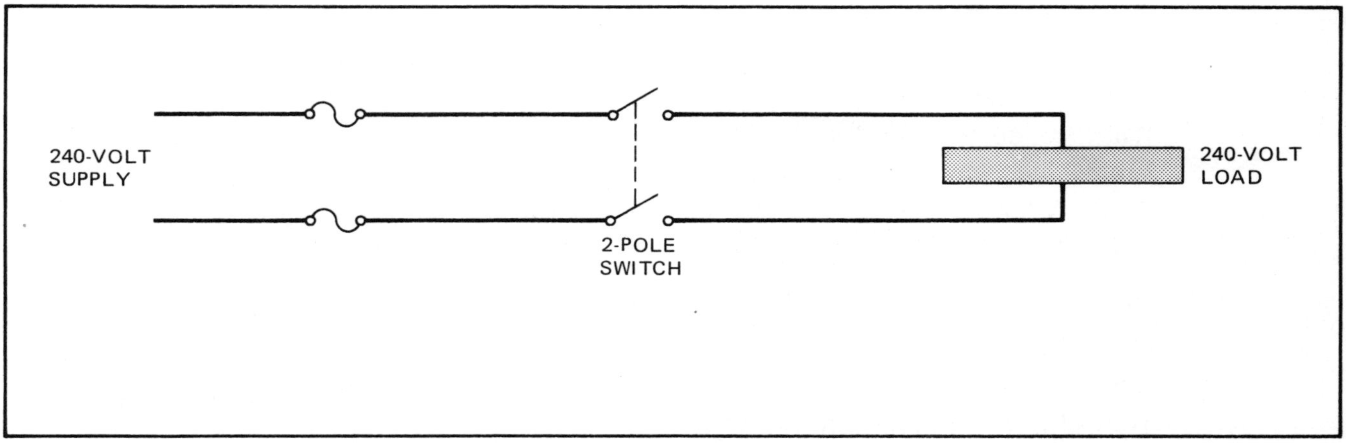

Fig. 5-14 Double-pole (2-pole) disconnect switch.

REVIEW

Note: Refer to the Code or the plans where necessary.

1. The identified grounded circuit conductor must be ____________ or ____________ in color.
2. Explain how lighting switches are rated. ____________
3. A T-rated switch may be used to its ____________ current capacity when controlling an incandescent lighting load.
4. What switch type and rating is required to control five 300-watt tungsten filament lamps on a 120-volt circuit? Show any calculations. ____________
5. List three types of lighting switches.
   a. ____________ b. ____________ c. ____________
6. To control a group of lights from one control point, what is the most practical type of switch to use? ____________
7. Single-pole switches are always connected into the ____________ wire.
8. Complete the connections in the following arrangement so that both ceiling light outlets are controlled from the single-pole switch. Assume the installation is in cable.

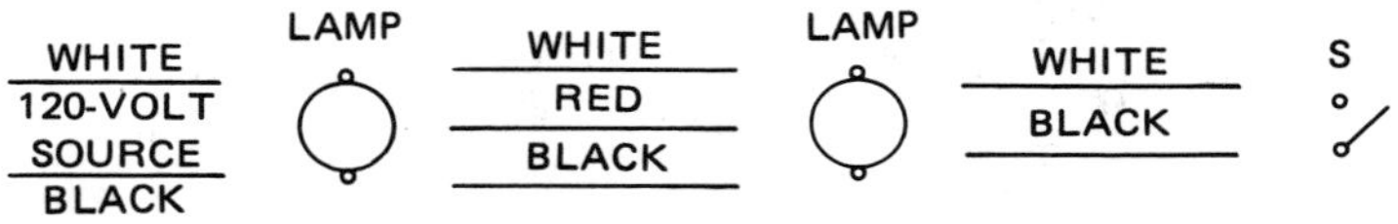

9. a. When installing a two-wire switch loop, which conductor of the cable feeds the switch? (See *Section 200-7.*) ____________
   b. Which conductor is used as the return wire? ____________
   c. From which wire does the switch feed tap? ____________
   d. What are the colors of the conductors connected to the fixture? ____________
10. A three-way switch may be compared to a ____________ switch.
11. What type of switch is installed to control a group of lights from two different control points? How many switches are needed? ____________
12. Complete the connections in the following arrangement so that the ceiling outlet may be controlled from either three-way switch.

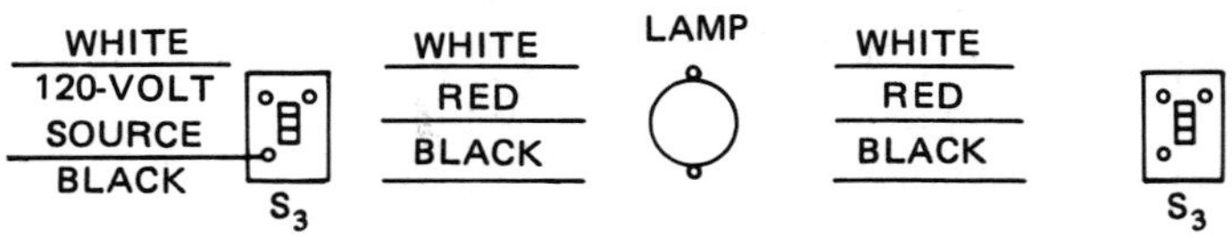

13. When connecting four-way switches, care must be taken to connect the travelers to the ____________ terminals.

14. Show the connections for a ceiling outlet which is to be controlled from any one of three switch locations. The 120-volt feed is at the light, *Section 200-7.* Label the conductor colors.

15. Match the following switch types with the correct number of terminals for each.

Three-way switch	Two terminals
Single-pole switch	Four terminals
Four-way switch	Three terminals

16. When connecting single-pole, three-way, and four-way switches, they must be wired so that all switching is done in the ______________________________ circuit conductor.

unit 6

Lighting Branch Circuit for Bedroom No. 2

OBJECTIVES

After studying this unit, the student will be able to

- explain the factors which influence the grouping of outlets into circuits.
- estimate wattages for the outlets of a circuit.
- draw a cable layout and a wiring diagram based on information given in the residence plans, the specifications, and Code requirements.
- select the proper wall box for a particular installation.
- explain how wall boxes can be grounded.
- list the requirements for the installation of fixtures in clothes closets.
- explain the operation and connections of ground-fault circuit interrupters (GFCI).

GROUPING OUTLETS

The grouping of outlets into circuits must conform to National Electrical Code standards and good wiring practices. There are many possible combinations or groupings of outlets. In most residential installations, circuit planning is usually done by the electrician who must insure that the circuits conform to Code requirements. In larger, more costly residences, the circuit layout may be completed by the architect and included on the plans.

Because many circuit arrangements are possible, there are few guidelines for selecting outlets for a particular circuit. An electrician plans circuits that are economical without sacrificing the quality of the installation. For example, some electricians do not place outlets in the same room on different circuits. However, one or more outlets in a room may be included on a separate circuit to economize on wiring materials. Some electricians consider it poor practice to include outlets on different floors on the same circuit. Here, too, the decision can be a matter of personal choice. Some local building codes limit this type of installation to lights at the head and foot of a stairway.

To summarize, the grouping of outlets into circuits must satisfy the requirements of the National Electrical Code, local building codes, good wiring practices, and common sense. A good wiring practice is to divide all loads as evenly as possible among the circuits as required by *Section 220-3(d).*

ESTIMATING WATTAGE FOR OUTLETS

Building plans typically do not specify the ratings in watts for the outlets shown. When planning circuits, the electrician must consider the types of fixtures which may be used at the various outlets. To do this, the electrician must know the general uses of the convenience outlets in the typical dwelling.

Unit 3 states that the general lighting load of a residence is determined by allowing a load of 3 watts for each square foot (0.093 m^2) of floor area, *Section 220-2(b)*. For the residence in the plans, it is shown that five lighting circuits meet the minimum standards set by the Code. However, to provide sufficient capacity, twelve general lighting circuits are to be installed.

The Code does not specify the maximum number of outlets permitted in dwellings. *Section 220-2(a)* states that the continuous load supplied by a branch circuit shall not exceed 80 percent of the branch-circuit rating. For a 15-ampere, 120-volt branch circuit, this means that the continuous load must be not more than 1440 watts.

15 amperes × 120 volts × 0.80 = 1440 watts
where 0.80 = 80 percent

Certain fixtures, such as recessed lights and fluorescent lights, are marked with their maximum lamp wattage and ballast current (for fluorescent fixtures only). Other fixtures, however, are not marked and the electrician does not know the exact load that will be connected to the convenience receptacles. Also unknown is the size of the lamps that will be installed in the lighting fixtures (other than the recessed and fluorescent types). In other words, it is difficult to anticipate what the homeowner may do after the installation is complete. The electrician should remember that the room in which the outlets are located does give some indications as to their possible uses. The circuits should be planned accordingly.

Estimating Wattage and Number of Outlets

The load in watts due to the lamps in the lighting fixtures and the number of outlets which can be included in a circuit can be found by estimating the wattages required for each living area. Then, an additional load is allowed of approximately 150 watts per convenience receptacle. The total load must not exceed the maximum load of 1440 watts for a 15-ampere circuit.

Although 20-ampere lighting circuits are not generally installed in residences, the maximum allowable load in watts for such a circuit is:

20 amperes × 120 volts × 0.80 = 1920 watts
where 0.80 = 80 percent

Estimating Number of Outlets by Assign... an Amperage Value to Each

Another method of determining the number of outlets to be included in one circuit requires that a value of 1 1/2 amperes be assigned to each outlet to a total of 15 amperes. Thus, a total of 10 outlets can be included in a 15-ampere circuit.

As stated previously, the National Electrical Code does not limit the number of outlets in one circuit. However, many local building codes do specify the maximum number of outlets per circuit. Before planning any circuits, the electrician must check the local building code requirements.

All outlets will not be required to deliver 1 1/2 amperes. For example, closet lights, night lights, and clocks will use only a small portion of the allowable current. A 60-watt closet light will draw less than 1 ampere.

$$I = \frac{W}{E} = \frac{60}{120} = 0.5 \text{ ampere}$$

For example, if low-wattage fixtures are connected to a circuit, the maximum number of outlets per circuit may be increased from 10 to 12 outlets. On the other hand, if the circuit supplies high-wattage lamps, the number of outlets must be reduced from 10 to perhaps 8 outlets.

SYMBOLS USED ON CIRCUIT DIAGRAMS

Symbols are used on circuit diagrams to indicate clearly the types of devices and outlets required. In the wiring diagram, the symbols are drawn in the locations corresponding with those on the plans. Both the residence specifications and plans should be reviewed for complete circuit information. Figure 6-1 illustrates some of the symbols which are frequently used.

DRAWING THE WIRING DIAGRAM OF A LIGHTING CIRCUIT

The electrician must take information from the building plans and convert it to the forms which will be most useful in planning an installation which is economical, conforms to all Code regulations, and follows good wiring practice. To do this, the electrician first makes a cable layout. Then, a wiring

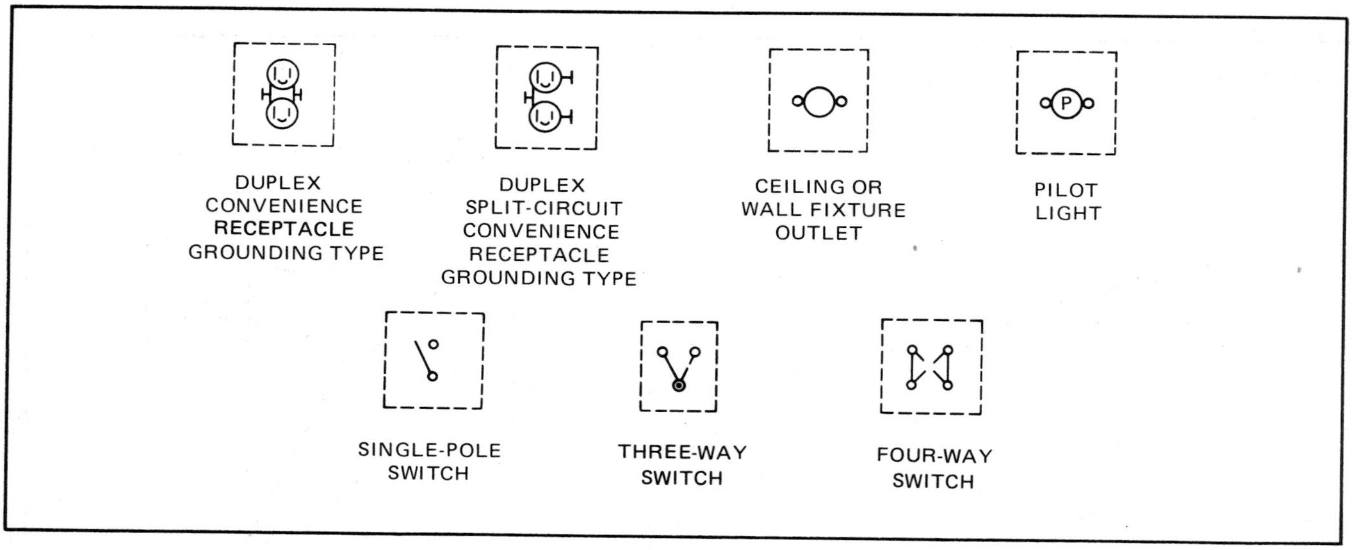

Fig. 6-1 Symbols used on circuit diagrams.

diagram is prepared to show clearly each wire and all connections in the circuit.

The following steps will guide the student in the preparation of wiring diagrams. (The student is required to draw wiring diagrams in later units of this text.)

1. Refer to the plans and make a cable layout of all lighting and convenience outlets. Show all switches and indicate the number of terminals for each, figure 6-2.
2. Draw a wiring diagram showing the traveler conductors for all three-way and four-way switches, if any, figure 6-3.
3. Draw a line between each switch and the outlet or outlets it controls. For three-way switches, do this for one switch only.
4. Draw a line from the grounded terminal on the lighting panel to each current-consuming outlet. This line may pass through switch boxes, but must not be connected to any switches.

   Note: An exception to step 4 may be made for double-pole switches. For these switches, all conductors of the circuit are opened simultaneously.
5. Draw a line from the ungrounded (hot) terminal on the panel to connect to each switch and unswitched outlet. Connect to one three-way switch only.

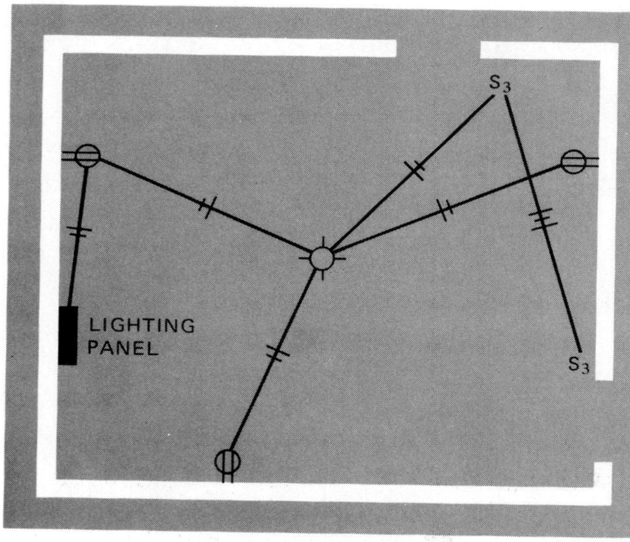

Fig. 6-2 Cable layout.

6. Show splices as small dots where the various wires are to be connected together. In the wiring diagram, the terminal of a switch or outlet may be used for the junction point of wires. In actual wiring practice, however, the Code does not permit more than one wire to be connected to a terminal unless the terminal is a type approved for more than one conductor. The standard screw-type terminal is NOT approved for more than one wire, *Section 110-14.*
7. The final step in preparing the wiring diagram is to mark the color of the conductors, figure

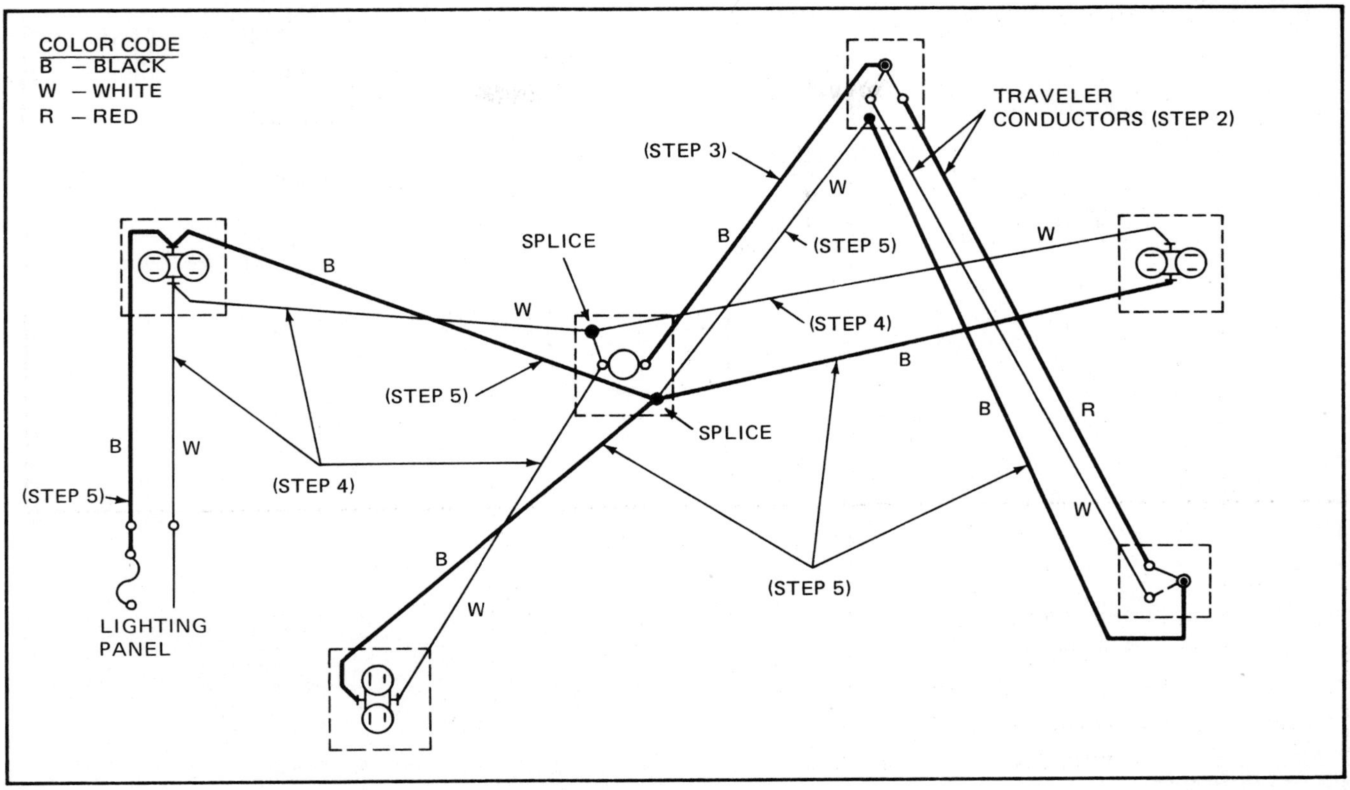

Fig. 6-3 Wiring diagram.

6-3. Note that the colors selected – black (B), white (W), and red (R), – are the colors of two- and three-conductor cables (refer to unit 5). It is suggested that the student use colored pencils for different conductors when drawing the wiring diagram. Yellow can be used to indicate white conductors to provide better contrast with white paper. A note to this effect should be placed on the diagram, *Section 200-7.*

BRANCH LIGHTING CIRCUIT FOR BEDROOM NO. 2

The outlets shown in the residence plans are grouped into 12 general lighting circuits. These circuits conform to the standards of the National Electrical Code and recommended wiring practices. It should be noted, however, that there are other possible groupings which are equally correct.

The circuit supplying Bedroom No. 2 originates in Panel A (unit 27) and is connected to Circuit No. 27. This circuit is a 15-ampere circuit and includes the following outlets and devices:

6 convenience receptacles (split circuit)
1 ceiling fixture
2 closet lights
1 weatherproof receptable (GFCI type)
Total: 10 Outlets

The estimated value of the lamp load in watts for a bedroom of this size is 200 watts. A minimum of 60 watts is added for each of the closet lights. In addition, an allowance of 150 watts must be provided for each convenience receptacle. Therefore, the approximate total connected load on this circuit is:

7 convenience outlets (150 watts each)	1050 watts
1 ceiling fixture	200 watts
2 closet lights (60 watts each)	120 watts
Total Estimated Wattage	1370 watts

The total load (1370 watts) is within the limits set for a 15-ampere circuit. Recall that a 15-ampere circuit at 120 volts is not to be subjected to a continuous load of more than 80 percent of its capacity or 1440 watts. It is assumed that all outlets will not be in use at full capacity at the same time.

DETERMINING THE WALL BOX SIZE

Several factors must be considered when the electrician determines the size of the wall box. These factors include:

- the number of conductors entering the box,
- the types of boxes available, and
- the space allowed for the installation of the box.

Box Size According to the Number of Conductors in a Box (*Section 370-6*)

The plans show that the branch-circuit feed enters the outlet box nearest Panel A. The cable is run from the top of Panel A and above the ceiling of the basement lavatory. It is then terminated in a wall box located approximately 2 1/2 feet (762 mm) from the outside bedroom wall, figure 6-4.

To determine the proper box size for each location, the total number of conductors entering the box must be determined. Eleven No. 12 AWG conductors enter the box on the bedroom wall (8 circuit conductors and 3 grounding conductors). However, all of the grounding wires are not counted when determining the box size, *Section 370-6*.

The following example shows how the proper wall box size is determined for a particular installation, figure 6-5.

1.	Add the circuit conductors: 2 + 3 + 3 =	8
2.	Add grounding wires (count one only)	1
3.	Add one conductor for the receptacle	1
	Total	10 Conductors

Once the total number of conductors is known, refer to *Table 370-6(a)* (page 13) to find the box that can hold the conductors. For example, a 4″ x 2 1/8″ square box with a suitable plaster ring can be used.

The volume of the box plus the space provided by plaster rings, extension rings, and raised covers may be used to determine the total available volume. (Refer to pages 12 and 13.) In addition, it is desirable to install boxes with external cable clamps. Remember that if the box contains one or more devices, such as cable clamps, fixture studs, or hickeys, the number of conductors permitted in the box shall be one less than shown in *Table 370-6(a)* for *each type* of device contained in the box. (See example on page 13.)

Box Size According to Space Allowances

The plans show that two switches are located next to the bedroom door and two switches are

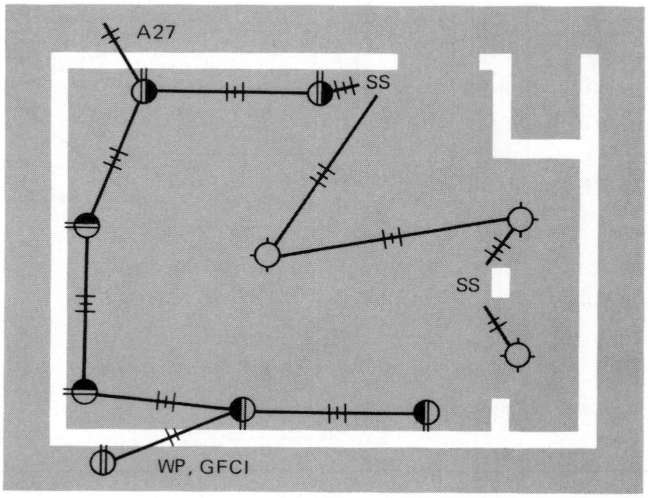

PANEL A
(CIRCUIT 27, 15 AMPERES)

Fig. 6-4 Bedroom No. 2.

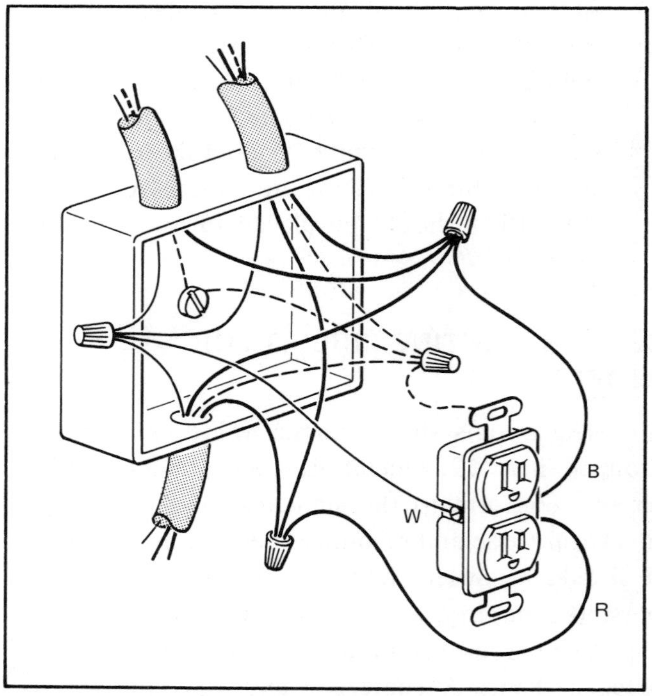

Fig. 6-5 Determining size of box according to number of conductors.

located between the sliding doors of the closets. There are a number of ways in which these switches can be arranged.

The electrician may install two ganged sectional switch boxes, each 2 3/4 inches deep. A maximum of six No. 12 AWG conductors is allowed in each box. This number is reduced by one conductor for the cable clamp and one conductor for the switch to be installed. Thus, a maximum of four No. 12 AWG conductors is permitted in each box. A total of eight No. 12 AWG conductors is permitted for the ganged assembly. One conductor can be added for a total of nine conductors if both cables are run into one box and one cable clamp is used to secure them. The electrician can then remove all unused cable clamps within the box to gain extra room. However, if another cable is to be added to the box in the future, there will be no cable clamps to secure it. The electrician must decide which is more important: gaining the additional wiring space or allowing for future additions.

The electrician may also install a 4-inch square, 1 1/2-inch deep outlet box with either a two-gang raised plaster cover or a single-gang raised plaster cover and an interchangeable line of switches. When the exposed portion of the wall between the two casings is narrow, a two-gang switch plate may not fit, since it is 4 9/16 inches (115.9 mm) wide, figure 6-6. A single-gang, two-hole switch plate is approximately 2 3/4 inches (69.85 mm) wide and can be used in the available space. (Standard wall plates, regardless of the width, are approximately 4 1/2 inches (114.3 mm) in height. Figure 6-6 gives standard wall plate dimensions.

If a 4-inch octagonal box is fastened to offset bar hangers on which lighting fixtures are to be mounted, a 4-inch octagonal raised plaster cover may be installed on the outlet box. This cover has a 2 3/4-inch opening through which connections may be made. This means that the electrician will complete the installation of the box and the plasterer can then finish the ceiling. Thus, the electrician does not have to worry about the plaster cracking at the edges of the box or the possibility that the fixture will not completely cover the space between the edge of the box and the plaster. This type of installation is preferred when 4-inch porcelain receptacles are to be mounted on finish surfaces.

No. of Gangs	Height	Width
1	4 1/2" (114.3 mm)	2 3/4" (69.85 mm)
2	4 1/2" (114.3 mm)	4 9/16" (115.9 mm)
3	4 1/2" (114.3 mm)	6 3/8" (161.9 mm)
4	4 1/2" (114.3 mm)	8 3/16" (207.9 mm)
5	4 1/2" (114.3 mm)	10" (254 mm)
6	4 1/2" (114.3 mm)	11 13/16" (300 mm)

Fig. 6-6 Height and width of standard wall plates.

Grounding of Wall Boxes

The specifications for the residence state that *all* metal boxes are to be grounded. The means of grounding is armored cable or nonmetallic-sheathed cable containing an extra grounding conductor. This conductor is used only to ground the metal box. This bare grounding conductor must NOT be used as a current-carrying conductor since severe shocks can result.

According to *Section 210-7*, grounding-type receptacles must be installed on 15-ampere and 20-ampere branch circuits. The methods of attaching the grounding conductor to the proper terminal on the convenience receptacle is covered in unit 16.

The six convenience receptacles shown in Bedroom No. 2 are called two-circuit or split-circuit receptacles. The top portion of such a receptacle is *hot* at all times and the bottom portion is controlled by the wall switch, figure 6-7. It is recommended that the electrician wire the bottom section of the receptacle as the switched section. As a result, when the attachment plug cap of a lamp is inserted into the bottom switched portion of the receptacle, the cord does not hang in front of the unswitched section. This unswitched section can be used as a receptacle for clock, vacuum cleaner, or radio cords, or for other appliances where a switch control is not necessary or desirable.

The receptacle outlets in the window corner of both bedrooms may not be installed above the electric baseboard unit. This requirement is related to NEC *Section 110-3(b)*. Such receptacles can be an integral part of the baseboard unit or the base-

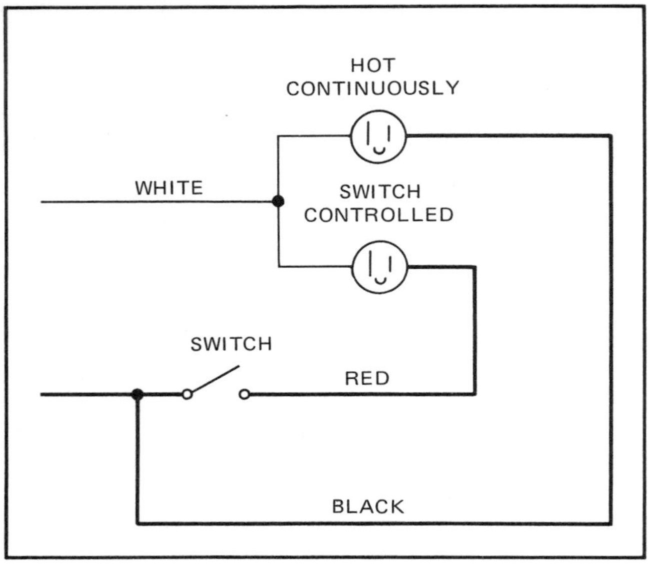

Fig. 6-7 Split circuit wiring for receptacles in bedroom.

board units can be moved out from the corner of the room to provide wall space for the receptacle. (These receptacles must NOT be connected to the heater circuit.)

See unit 23, "Electric Heating," for a full discussion of electric baseboard heaters and the location of receptacle outlets.

GROUND-FAULT CIRCUIT INTERRUPTERS (*SECTION 210-8*)

Many lives have been lost because of electrical shock from an appliance or a piece of equipment that is *hot*. This means that the hot circuit conductor in the appliance is contacting the metal frame of the appliance. This condition may be due to the breakdown of the insulation because of defective construction or accidental misuse of the equipment. **The shock hazard exists whenever the user can touch both the defective equipment and grounded surfaces such as water pipes, metal rims of sinks, grounded metal lighting fixtures, earth, and concrete in contact with the earth, water, or any other grounded surface.**

To protect against shocks, the National Electrical Code requires that ground-fault circuit interrupters (GFCI) (also called ground-fault interrupter, GFI) be provided for the receptacle outlets of dwellings as follows:

- on *all* 125-volt, single-phase, 15- and 20-ampere receptacles in bathrooms, and outdoors where there is direct grade access to the dwelling. Thus, it is not required to provide a GFCI-protected receptacle on a second floor balcony, such as in condominium and apartment wiring.
- on *all* 125-volt, single-phase, 15- and 20-ampere receptacles in garages. GFCI protection, however, is not required if the receptacle is not readily accessible, such as one which is installed on the ceiling for the overhead door operator. GFCI protection is not required if the receptacle is installed for a cord-connected appliance occupying a dedicated space, such as a freezer.
- on *all* 125-volt, single-phase, 15- and 20-ampere receptacle outlets used for temporary power while the building is under construction. An exception to this requirement is made when the receptacle is part of small portable generators rated at 5 kW or less. Plug-in GFCI devices may be used in such a case.

NEC *Article 680* gives the electrical requirements for swimming pools and the use of GFCI devices. (See unit 30.)

The Code requirements for ground-fault circuit protection can be met in many ways. Figure 6-8 shows how a ground-fault circuit interrupter (GFCI) may be combined with the branch-circuit breaker, the convenience receptacle outlet, or the feeder overcurrent device. Regardless of the location of the GFCI in the circuit, it must open the circuit when a current to ground exceeds 6 milliamperes (0.006 ampere).

The GFCI monitors the current balance between the hot conductor and the neutral conductor. As soon as the current in the neutral conductor is less than the current in the hot conductor, the GFCI senses this imbalance and opens the circuit. The imbalance indicates that part of the current in the circuit is being diverted to some path other than the normal return path along the neutral conductor. Thus, if the GFCI trips off, it is an indication of a possible shock hazard from a line-to-ground fault.

Nuisance tripping of GFCIs is known to occur. Sometimes, this can be attributed to extremely long runs of cable for the protected circuit. Consult the manufacturer's instructions to determine if maximum lengths for a protected circuit are recommended. Some electricians advise the use of nonmetallic staples (see figure 14-4C) or nonmetallic straps, instead of metallic, as a means of preventing nuisance tripping of GFCIs.

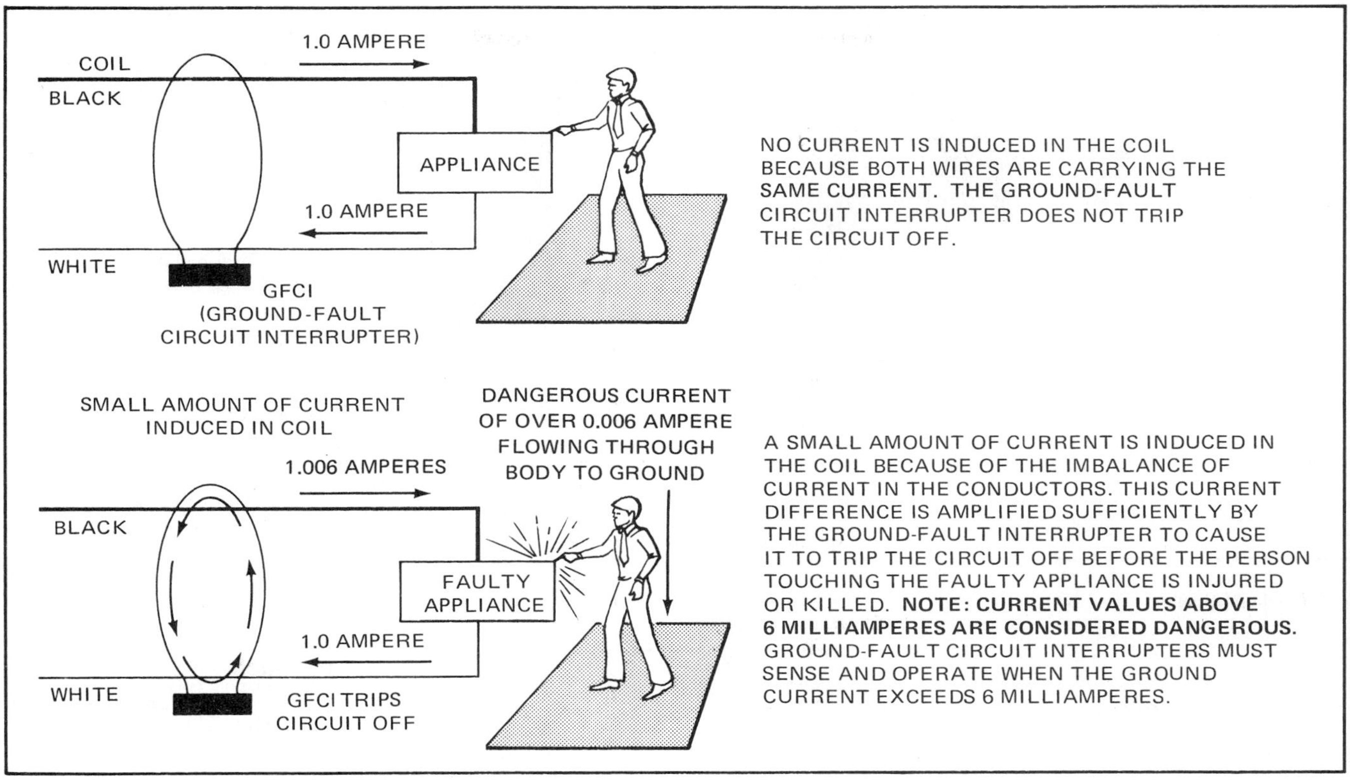
1.0 AMPERE
COIL
BLACK
APPLIANCE
1.0 AMPERE
WHITE
GFCI
(GROUND-FAULT
CIRCUIT INTERRUPTER)
NO CURRENT IS INDUCED IN THE COIL BECAUSE BOTH WIRES ARE CARRYING THE SAME CURRENT. THE GROUND-FAULT CIRCUIT INTERRUPTER DOES NOT TRIP THE CIRCUIT OFF.
SMALL AMOUNT OF CURRENT INDUCED IN COIL
DANGEROUS CURRENT OF OVER 0.006 AMPERE FLOWING THROUGH BODY TO GROUND
1.006 AMPERES
BLACK
FAULTY APPLIANCE
1.0 AMPERE
WHITE
GFCI TRIPS CIRCUIT OFF
A SMALL AMOUNT OF CURRENT IS INDUCED IN THE COIL BECAUSE OF THE IMBALANCE OF CURRENT IN THE CONDUCTORS. THIS CURRENT DIFFERENCE IS AMPLIFIED SUFFICIENTLY BY THE GROUND-FAULT INTERRUPTER TO CAUSE IT TO TRIP THE CIRCUIT OFF BEFORE THE PERSON TOUCHING THE FAULTY APPLIANCE IS INJURED OR KILLED. NOTE: CURRENT VALUES ABOVE 6 MILLIAMPERES ARE CONSIDERED DANGEROUS. GROUND-FAULT CIRCUIT INTERRUPTERS MUST SENSE AND OPERATE WHEN THE GROUND CURRENT EXCEEDS 6 MILLIAMPERES.

Fig. 6-8 Basic principle of how a ground-fault circuit interrupter operates.

GROUND-FAULT CIRCUIT INTERRUPTER IN RESIDENCE CIRCUITS

In this residence, receptacle outlets installed outdoors, in the garage, and in the bathrooms are connected to the circuits listed in Table 6-1. The National Electrical Code requires that the convenience receptacle outlets at these locations be protected by ground-fault circuit interrupters, figure 6-9. These interrupters may be installed in Panel A on each of the specified circuits, or they may be installed as an integral part of the receptacle.

When interrupters are installed on the specified circuits at Panel A, figure 6-10, a fault or current in excess of 6 milliamperes shuts off the entire circuit. For example, a ground fault at any point on Circuit A27 will shut off the entire bedroom circuit, including the wall convenience receptacles, ceiling light, and closet lights.

When a ground-fault circuit interrupter is installed as an integral part of the convenience receptacle, then only that receptacle is shut off when a ground fault greater than 6 milliamperes occurs in a device plugged into the receptacle, figure 6-11.

Circuit	Receptacle Location
A26	Bathroom and passage
A27	Front of residence outside of Bedroom No. 2
A29	Front entry
A30	Garage
A32	Rear porch
A34	Rear terrace
A35	Lavatory in basement

Table 6-1.

Figure 6-12 shows the effect of an interrupter installed as part of a main feeder.

Feedthrough Ground-fault Circuit Interrupter

Feedthrough ground-fault circuit interrupter receptacles are also available. Figure 6-13 shows how the feedthrough GFCI receptacle is connected into the circuit. When connected in this manner, the feedthrough GFCI receptacle and all downstream outlets on the same circuit will have ground-fault protection, figure 6-14.

Several factors influence the choice of individual ground-fault circuit interrupter devices, feedthrough GFCI devices, or GFCI breakers for installation in the panel. The electrician must determine

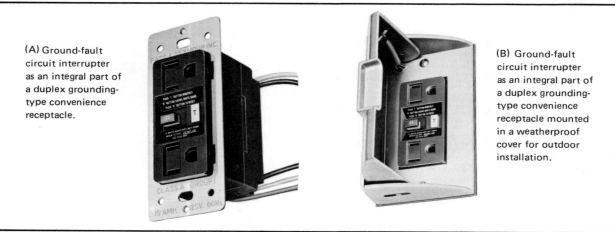

(A) Ground-fault circuit interrupter as an integral part of a duplex grounding-type convenience receptacle.

CLASS A GROUP

(B) Ground-fault circuit interrupter as an integral part of a duplex grounding-type convenience receptacle mounted in a weatherproof cover for outdoor installation.

Fig. 6-9 Ground-fault circuit interrupters.

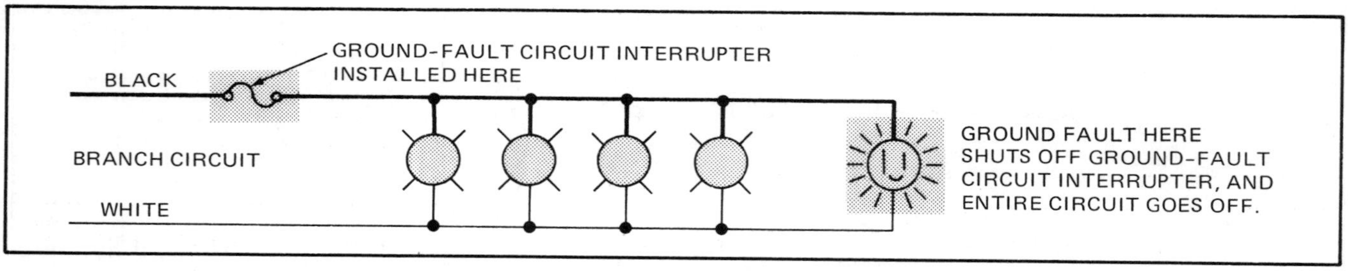
GROUND-FAULT CIRCUIT INTERRUPTER
INSTALLED HERE
BLACK
BRANCH CIRCUIT
WHITE
GROUND FAULT HERE
SHUTS OFF GROUND-FAULT
CIRCUIT INTERRUPTER, AND
ENTIRE CIRCUIT GOES OFF.

Fig. 6-10 Ground-fault circuit interrupter as a part of the branch-circuit overcurrent device.

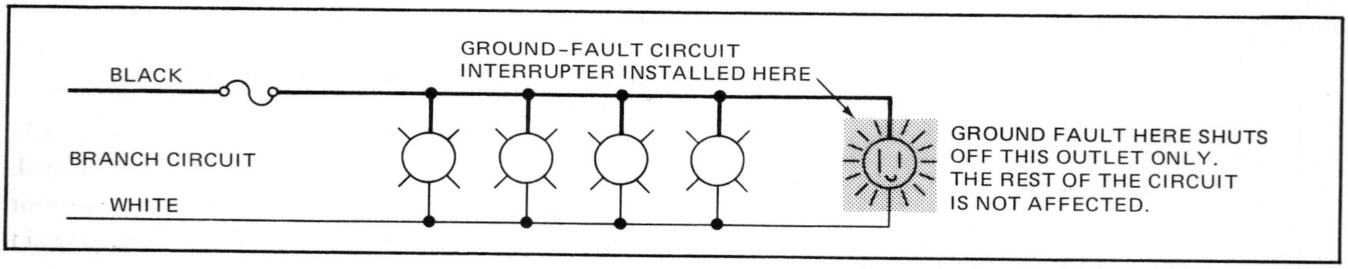
GROUND-FAULT CIRCUIT
INTERRUPTER INSTALLED HERE
BLACK
BRANCH CIRCUIT
WHITE
GROUND FAULT HERE SHUTS
OFF THIS OUTLET ONLY.
THE REST OF THE CIRCUIT
IS NOT AFFECTED.

Fig. 6-11 Ground-fault circuit interrupter as an integral part of a convenience receptacle outlet.

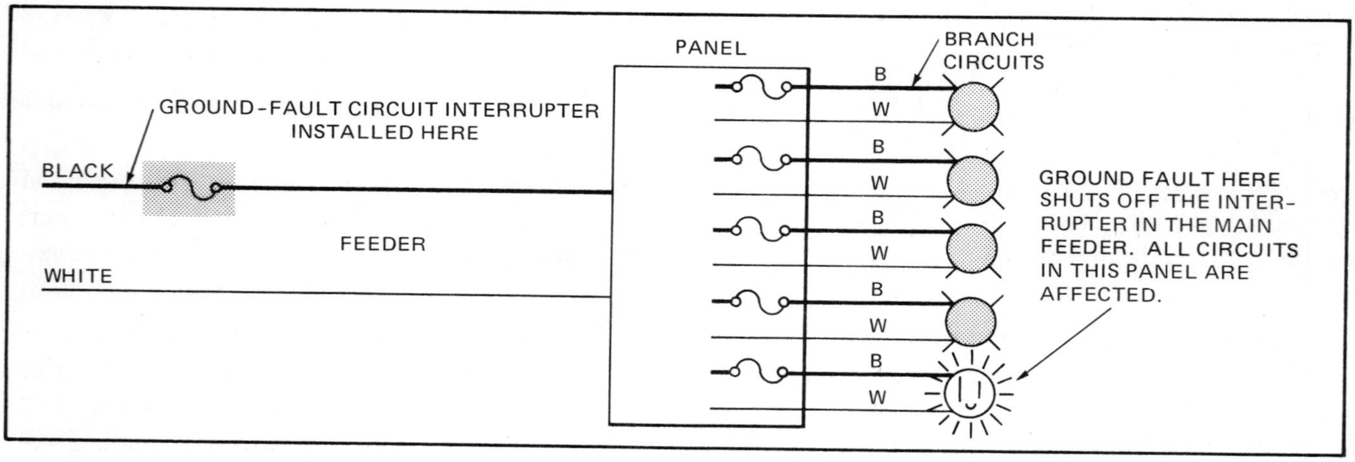
PANEL
BRANCH
CIRCUITS
GROUND-FAULT CIRCUIT INTERRUPTER
INSTALLED HERE
BLACK
FEEDER
WHITE
B
W
B
W
B
W
B
W
B
W
GROUND FAULT HERE
SHUTS OFF THE INTER-
RUPTER IN THE MAIN
FEEDER. ALL CIRCUITS
IN THIS PANEL ARE
AFFECTED.

Fig. 6-12 Ground-fault circuit interrupter as part of the main feeder, *Section 215-9*.

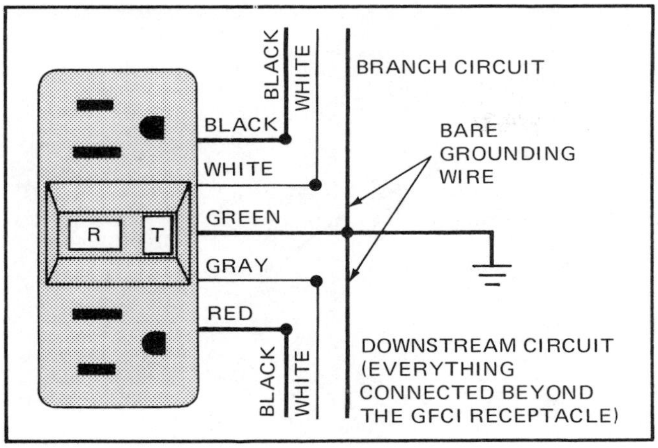

Fig. 6-13 Connecting feedthrough ground-fault circuit interrupter into circuit.

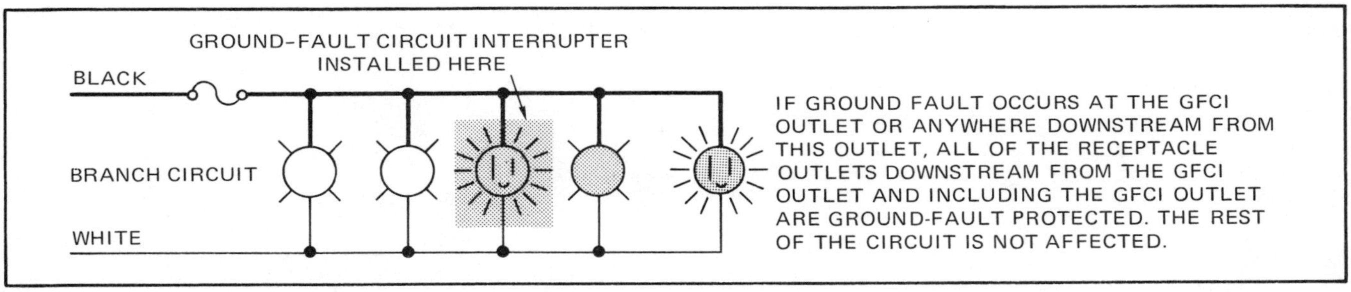

Fig. 6-14 Feedthrough ground-fault circuit interrupter as an integral part of the convenience receptacle outlet.

if it is more desirable to lose power completely because of a ground fault in one location, or to lose power only in that portion of the electrical circuit containing the ground fault. The electrician must also decide if all the outdoor, garage, and bathroom receptacle outlets are to be connected to one or more than one GFCI-protected circuit. Such an installation uses fewer GFCI devices and more cable. The most important factors to be considered are the continuity of electrical power and the economy of the installation. The decision must be made separately for each installation.

The electrician plans the installation carefully to insure that it conforms to all Code regulations and is economical as well. An economical installation is important because GFCI devices are expensive. Although many circuit arrangements are possible, it is suggested that the connections follow the cable layout for each circuit. It is assumed that a ground-fault circuit interrupter receptacle is to be installed at each location where required by the Code.

Figure 6-15 illustrates one circuit arrangement where the outdoor, garage, and bathroom receptacles are connected to provide ground-fault protection using two GFCI receptacle outlets.

FIXTURES IN CLOTHES CLOSETS

Clothing, boxes, and other material normally stored in clothes closets are a potential fire hazard. These items may ignite on contact with the hot surface of an exposed light bulb. The bulb, in turn, may shatter and spray hot sparks and hot glass onto other combustible materials. NEC *Section 410-8* gives the following rules for the installation of lighting fixtures in clothes closets.

- Pendants shall *not* be installed in clothes closets, figure 6-16.
- Surface-mounted clothes closet fixtures must be placed so that there is an 18-inch (457 mm) clearance between the fixtures and combustible material storage areas, figure 6-17.
- Flush, recessed fixtures having a solid lens, or ceiling-mounted fluorescent fixtures, must be placed so that there is at least a 6-inch (152 mm) clearance between the fixtures and the combustible material storage area, figure 6-18.

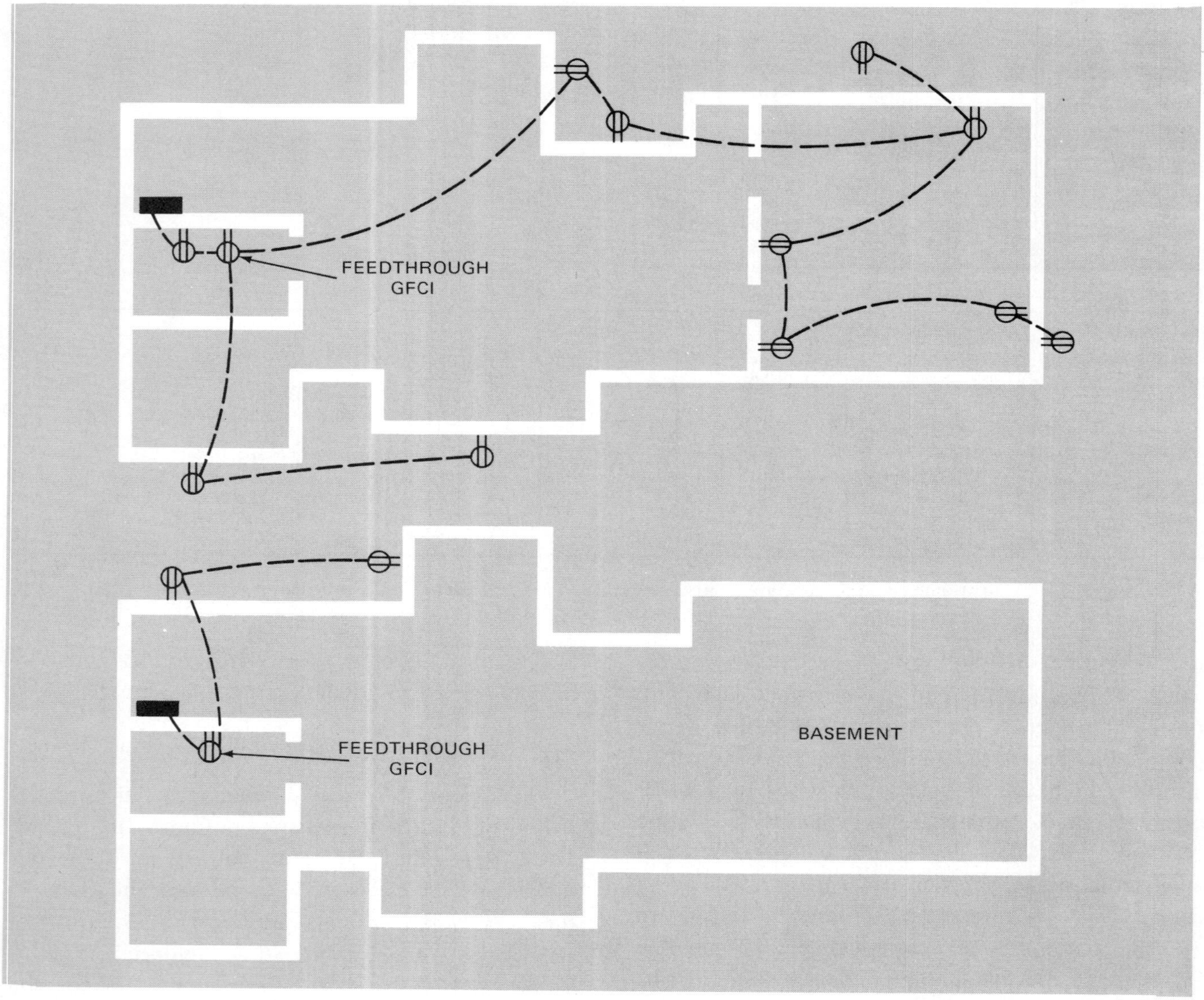

Fig. 6-15 One possible method of connecting bathroom, garage, and outdoor outlets using GFCI devices.

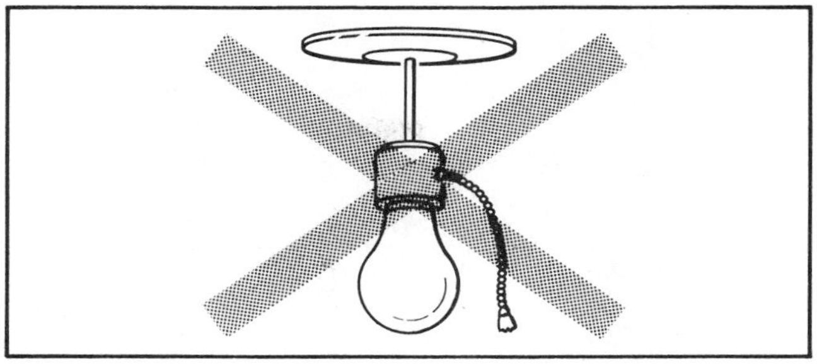

Fig. 6-16 Pendants are *not permitted* in closets.

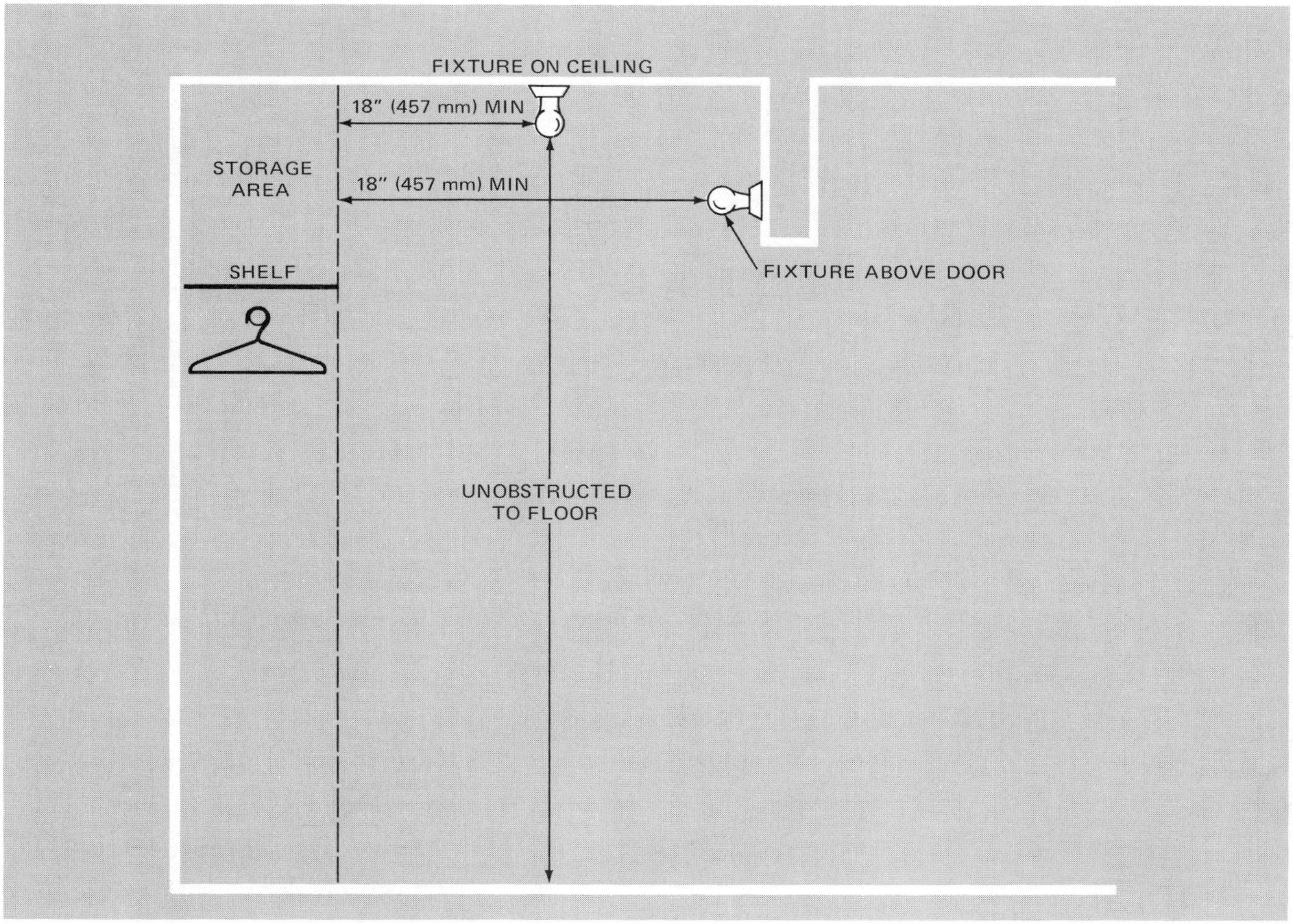

Fig. 6-17 Clearances for mounting fixtures in clothes closets.

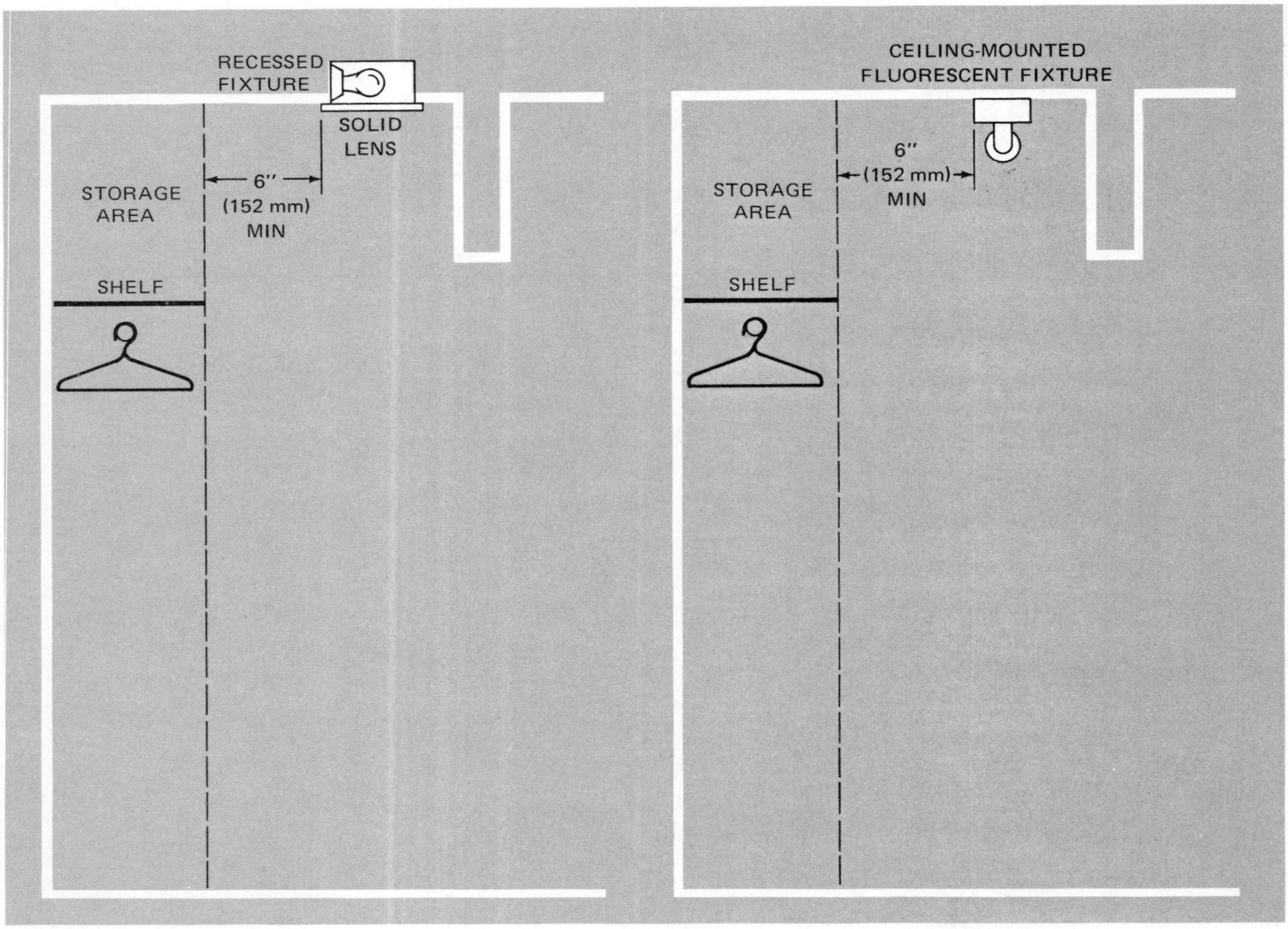

Fig. 6-18 Location of flush, recessed fixtures or ceiling-mounted fluorescent fixtures.

REVIEW

Note: Refer to the Code or the plans where necessary.

1. Can the outlets in a circuit be arranged in different groupings to obtain the same result? Why? ______________________________
____
____

2. Is it good practice to have outlets on different floors on the same circuit? Why?
____
____

3. What usually determines the grouping of outlets into a circuit? ______________________________
____

4. The continuous load on a lighting branch circuit must not exceed ______________ percent of the branch-circuit rating.

5. To determine the maximum number of outlets in a circuit, ______________ amperes per outlet are allowed. For a 15-ampere circuit, this results in a maximum of ________ outlets.

6. For residential installations, what are the estimated typical wattages used in determining the loading of branch circuits which supply lighting outlets and convenience receptacles?
ceiling fixtures ________ watts convenience receptacles ________ watts
closet fixtures ________ watts recessed fixtures ________ watts

7. To what branch circuit is Bedroom No. 2 connected? ________

8. What is the ampere rating of this circuit? ________

9. What size wire is used for the lighting circuit in Bedroom No. 2? Why? ________

____

10. a. How many convenience receptacles are connected to this circuit? ________
b. How many lighting fixtures? ________

11. What main factor influences the choice of wall boxes? ________

____

12. How is a wall box grounded? ________

____

13. What is the width of two-gang wall plate? ________

14. What is meant by a split-circuit receptacle? ________

15. Is the switched portion of a convenience outlet mounted toward the top or bottom? Why? ________

____

____

16. When an electrician desires to install a surface-mounted fixture in a clothes closet, a(an) ________ -inch (________ -millimeter) clearance must be maintained between the fixture and the storage area above the closet shelving. If this clearance is not possible, it is permissible to install a(an) ________ fixture equipped with a(an) ________ lens, by maintaining at least a(an) ________ -inch (________ -millimeter) clearance between the fixture and the storage area.

17. How many switches are there in the bedroom circuit and of what type are they?

____

____

18. Explain the operation of a ground-fault circuit interrupter. Why are GFCI devices used? Where are GFCI receptacles required? ________

____

____

____

____

____

19. How far above grade line is the weatherproof outlet on this circuit to be mounted?

____

20. The following is a layout of the lighting circuit for Bedroom No. 2. Using the cable layout shown in figure 6-4, make a complete wiring diagram of this circuit. Indicate the color of each conductor.

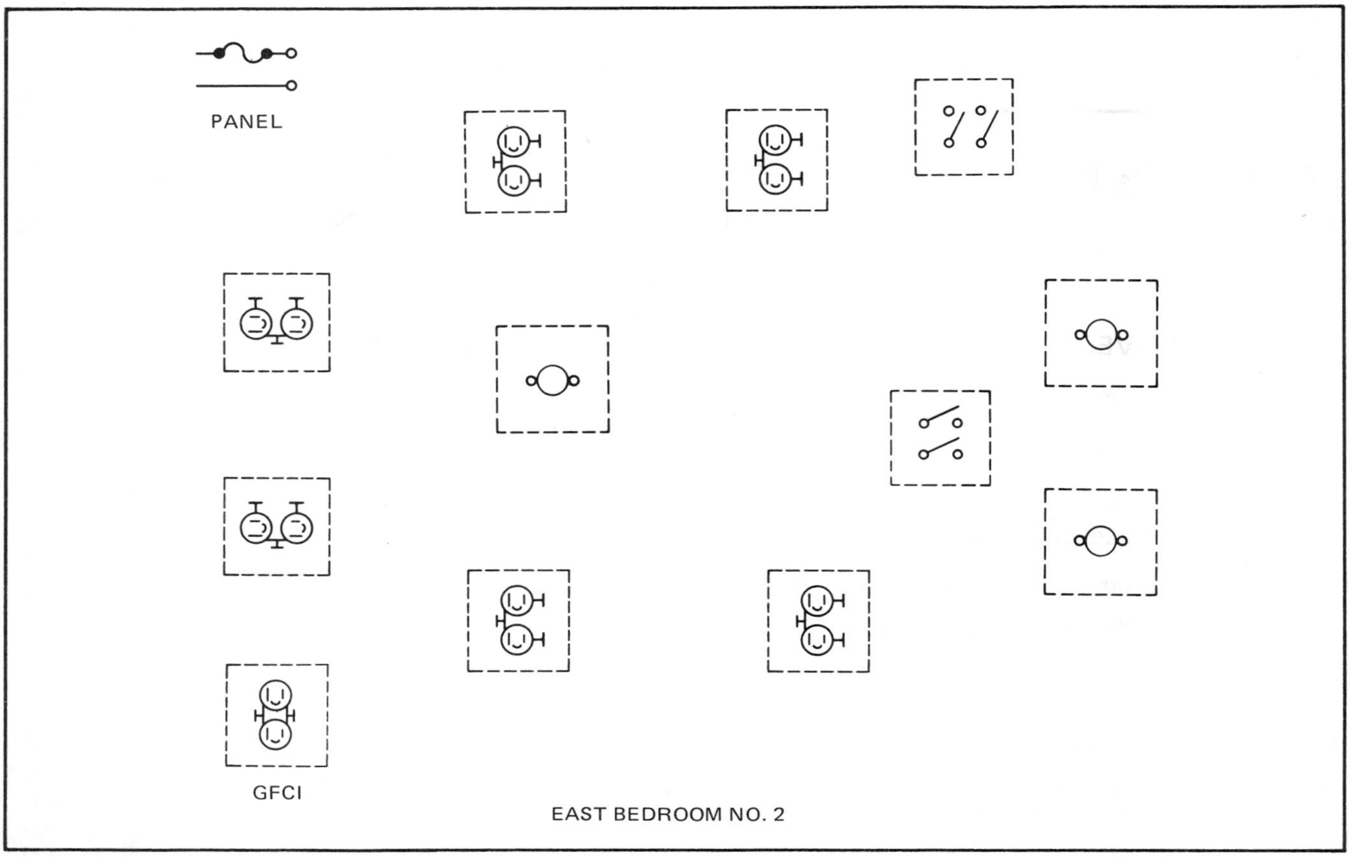

21. When planning circuits, what common practice is followed regarding the division of loads? ______________________________

____

22. Is it permissible to install an electric baseboard heater below a receptacle outlet? Explain. ______________________________

____

23. Residential GFCI devices are set to trip at ground-fault currents above __________ milliamperes.

24. Where must GFCI receptacles be installed in garages? ______________________________

____

unit 7

Lighting Branch Circuit for Bedroom No. 1

OBJECTIVES

After studying this unit, the student will be able to

- draw the wiring diagram of the cable layout for Bedroom No. 1.
- explain how outlets in one room can be connected to a circuit in another room to shift the load from one circuit to another.
- estimate the probable connected wattage for a room based on a number of fixtures and outlets included in the circuit supplying the room.

The discussion in unit 6 of the grouping of outlets, estimating loads, selecting wall box sizes, and drawing wiring diagrams can also be applied to the circuit for Bedroom No. 1.

The residence plans and specifications show that the bedroom is supplied by Circuit 25, a 15-ampere circuit from Panel A. The feed from this panel enters the bedroom at an outlet box located on the northeast bedroom wall 2 1/2 feet from the outside wall. This convenience receptacle is located almost directly over the panel in the workshop.

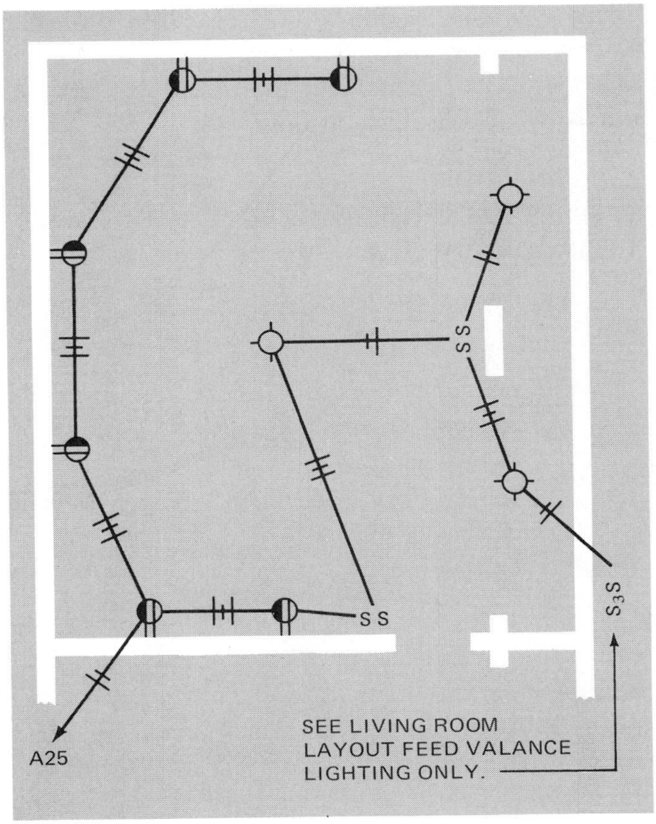

Fig. 7-1 Cable layout for Bedroom No. 1.

CIRCUIT DESIGN

The cable layout shown in figure 7-1 differs from the layout given in unit 6 for Bedroom No. 2. This difference gives the student the opportunity to become familiar with a variety of methods for making connections in these areas.

Figure 7-1 shows that the bedroom circuit also supplies the valance lighting in the living room. The convenience receptacles in the living room must supply all of the lamps (floor lamps and table lamps) used in that area. Because of the lamp load, it was decided to put the valance lighting load on a different circuit. Therefore, the valance lighting is connected to the bedroom circuit. In general, this circuit is less heavily loaded than the living room circuit. The connection for the valance lighting is made at the closet ceiling outlet box nearest the switch controlling the valance lighting.

Figure 7-1 shows that the valance lighting connection is made by running a two-wire cable from the closet ceiling outlet box to the switch box on the living room wall. The actual connection of the valance lighting is covered in unit 11. The cost of the few extra feet of cable required to connect the valance lighting to the bedroom circuit is offset by the fact that there is less load on the living room receptacle circuit. In addition, the use of the bedroom circuit for the valance lighting means that some form of light is always available in the living room if the living room receptacle circuit fails.

BASEBOARD ELECTRIC HEATERS AND RECEPTACLE OUTLETS

Care must be used in wiring Bedroom No. 1 because of the receptacle outlet located in the corner by the windows. See page 55 and unit 23 for the details of installing receptacle outlets and electric baseboard heater units.

OUTLETS AND WATTAGE FOR BEDROOM CIRCUIT

The circuit for Bedroom No. 1 consists of the following devices:

6 convenience (split-current) receptacles
1 ceiling fixture
2 closet lights
1 twelve-foot section of valance lighting (in living room)

Total: 10 Outlets

The probable connected load in watts must be estimated for this circuit. The following values can be assigned: 150 watts for each convenience receptacle, 200 watts as the general lamp load for a bedroom of this size, and a minimum of 60 watts for each closet light. The three ballasts required for the fluorescent valance lighting contribute 56 watts each to the load. The total approximate connected load in watts on the bedroom circuit is:

1 ceiling fixture	200 watts
2 closet lights (60 watts each)	120 watts
6 convenience receptacles (150 watts each)	900 watts
3 fluorescent ballasts (56 watts each)	168 watts
Total	1388 watts Estimated Load

This load is within the 1440-watt limit for the load on a 15-ampere circuit. Note that it is unlikely that all of these outlets will be loaded to capacity at the same time.

SELECTION OF BOXES

As discussed in unit 6, the selection of outlet boxes and switch boxes is made by the electrician. These decisions are based on Code requirements, space allowances, good wiring practices, and common sense.

For example, in Bedroom No. 1, the electrician may decide to install a 4-inch square box with a two-gang raised plaster cover at the box location next to the door. Or, two sectional switch boxes ganged together may be installed at that location.

The type of outlet box to be installed on the ceiling depends upon the number of conductors entering the box. As indicated in figure 7-1, two cables enter the box for a total of seven conductors (five circuit conductors and two grounding conductors). Since only one of the two grounding conductors in the cables is counted, the total number of conductors is five circuit conductors and one grounding conductor or six conductors.

Assume that the box to be used for this installation has external cable clamps and a fixture stud. NEC *Table 370-6(a)* indicates that the following boxes may be used for the installation.

	Maximum number of conductors permitted
4" x 1 1/2" octagonal box	6-1 (for stud) = 5 No. 12 conductors
4" x 2 1/8" octagonal box	9-1 (for stud) = 8 No. 12 conductors

The 4" x 2 1/8" octagonal box is preferred because it provides enough space for the necessary conductors.

As in Bedroom No. 2, the six receptacles are of the split-circuit type. Part of the receptacle is *hot* at all times and the other part of the receptacle is controlled through the wall switch located near the bedroom door.

ESTIMATING CABLE LENGTHS

The length of cable needed to complete an installation can be estimated roughly. It may be possible to run the cable in a straight line directly from one wall outlet to another. In some cases, obstacles such as steel columns, sheet metal ductwork, and plumbing may require the cable routing to follow a longer path.

To insure that the estimate is not short, all measurements are to be made "square." For example, measure from the ceiling outlet straight to the wall, then measure straight down the wall, and finally measure straight over to the wall outlet. Add one foot (305 mm) of cable at each cable termination. This is an allowance to permit the outer jacket of the cable to be stripped at all junction and outlet boxes. Check the plans carefully as an aid in determining where the cables can be routed.

REVIEW

Note: Refer to the Code or the plans where necessary.

1. What circuit supplies Bedroom No. 1? ______
2. What lighting outside of the bedroom is supplied by this circuit? ______
3. What type of convenience receptacles are provided in the bedroom? How many receptacles are there? ______
4. How many ceiling outlets are included in this circuit? ______
5. What is the recommended lamp load in watts for the bedroom ceiling fixture? ______
6. What is the current draw for the bedroom ceiling fixture? ______
7. What is the estimated load in watts for the circuit supplying Bedroom No. 1? ______
8. Which is installed first, the switches and outlet boxes or the cable runs? ______
9. How many conductors enter the bedroom ceiling box? ______
10. What type and size of box may be used for the convenience receptacle located at the point where the feed connects? ______
11. What type of covers are used with 4-inch square outer boxes? ______

    ______
12. a. Does the circuit in Bedroom No. 1 have a grounded conductor? ______
    b. Does it have a grounding conductor? ______
    c. Explain the answers to (a) and (b). ______

    ______
13. Approximately how many feet (meters) of two-wire cable and three-wire cable are needed to complete the circuit supplying Bedroom No. 1? Two-wire cable ______ feet (______ meters). Three-wire cable ______ feet (______ meters).
14. Is there any obstruction in the south corner of Bedroom No. 1 that can hinder attempts to run the three-wire cable between the two convenience outlets near that corner? Explain. ______

    ______

15. If the cable is laid in the notches cut in the corner studs, what protection for the cable must be provided? ______________________________

____

16. How high are the convenience receptacles mounted above the finish floor? __________

17. Approximately how far from the door opening is the first convenience receptacle mounted? (See the plans.) ______________________________

18. What is the distance from the finish floor to the center of the wall switches in this bedroom? ______________________________

19. What is the distance from the rough floor to the center of the wall switches in this bedroom? ______________________________

____

20. It may be necessary to furnish a receptacle as part of the electric baseboard heater. Which receptacle may be provided with a baseboard heater and why? ______________

____

21. The following is a layout of the lighting circuit for Bedroom No. 1. Using the cable layout shown in figure 7-1, make a complete wiring diagram of this circuit. Indicate the color of each conductor.

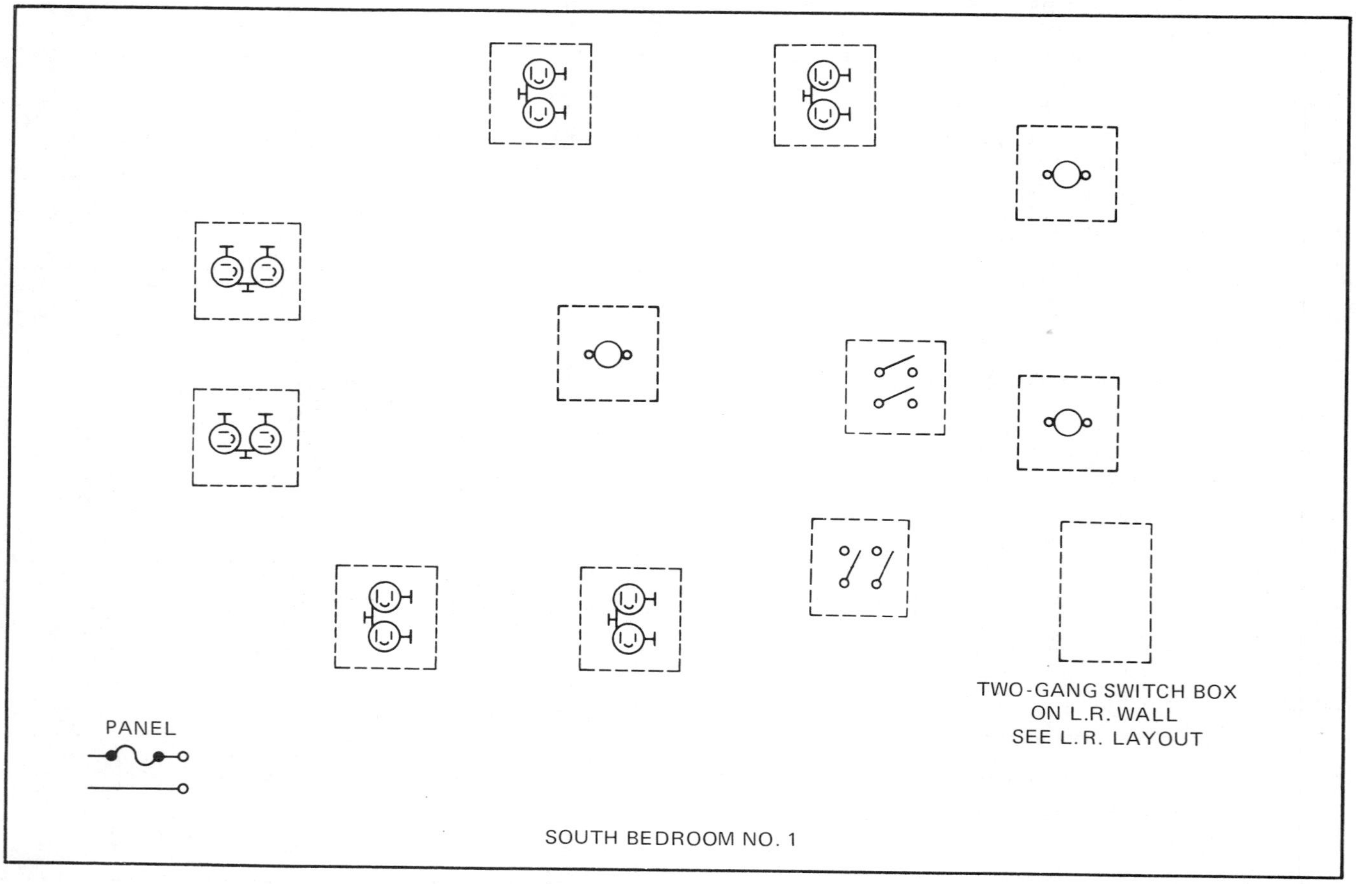

unit 8

Lighting Branch Circuit for Bathroom and Passage

OBJECTIVES

After studying this unit, the student will be able to

- connect recessed fixtures, both prewired and nonprewired types, according to Code requirements.
- discuss the importance of temperature effects when planning recessed installations.
- list equipment grounding requirements for bathroom installations.
- draw the wiring diagram of the bathroom and passage branch lighting circuit.

BATHROOM FIXTURES AND OUTLETS

Surface-Mounted Fixtures

The bathroom lighting system is designed to minimize shadows in the bathroom, particularly at the mirror location for shaving and makeup purposes. As shown in figure 8-1, one 20-watt, trigger-start fluorescent fixture is mounted on each side of the mirror. Two 40-watt, rapid-start fluorescent fixtures are installed within the soffit (enclosure) above the mirror. When the proper fluorescent lamps are installed in these fixtures, there will be a good color rendition. For example, deluxe warm white (WWX) or 3000-K (kelvin) fluorescent lamps can be used in these fixtures to provide *warm* lighting very similar to that provided by incandescent lamps. These fluorescent lamps do not have the *cold* appearance of regular fluorescent lamps but, rather, they bring out the warm (or red) tones in lighting.

The two 40-watt fluorescent fixtures mounted in the soffit, figure 8-2, are installed by the electrician. The carpenter and plasterer are responsible for the framing, wood trim, and plastering of the soffit. The diffusing or translucent plastic panel installed in the soffit shields the fixtures. Plastic sheets are available in many colors and patterns to complement the bathroom decorating scheme.

A three-hole, single-gang switch plate is centered below the 20-watt fixture to the right of the

Fig. 8-1 Cabinet lavatory, first floor bathroom. Note soffit lighting, side lighting, outlets, and electrical baseboard heating unit.

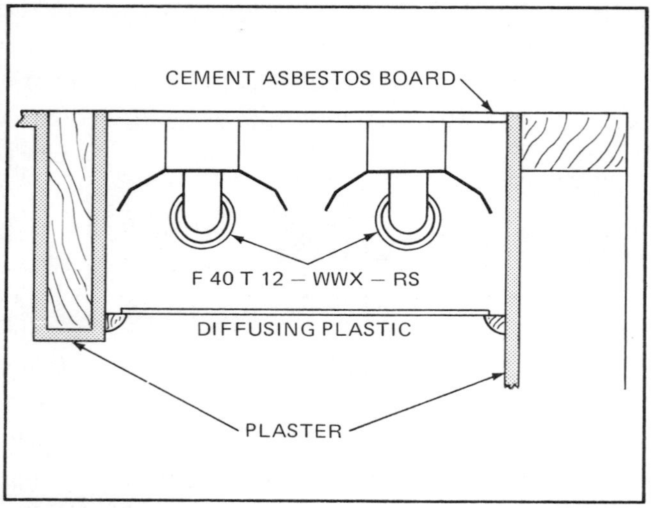

Fig. 8-2 Soffit detail.

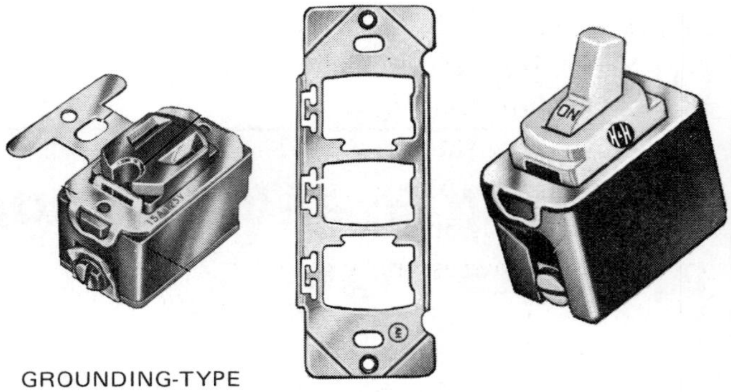

Fig. 8-3 Interchangeable devices used in the bathroom.

mirror. This plate covers a line of interchangeable devices: one switch for the side lights, one switch for the soffit lights, and one grounding-type convenience receptacle, figure 8-3.

Bathroom receptacle outlets must have ground-fault circuit protection according to NEC *Section 210-8*. The NEC defines a bathroom as a room that contains a basin, plus one or more of the following: a toilet, a tub, or a shower, figure 8-4.

Recessed Fixtures

The bathroom also contains several recessed fixtures. One fixture is installed in the center of the bathroom ceiling to provide general lighting. This fixture is controlled by a switch located just inside the bathroom door.

A second recessed fixture is installed in the ceiling in the shower and is controlled by a switch located outside the shower stall. This fixture must be approved for installation in a wet location, *Section 410-4(a)*.

The first floor electrical plans indicate that there are two switches outside the shower stall. One switch box and faceplate are provided for the recessed fixture. The second switch controls a combination heater, light, and exhaust fan which is mounted in the bathroom ceiling. The fixture is supplied with its own two-gang faceplate containing a pilot light and special switch. (Unit 21 covers the installation of this special fixture.)

Code Requirements for Installing Recessed Fixtures

The Code requirements for the installation and construction of recessed fixtures are given in *Sections 410-64* through *410-72*. Of particular importance are the Code restrictions on conductor temperature ratings, fixture clearances from combustible materials, and maximum lamp wattages. Recessed fixtures generate a great deal of heat within the enclosure. **Thus, these fixtures are a fire hazard if they are not wired and installed properly, figure 8-5.**

The branch-circuit conductors are run to the junction box for the recessed fixture. Here they are connected to conductors whose insulation can handle the temperature at the fixture lampholder. The junction box is placed at least one foot (305 mm) from the fixture. Thus, heat radiated from the fixture cannot overheat the wires in the junction box. Conductors rated for higher temperatures must run through at least four feet (1.22 m) of metal raceway, but not to exceed six feet (1.83 m), between the fixture and the junction box. As a result, any heat conducted from the fixture along the metal raceway will be reduced before reaching the junction box. Many recessed fixtures are factory equipped with a flexible metal raceway containing type AF (asbestos fixture) wires that meet the requirements of *Section 410-67*. This is true of adjustable-type recessed fixtures in particular.

A factor to be considered by the electrician when installing recessed fixtures is the necessity of working with the installer of the insulation to be sure that the clearances, as required by the Code, are maintained.

A boxlike device, figure 8-6(A), is available which snaps together around the recessed fixture, thus preventing the insulation from coming into contact with the fixture, as required by the Code. The material used in these boxes is fireproof. Figure 8-6(B) indicates the clearances for a recessed lighting fixture installed near thermal insulation (*Section 410-66*).

Recessed fixtures of the type installed in the shower and ceiling of the bathroom usually have a box mounted on the side of the fixture. The branch-circuit conductors can be run directly into this box where they are connected to the conductors entering the fixture.

Prewired fixtures do not require additional wiring, figure 8-7. *Section 410-11* states that branch-circuit wiring shall not be passed through an outlet box that is an integral part of an incandescent fixture unless the fixture is identified for through wiring.

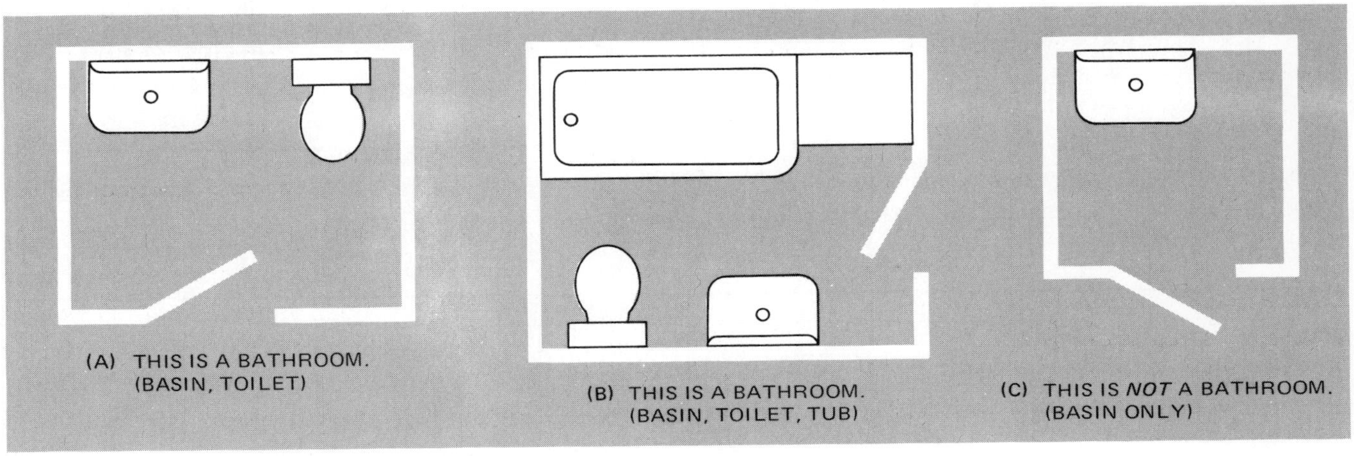
(A) THIS IS A BATHROOM.
(BASIN, TOILET)
(B) THIS IS A BATHROOM.
(BASIN, TOILET, TUB)
(C) THIS IS NOT A BATHROOM.
(BASIN ONLY)

Fig. 8-4 Definition of a bathroom.

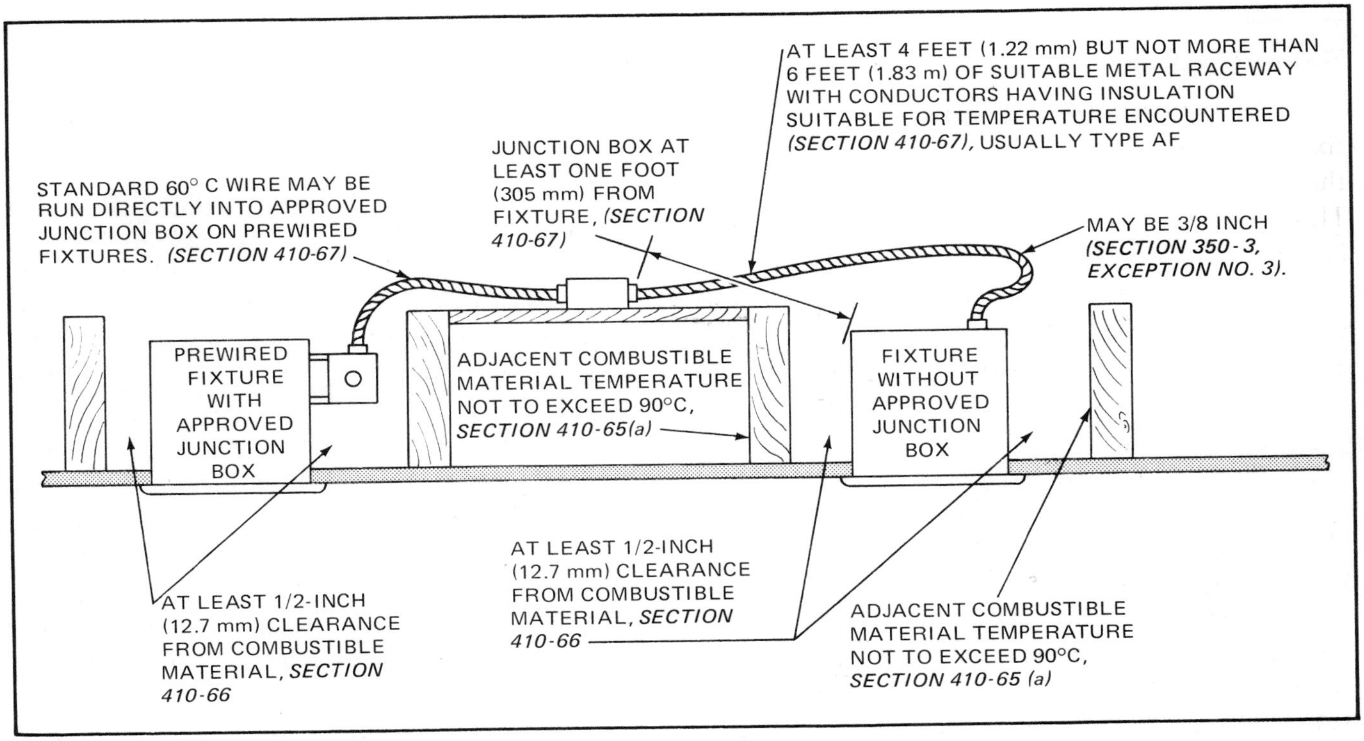
AT LEAST 4 FEET (1.22 mm) BUT NOT MORE THAN 6 FEET (1.83 m) OF SUITABLE METAL RACEWAY WITH CONDUCTORS HAVING INSULATION SUITABLE FOR TEMPERATURE ENCOUNTERED (SECTION 410-67), USUALLY TYPE AF
JUNCTION BOX AT LEAST ONE FOOT (305 mm) FROM FIXTURE, (SECTION 410-67)
STANDARD 60° C WIRE MAY BE RUN DIRECTLY INTO APPROVED JUNCTION BOX ON PREWIRED FIXTURES. (SECTION 410-67)
MAY BE 3/8 INCH (SECTION 350-3, EXCEPTION NO. 3).
PREWIRED FIXTURE WITH APPROVED JUNCTION BOX
ADJACENT COMBUSTIBLE MATERIAL TEMPERATURE NOT TO EXCEED 90°C, SECTION 410-65(a)
FIXTURE WITHOUT APPROVED JUNCTION BOX
AT LEAST 1/2-INCH (12.7 mm) CLEARANCE FROM COMBUSTIBLE MATERIAL, SECTION 410-66
AT LEAST 1/2-INCH (12.7 mm) CLEARANCE FROM COMBUSTIBLE MATERIAL, SECTION 410-66
ADJACENT COMBUSTIBLE MATERIAL TEMPERATURE NOT TO EXCEED 90°C, SECTION 410-65 (a)

Fig. 8-5 Code requirements for installing recessed lighting fixtures.

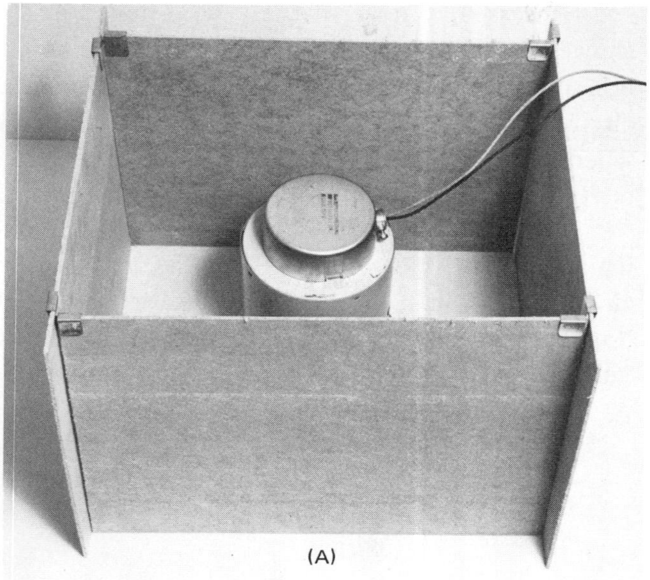

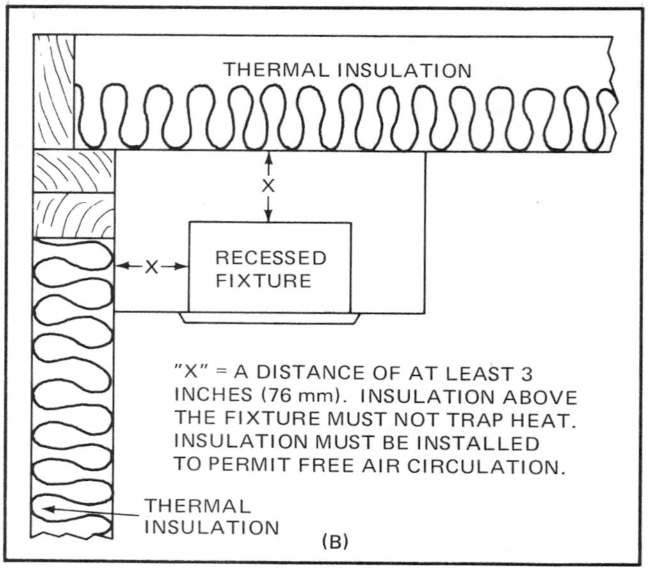

Fig. 8-6 (A) Boxlike device prevents insulation from coming into contact with fixture. (B) Clearances for recessed lighting fixture installed near thermal insulation (*Section 410-66*).

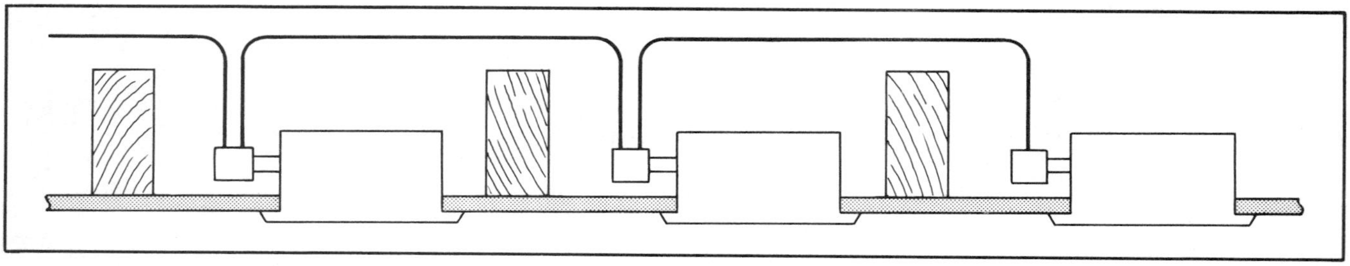

Fig. 8-7 Installation permissible only with prewired recessed fixtures with approved junction box.

For a recessed fixture that is not prewired, the electrician must check the fixture for a label indicating that the conductor used must have an insulation temperature rating of more than 60°C (140°F).

The cables to be installed in the residence have Type T or Type TW insulation. This insulation has a maximum operating temperature rating of 60°C (140°F). If the temperature will exceed this value, conductors with other types of insulation must be installed, *Table 310-13.*

Thermal Protection (*Section 410-65*)

Recessed incandescent fixtures installed after April 1, 1982, must have some type of integral thermal protection, NEC *Section 410-65.* Thermal protection will be provided in or on the fixture. The purpose of thermal protection is to open the circuit supplying the fixture should the temperature exceed the limit for the specific fixture.

Thermal protection is not required where the recessed fixture is identified for use and installation in poured concrete, or if the recessed fixture is identified for installations where the thermal insulation will come into direct contact with the fixture. Fixtures must be so marked by the manufacturer in order to obtain UL listing.

PASSAGE CLOSET FIXTURES

The residence plans show that a door switch, figure 8-8, is mounted in the door jamb of the closet adjacent to the front hall. This switch turns the closet light on when the door is opened and off when the door is closed.

Prior to installing the switch, the cable can be brought out through the rough door framing on the hinge side of the door at a point about six feet (1.83 m) from the floor. The electrician must verify the height of the switch when roughing-in the cable. This step insures that the switch will not interfere with the installation of the door hinges. To complete the installation, the carpenter will cut the proper size opening for the box in the door jamb. The electrician will then finish installing the switch.

Fig. 8-8 Door switch.

LIGHTING BRANCH CIRCUIT FOR BATHROOM AND PASSAGE

Estimating Wattage

The cable supplying the bathroom and passage area is connected to Circuit 26. This is a 15-ampere circuit from Panel A. The cable is fed into the outlet box adjacent to the cabinet lavatory, figure 8-9. It is recommended that a 4-inch square box be used. A single-gang, 4-inch square raised plaster cover can be mounted on the box to receive the grounded convenience receptacle.

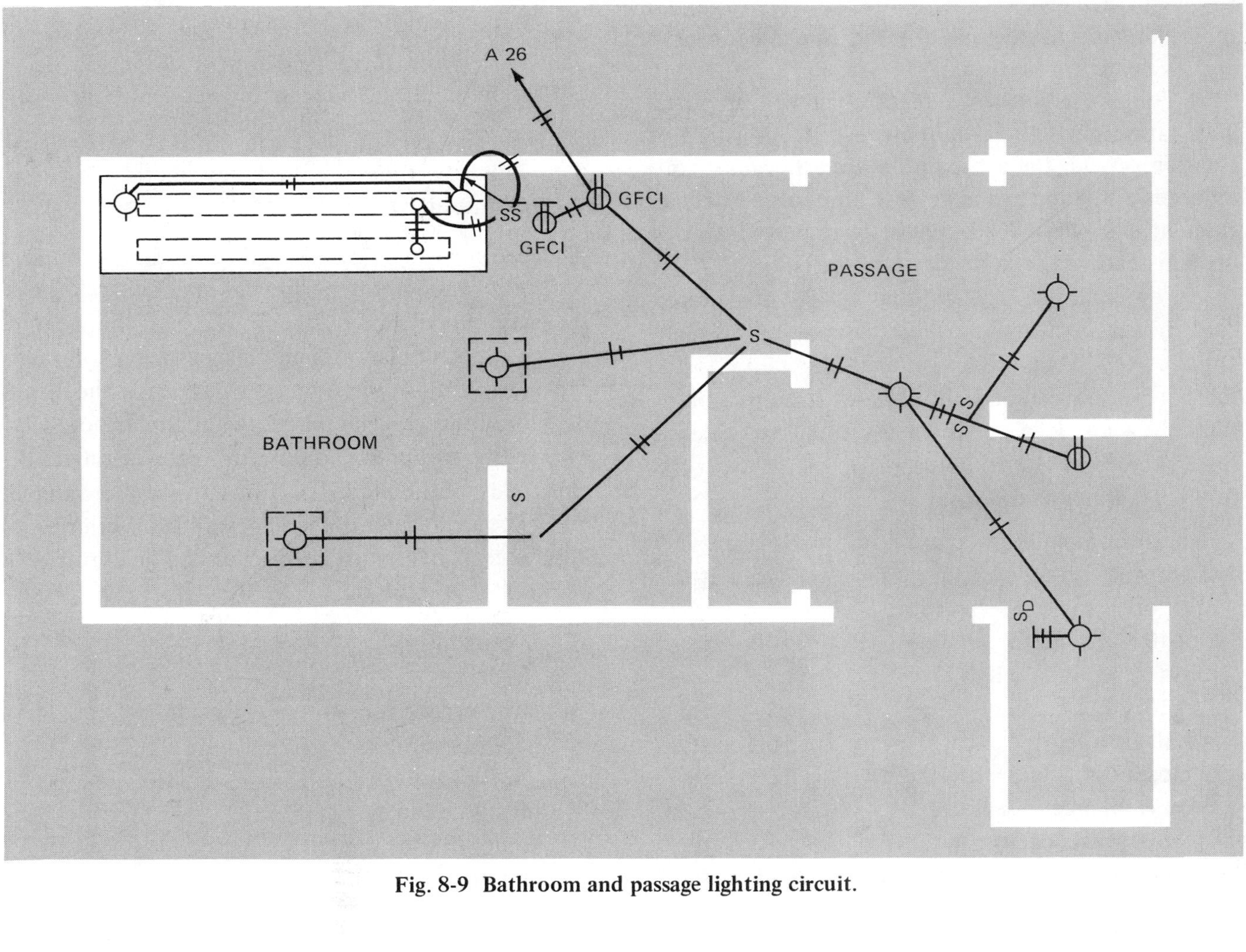

Fig. 8-9 Bathroom and passage lighting circuit.

The bathroom and passage branch lighting circuit consists of the following devices:

3 convenience receptacles
1 ceiling fixture (bathroom)
1 ceiling fixture (shower)
1 pair fluorescent side lights
1 pair fluorescent soffit lights
1 ceiling fixture (hall)
2 closet lights
Total: 10 Outlets

The estimate of the probable connected load for this circuit is fairly accurate because of the fixed maximum loads in watts for the recessed fixtures and the fluorescent fixtures. A value in watts is assumed for those fixtures which do not have rated loads. Thus, values are assumed for the hall, closet lighting, and convenience receptacles. The total rating for a fluorescent lamp consists of the lamp load plus the load of the ballast. To correct for the ballast load, the lamp wattage is multiplied by a factor of 1.25. The resulting value is a reasonably accurate load for the fluorescent lamp and ballast combination.

The ballast may be marked in volt-amperes. This rating gives the true picture of the wattage of the lamp and ballast combination because volt-amperes actually means volts x amperes. For load calculations, therefore, the volt-ampere rating is used instead of the lamp value in watts.

For example, assume that a 40-watt fluorescent fixture contains a ballast rated at 50 volt-amperes (50 VA). This value is the same as the total power consumed by a 40-watt lamp using a factor: 40 x 1.25 = 50 watts. In a load calculation, therefore, the value used for the 40-watt fixture is 50 watts or 50 VA.

1 bathroom ceiling fixture	150 watts
1 shower fixture	75 watts
2 fluorescent fixtures in soffit (40 x 2 x 1.25)	100 watts
2 fluorescent side lights (20 x 2 x 1.25)	50 watts
1 hall fixture	100 watts
2 closet lights (60 watts each)	120 watts
3 convenience receptacles (150 watts each)	450 watts
Total Estimated Load	1045 watts

The total load for Circuit 26 is well within the 1440-watt limit for a 15-ampere branch circuit (15 x 120 x 0.80 = 1440 watts).

EQUIPMENT GROUNDING REQUIREMENTS FOR A BATHROOM CIRCUIT

All exposed metal equipment including fixtures, electric heaters, faceplates, and similar items must be grounded. The equipment grounding requirements for bathroom circuits and any other electrical equipment in a residential building are contained in *Section 250-42*.

In general, all exposed noncurrent-carrying metal parts of electrical equipment must be grounded:

- if they are within eight feet (2.44 m) vertically or five feet (1.52 m) horizontally of the ground or other grounded metal objects.
- if the "ground" and the electrical equipment can be touched at the same time, then the equipment must be grounded.
- if they are located in wet or damp locations, such as in bathrooms, showers, and outdoors.
- if they are in electrical contact with metal. This requirement includes metal lath and aluminum foil insulation.

Equipment is considered grounded when it is properly and permanently connected to metal raceway, the armor of armored cable, the grounding conductor in nonmetallic-sheathed cable, or a separate grounding conductor. Of course, the means of grounding (the metal raceway, the armored cable, or the grounding conductor in the nonmetallic-sheathed cable) must itself be properly grounded.

Those residential electrical cord-connected appliances that must be grounded according to *Section 250-45(c)* are:

refrigerators
freezers
air conditioners
clothes washers
clothes dryers
dishwashers
sump pumps
aquarium equipment

wet scrubbers
hand-held motor operated tools
electric motor operated hedge trimmers
electric motor operated lawn mowers
electric motor operated snow blowers
portable hand lamps

When any of the appliances in this list have special "double insulation" or equivalent, then grounding is not required. Such equipment is clearly marked to indicate the use of double insulation. Generally, such equipment is furnished with a two-wire cord.

REVIEW

Note: Refer to the Code or the plans where necessary.

1. List the number and types of all switches and convenience receptacles used in Circuit 26. ______________________________
____
____
2. What type of fluorescent lamps are to be installed in the soffit? ______________
3. Two switches and one convenience receptacle are located below the side light to the right of the mirror. What type of switches and receptacle are used? ______________
4. What size boxes can be used at the location given in question 3? ______________
5. How high to center from the finish floor should this box be mounted? ______________
6. What type of fixture is used in the shower? ______________
7. Is it permitted to mount the switch for the shower fixture close to the end of the partition that extends toward the center ceiling fixture in the bathroom? Explain.
____
____
8. Is it permissible to place the ceiling or shower recessed fixtures directly against the wood ceiling joists? Why? ______________________________
____
9. If a recessed fixture without an approved junction box is installed, what extra wiring must be provided? ______________________________
____
____
10. How many closet lights are supplied by this circuit? ______________
11. What special type of switch is used to control one of the passage closet lights?
____
12. How much current does the bathroom ceiling fixture draw? ______________
13. Thermal insulation shall not be installed within ____________ inches (millimeters) of the top, or ____________ inches (millimeters) of the side of a recessed fixture.
14. Exposed noncurrent-carrying metallic parts of electrical equipment must be grounded if installed within ____________ feet (meters) vertically or ____________ feet (meters) horizontally of bathtubs, plumbing fixtures, pipes, or other grounded metal work or grounded surfaces.

15. Are there any metal faceplates or lighting fixtures in the bathroom that are *not* required to be grounded? (Refer to *Sections 250-42, 410-18,* and *410-21.*) ____________

____

16. The top of the soffit is lined with what type of material? ____________

17. The faceplates to be installed in the bathroom are made of what material? ____________

18. Most electrical appliances, such as hedge trimmers, motorized barbecue grills, lawn mowers, among others, are furnished with a three-wire cord and a three-wire grounding-type attachment plug cap. If the appliance has a two-wire cord it is probably ________ insulated.

19. The following is a layout of the lighting circuit for the Bathroom and Passage. Using the cable layout shown in figure 8-9, make a complete wiring diagram of this circuit. Use colored pencils to indicate the conductors.

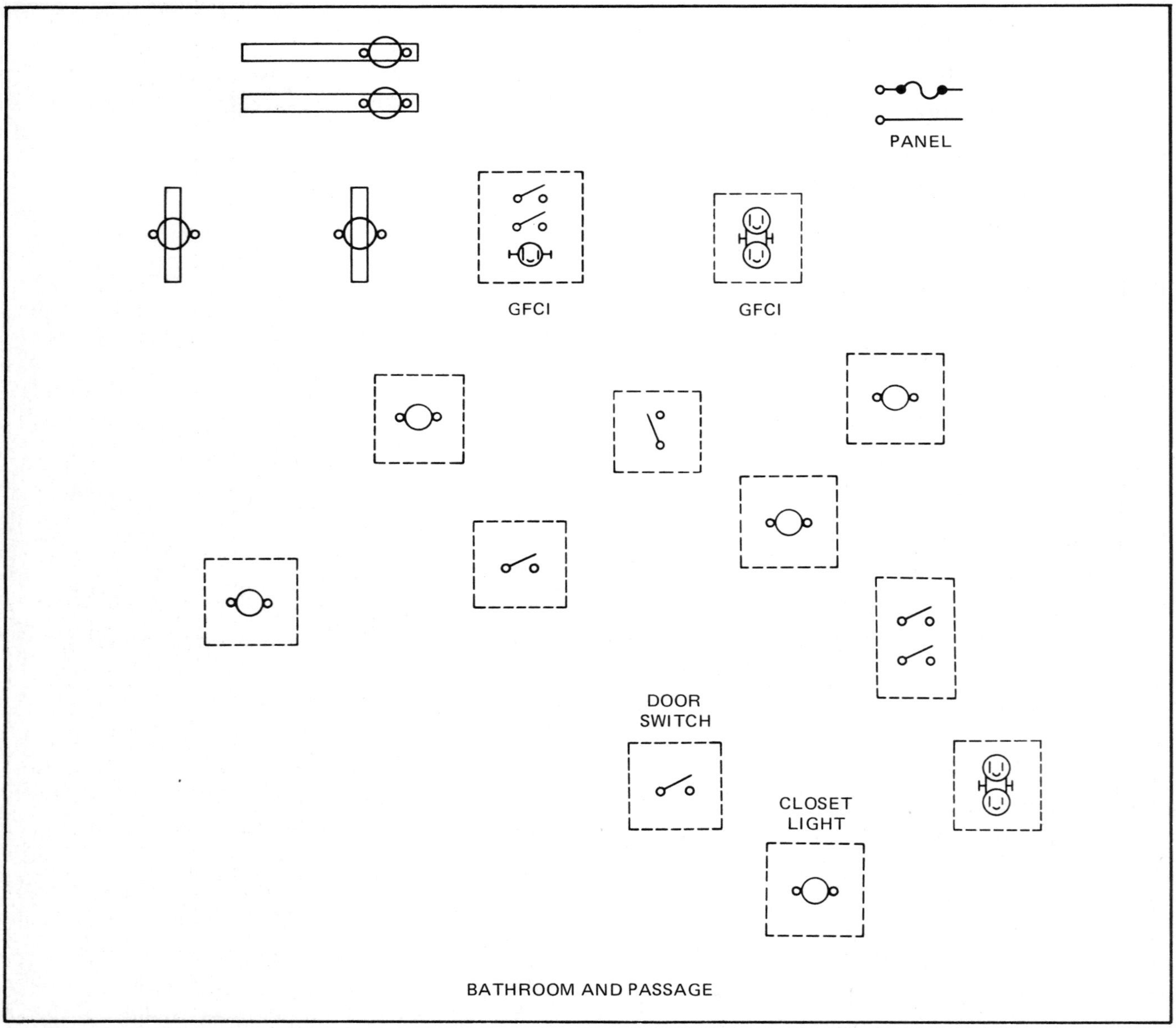

unit 9

Lighting Branch Circuit for the Hall and Front Entrance

OBJECTIVES

After studying this unit, the student will be able to

- install three-wire circuits, making the proper connections to prevent overloads.
- demonstrate the proper methods of installing pilot light switches in circuits.
- explain how outdoor lighting can be installed using type UF underground cable.
- describe two methods of bringing conduit into a residence from an external location.
- define a wet location.

WIRING CONSIDERATIONS FOR THE CIRCUIT

The hall, front entrance, shrub lights, and post lights are grouped to form Circuit A29. (Remember, this is only one of many possible groupings.) Figure 9-1 shows the outlets that are part of Circuit A29.

The cable layout of figure 9-1 differs from those in previous units because a three-wire cable is connected to Panel A (Circuits A29 and A31). The cable runs to the front hall ceiling outlet box. The black conductor of the cable (Circuit A29) supplies the hall and front entrance area. The red conductor (Circuit A31) is spliced to the black conductor of the two-wire cable supplying the kitchen lighting.

Note: A three-wire cable installation is an economical method of carrying two circuits to a given point and then dividing them so that one circuit feeds one area and the other circuit feeds another area.

The white conductor of the three-wire cable is common to both the hall and kitchen circuits. Care must be used when connecting a three-wire circuit to the power panel. The black and red conductors of the three-wire cable must be connected to opposite phases in Panel A to prevent heavy overloading of the neutral grounded (white) wire, figure 9-2.

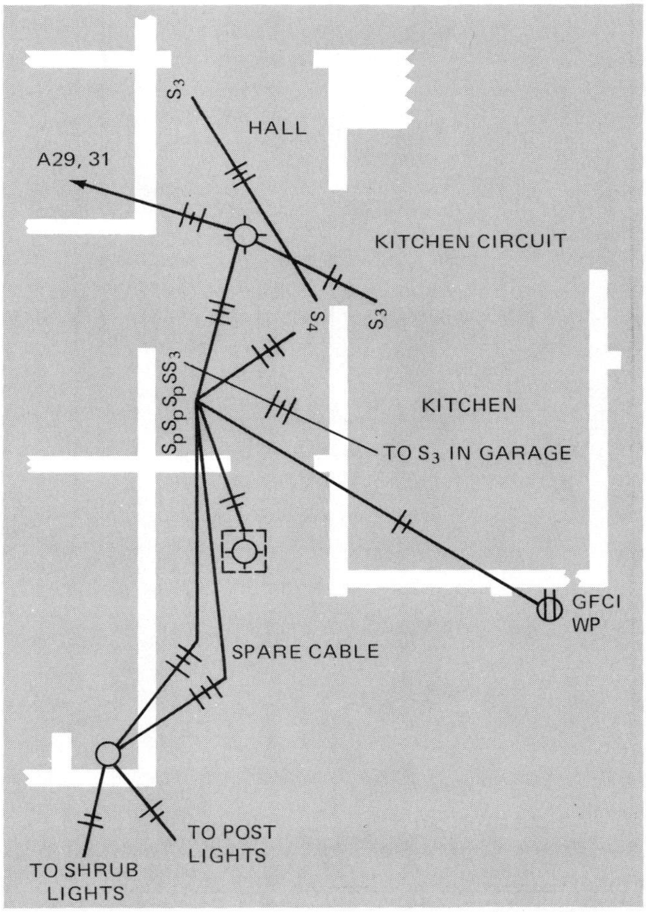

Fig. 9-1 Hall and front entrance lighting branch circuit.

The neutral (grounded) conductor of the three-wire cable carries the unbalanced current. This current is the difference between the current in the black wire and the current in the red wire. For example, if the kitchen load is 12 amperes and the hall front entrance load is 10 amperes, the neutral current is the difference between these loads, or 2 amperes, figure 9-3.

If the black and red conductors of the three-wire cable are connected to the same phase in Panel A, figure 9-4, the neutral conductor must carry the total current of both the red and black conductors rather than the unbalanced current. As a result, the neutral conductor will be overloaded. All single-phase, 120/240-volt panels are clearly marked to help prevent an error in phase wiring. The electri-

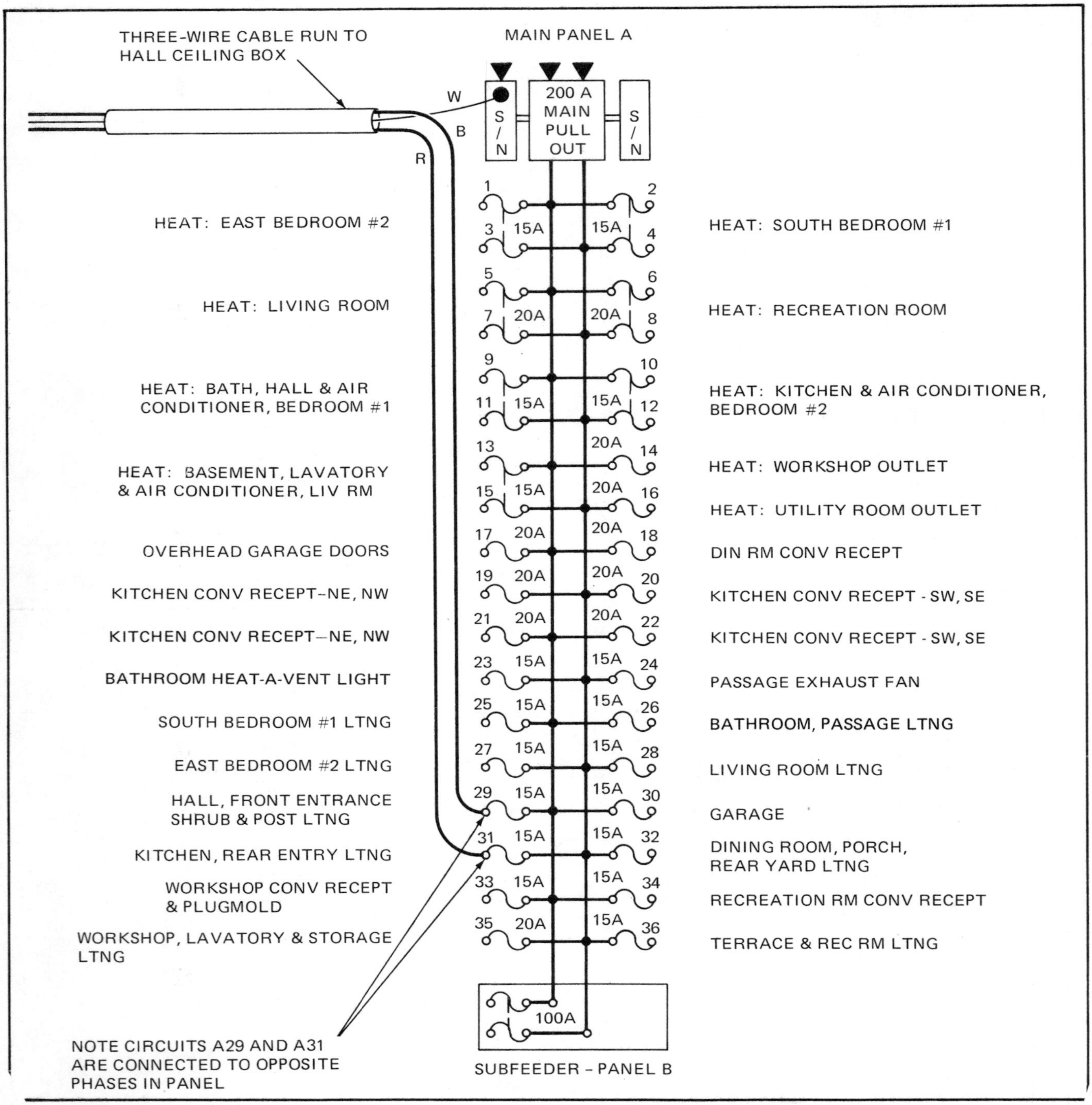

Fig. 9-2 Connection of circuits A29 and A31 to main Panel A.

cian must check all panels for the proper wiring diagrams before beginning an installation.

Figure 9-4 shows how an improperly connected three-wire circuit results in an overloaded neutral conductor.

An advantage of the three-wire circuit is that the voltage drop and wattage loss in the circuit are less than would occur if a similar load were connected to separate two-wire circuits. If an open neutral occurs on a three-wire circuit, some of the electrical appliances in operation may experience voltages higher than the rated voltage at the instant the neutral opens.

For example, figure 9-5 shows that for an open neutral condition, the voltage across load A decreases and the voltage across load B increases. If the load on each circuit changes, the voltage on each circuit also changes. According to Ohm's

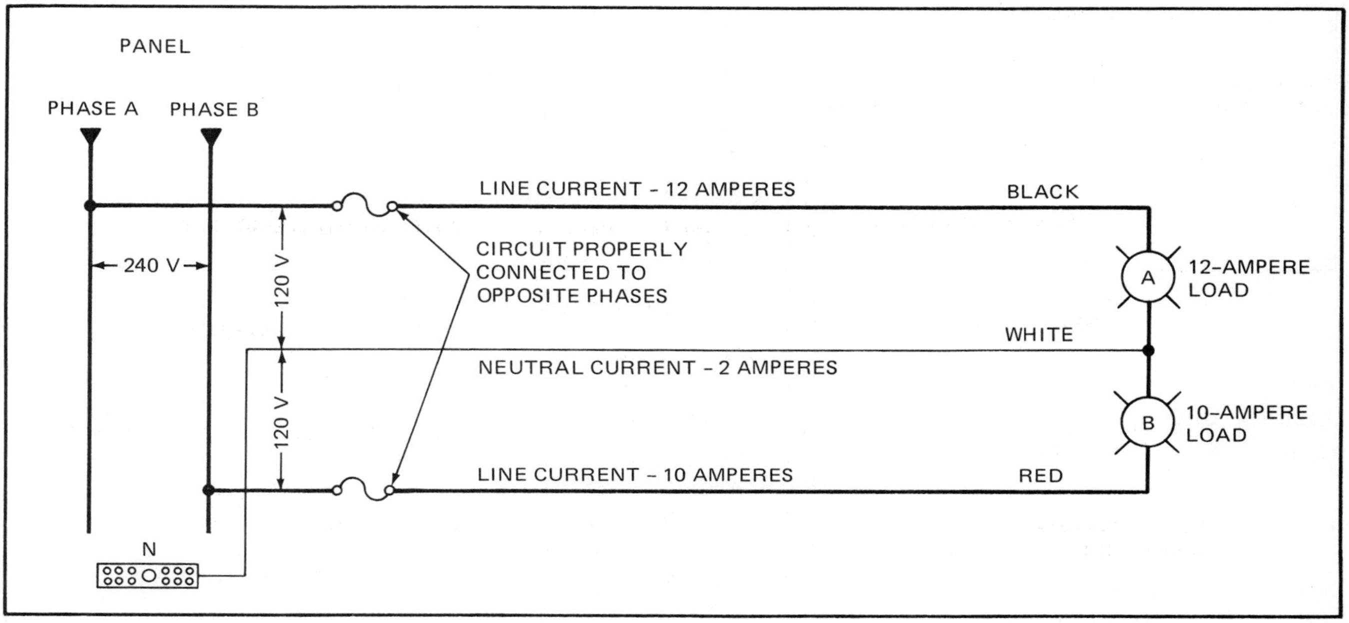

Fig. 9-3 Correct wiring connections for three-wire cable.

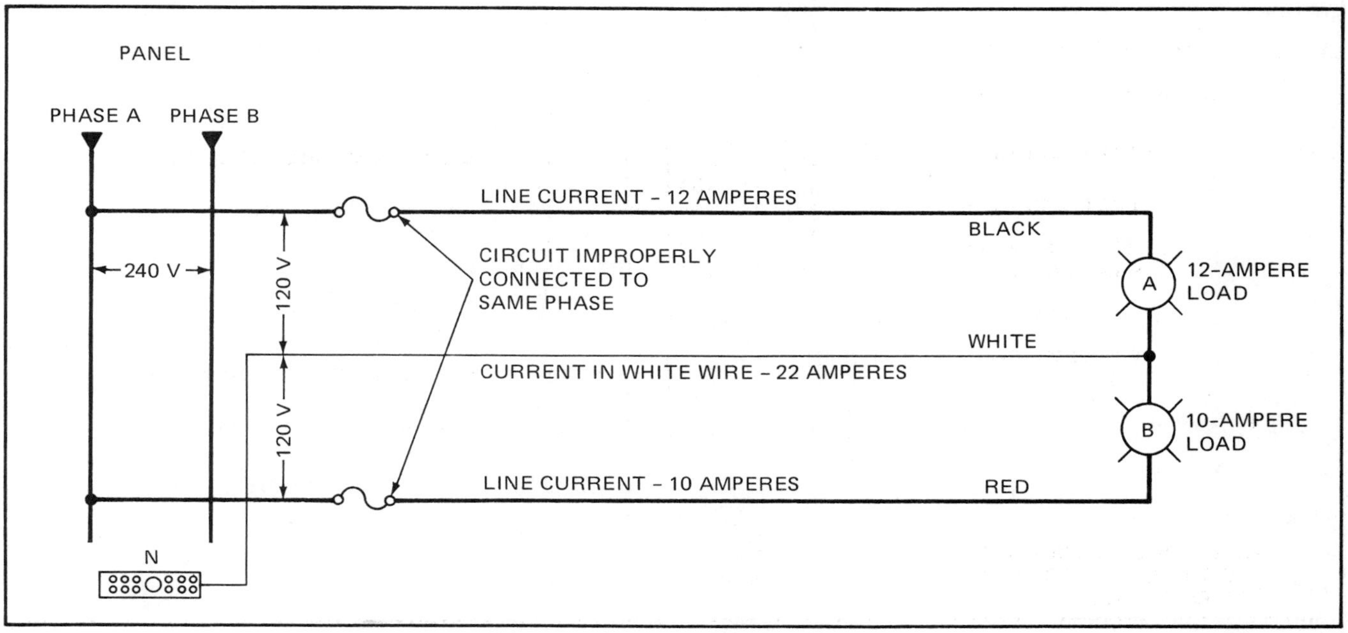

Fig. 9-4 Improperly connected three-wire cable.

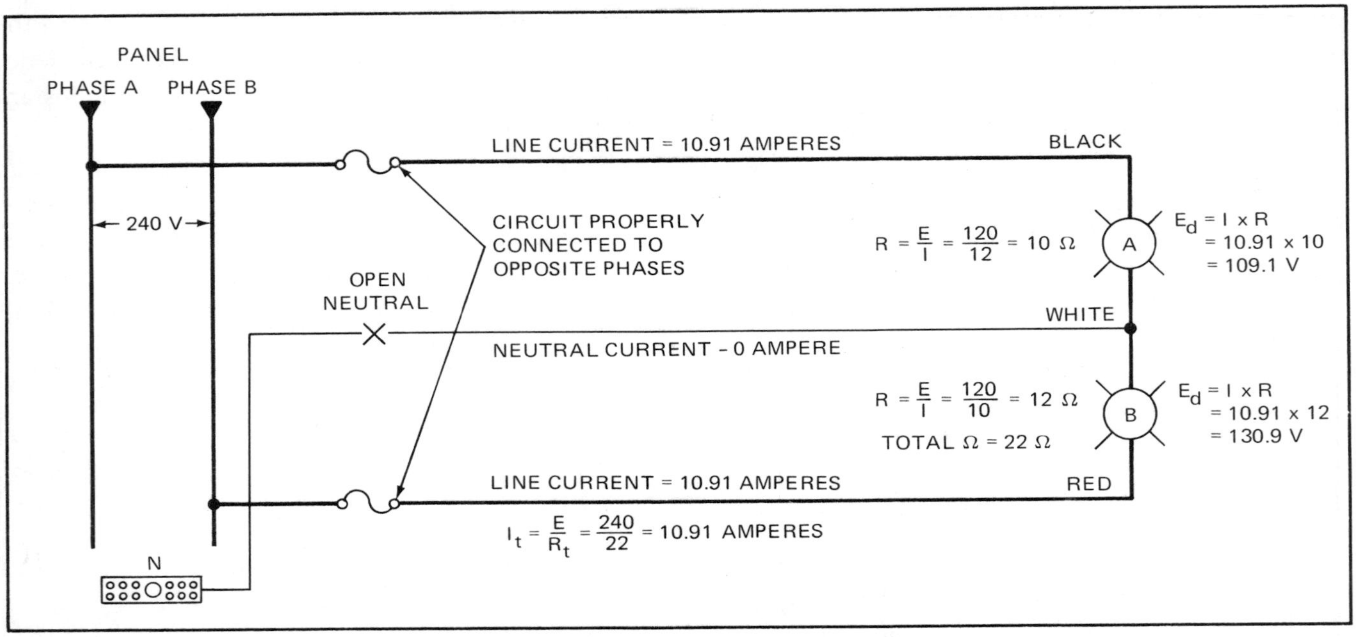

Fig. 9-5 Example of an open neutral conductor.

Law, the voltage drop across any device in a series circuit is directly proportional to the resistance of that device. In other words, if load B has twice the resistance of load A, then load B will be subjected to twice the voltage of load A for an open neutral condition. To insure the proper connection, care must be used when splicing the conductors.

An example of what can occur should the neutral of a 3-wire multiwire circuit open is shown in figure 9-6. Trace the flow of current from Phase A, through the television set, then through the toaster, then back to Phase B, thus completing the circuit. The following simple calculations show why the television set (or stereo or home computer) can be expected to burn up.

$$R_t = 8.45 + 80 = 88.45 \text{ ohms}$$

$$I = \frac{E}{R} = \frac{240}{88.45} = 2.71 \text{ amperes}$$

Voltage appearing across the toaster:

$$I_R = 2.71 \times 8.45 = 22.9 \text{ volts}$$

Voltage appearing across the television:

$$I_R = 2.71 \times 80 = 216.8 \text{ volts}$$

This example illustrates the problems that can arise with an open neutral on a 3-wire, 120/240-volt multiwire branch circuit.

The same problem can arise when the neutral of the utility company's incoming service-entrance conductors (underground or overhead) opens. The problem is minimized because the neutral of the service is solidly grounded to the water piping system within the building. However, there are cases on record where poor service grounding has resulted in serious and expensive damage to appliances within the home because of an open neutral in the incoming service-entrance conductors. For example, poor service grounding results when relying on a driven iron pipe (that rusts in a short period of time) as the only means of obtaining the service equipment ground.

The integrity of using grounds rods *only* is highly questionable, whether the rod is an iron pipe, a galvanized pipe, or a copper-clad rod. The Code prohibits the use of ground rods as the only means of attaining a good service-entrance equipment ground.

See unit 27 for details on the grounding and bonding of service-entrance equipment.

In summary, three-wire circuits are used frequently in residential installations. The two hot conductors of three-wire cable are connected to opposite phases. One white grounded neutral conductor is common to both hot conductors. In the single-phase systems used in residences, only two hot conductors can be connected with a common neutral. Thus, a four-wire cable cannot be used for the cir-

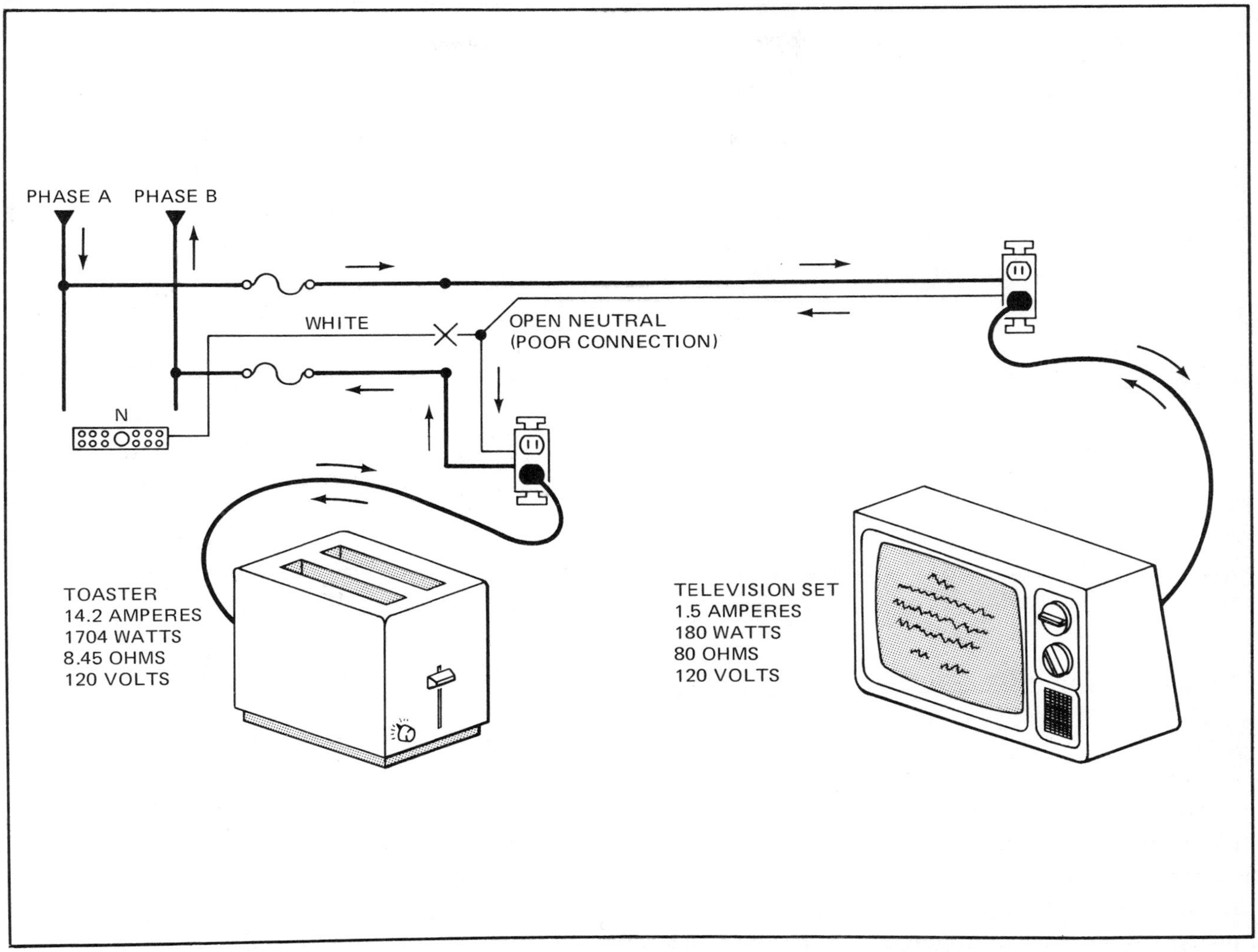

Fig. 9-6 Problems that can occur with an open neutral on a 3-wire multiwire circuit.

cuit since the white wire would then be common to three hot conductors, resulting in overloading of the neutral conductor. However, four-wire cable can be used for switch legs and as travelers of three-way switches.

OUTDOOR LIGHTING

Most adequately wired homes have some form of outdoor lighting. The residence plans show two post lights located on the front lawn and three 150-watt PAR reflector flood lamps located under, in front of, or behind certain shrubs or trees. The electrician must verify the exact locations of these lights with the owner. To provide for outdoor lighting, weatherproof receptacle outlets may be stubbed out of the ground in various locations, figure 9-7. Either 120-volt or low-voltage lighting devices can then be plugged into these outlets. Figure 9-8 shows how these outlets or threaded conduit bodies can be supported.

PILOT LIGHT SWITCHES IN CIRCUIT

As explained in unit 6, the grounding-type convenience receptacle provided on the front of the house must be weatherproof and must have ground-fault protection. The receptacle is controlled by a pilot light switch. The pilot light may be located in the handle of the switch or it may be separately mounted. The pilot light indicates when the switch is on. Thus, when any load is plugged

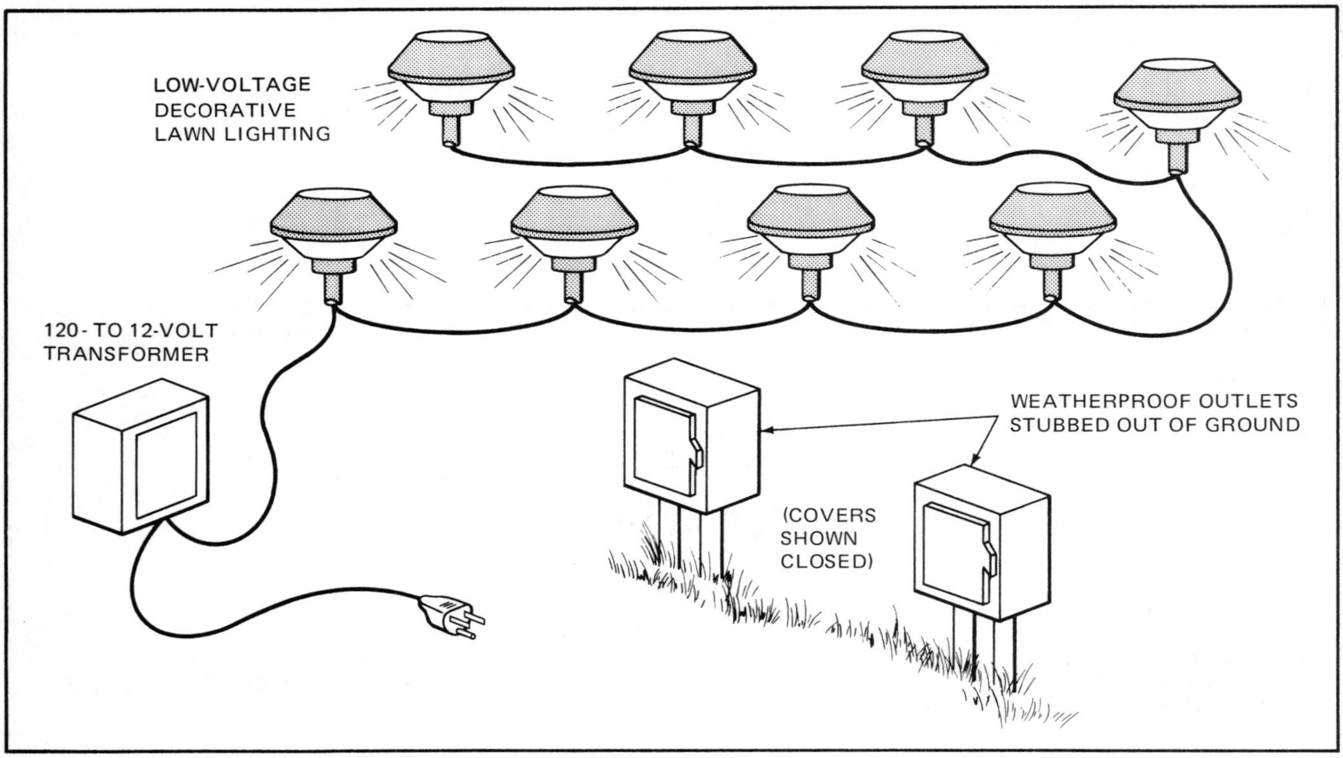

Fig. 9-7 Weatherproof receptacle outlets stubbed out of the ground; either low-voltage decorative lighting or 120-volt PAR lighting fixtures may be plugged into these outlets.

into the receptacle and the pilot light is on, the load is also on. Pilot light switches are also used to control the shrub lighting and post lights. Figure 9-9 shows how pilot lamps are connected in circuits containing either single-pole or three-way switches.

(*Note:* Unit 12 covers the installation and grounding requirements of outdoor convenience receptacles. Unit 6 covers the installation of ground-fault circuit interrupters for such receptacles.)

WIRING WITH TYPE UF CABLE (*ARTICLE 339*)

The plans show that Type UF underground cable, figure 9-10, is used to connect the outdoor lighting. According to *Article 339*, Type UF cable

- is marked "Underground Feeder Cable."
- is available in sizes from No. 14 AWG through No. 4/0 (for copper conductors) and from No. 12 AWG through No. 4/0 (for aluminum conductors). See *Table 310-16* for the ampacity ratings.
- may be used in direct exposure to sun if the cable is marked "Sunlight Resistant."
- may be used with nonmetallic-sheathed cable fittings.
- is flame-retardant.
- is moisture-, fungus-, and corrosion-resistant.
- may be buried directly in the earth.
- may be used for branch-circuit and feeder wiring.
- may be used in interior wiring for wet, dry, or corrosive installations.
- is installed by the same methods as nonmetallic-sheathed cable. (*Article 336*)
- must *not* be used as service-entrance cable.
- must *not* be embedded in concrete, cement, or aggregate.
- must be buried in the same trench where single-conductor cables are installed.
- shall be installed according to NEC *Section 300-5*.

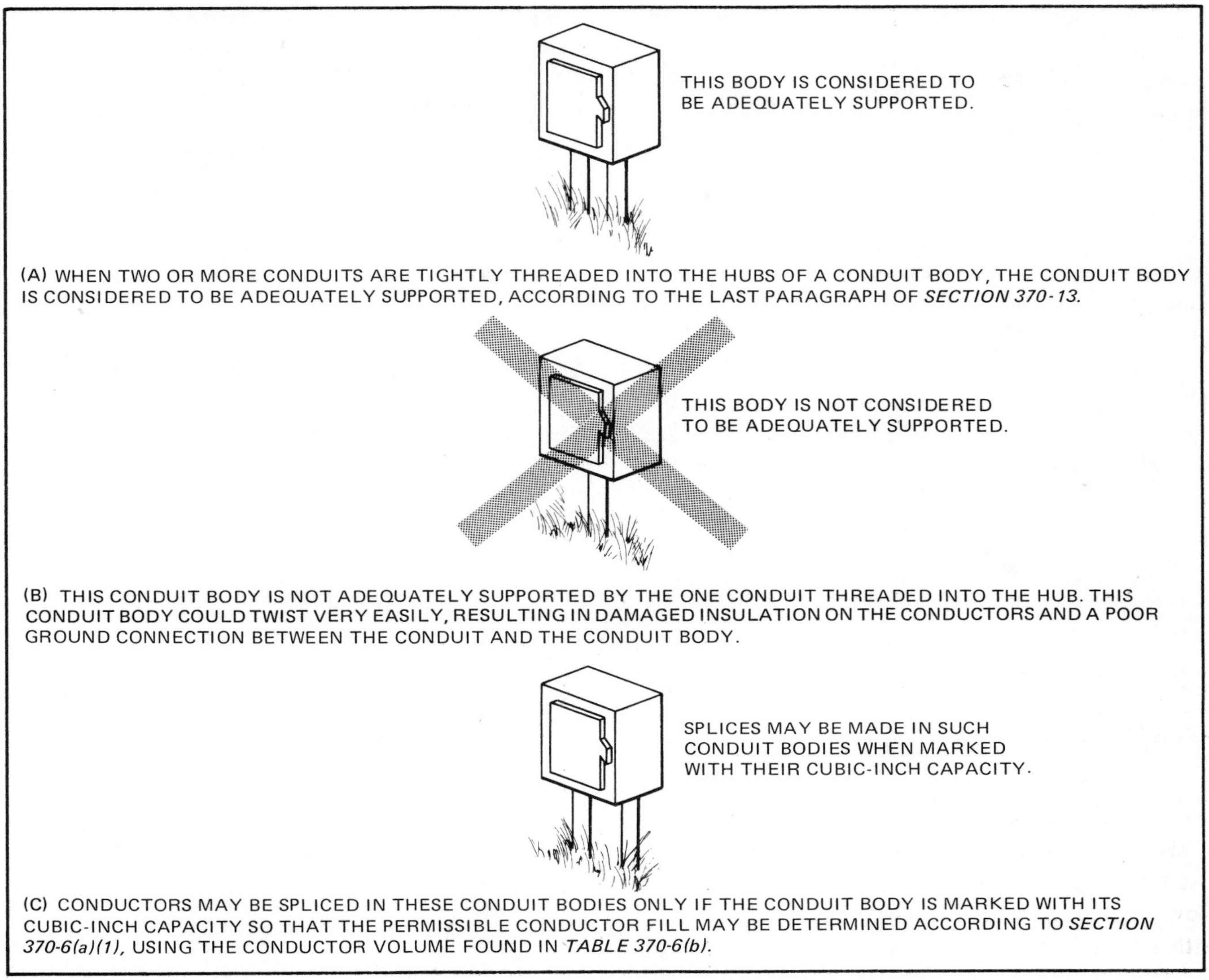

THIS BODY IS CONSIDERED TO BE ADEQUATELY SUPPORTED.
(A) WHEN TWO OR MORE CONDUITS ARE TIGHTLY THREADED INTO THE HUBS OF A CONDUIT BODY, THE CONDUIT BODY IS CONSIDERED TO BE ADEQUATELY SUPPORTED, ACCORDING TO THE LAST PARAGRAPH OF *SECTION 370-13.*
THIS BODY IS NOT CONSIDERED TO BE ADEQUATELY SUPPORTED.
(B) THIS CONDUIT BODY IS NOT ADEQUATELY SUPPORTED BY THE ONE CONDUIT THREADED INTO THE HUB. THIS CONDUIT BODY COULD TWIST VERY EASILY, RESULTING IN DAMAGED INSULATION ON THE CONDUCTORS AND A POOR GROUND CONNECTION BETWEEN THE CONDUIT AND THE CONDUIT BODY.
SPLICES MAY BE MADE IN SUCH CONDUIT BODIES WHEN MARKED WITH THEIR CUBIC-INCH CAPACITY.
(C) CONDUCTORS MAY BE SPLICED IN THESE CONDUIT BODIES ONLY IF THE CONDUIT BODY IS MARKED WITH ITS CUBIC-INCH CAPACITY SO THAT THE PERMISSIBLE CONDUCTOR FILL MAY BE DETERMINED ACCORDING TO *SECTION 370-6(a)(1)*, USING THE CONDUCTOR VOLUME FOUND IN *TABLE 370-6(b)*.

Fig. 9-8 Supporting threaded conduit bodies.

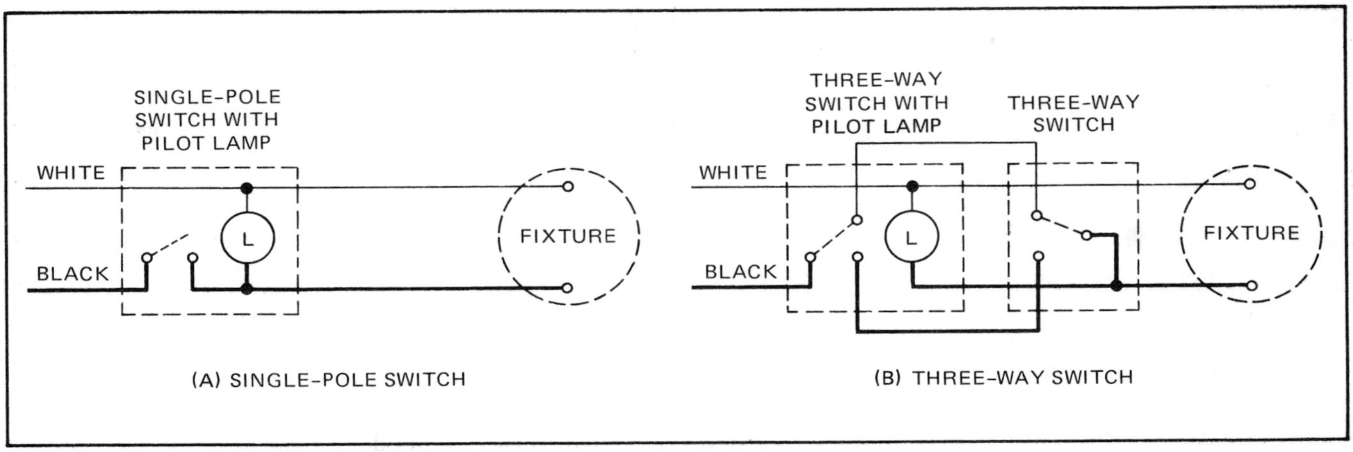

Fig. 9-9 Pilot lamp connections.

The underground cable is run from the shrub lights and post lights in a conduit which passes through the basement wall to a junction box in the basement storage room. Standard nonmetallic-sheathed cable runs from the junction box to the switches in the front hall. The plans show that a spare cable is specified to accommodate any additional switch-controlled lighting that may be installed. To complete the installation, the electrician must use one of the three-wire cables between the switch location and the junction box. The white wire from this cable must be spliced to the white wires of the underground cables from the post and shrub lights. The black wire must be spliced to the black wire of the post light underground cable. The red wire must be spliced to the black wire of the shrub light underground cable.

The connections are made properly if 10 amperes are carried by the *hot* conductors and 10 amperes return through the white conductor of the same cable. All conductors of a circuit must be run in the same conduit, in the same trench, and through the same opening in a metal switchbox. This requirement prevents "inductive heating" produced by the alternating current, *Section 300-20.* The larger the current value, the more heating will take place in the metal.

In many cases, unusual methods are used to connect three-way switches and other circuitry in an attempt to save wiring materials. For example,

Fig. 9-10 Type UF underground cable.

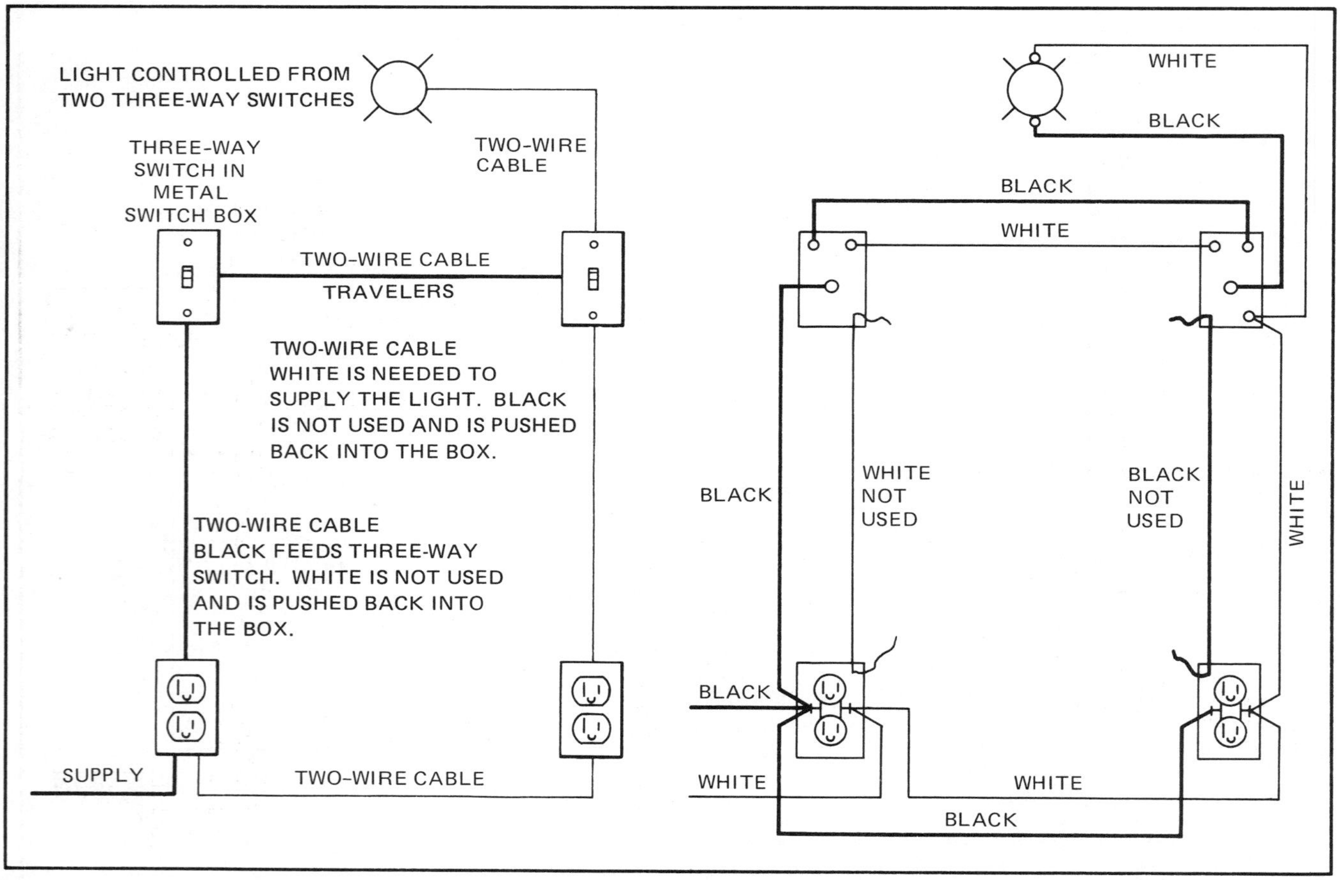

Fig. 9-11 Example of violation of *Section 300-20* when metal switch boxes are used. When nonmetallic boxes are used, there is no Code violation.

someone may use two-wire cable between three-way switches and then pick up the grounded conductor at the other end, figure 9-11. This arrangement is in violation of NEC *Section 300-20* and causes induced heat where the conductors enter the metal boxes. (See unit 5 for the correct way to connect switches.)

When two three-wire cables are installed between the switch location and the junction box, it is possible to make improper connections. In a correct installation, the two switch legs and the white conductor must all be within the same cable. (At present, none of the conductors in the spare three-wire cable is to be used.)

Depth of Installation of UF Cable

Figure 9-12 illustrates the recommended depth of installation of UF cable for residential uses (*Section 300-5*).

INSTALLATION OF CONDUIT

Many local building codes require the installation of rigid metal conduit with Type RW or TW conductors for all underground wiring. When conduit is used in the residence in the plans, the post and shrub lights are grounded by connecting them to the conduit. The Code requirements for grounding to minimize the shock hazard are discussed in unit 8.

Figure 9-12 shows two methods of bringing the conduit into the basement: (1) it can be run below ground level and then can be brought through the basement wall, or (2) it can be run up the side of the building and through the basement wall at the ceiling joist level. When the conduit is run through the basement wall, the opening must be sealed to prevent moisture from seeping into the basement. The electrician must decide which of the two methods is more suitable for each installation.

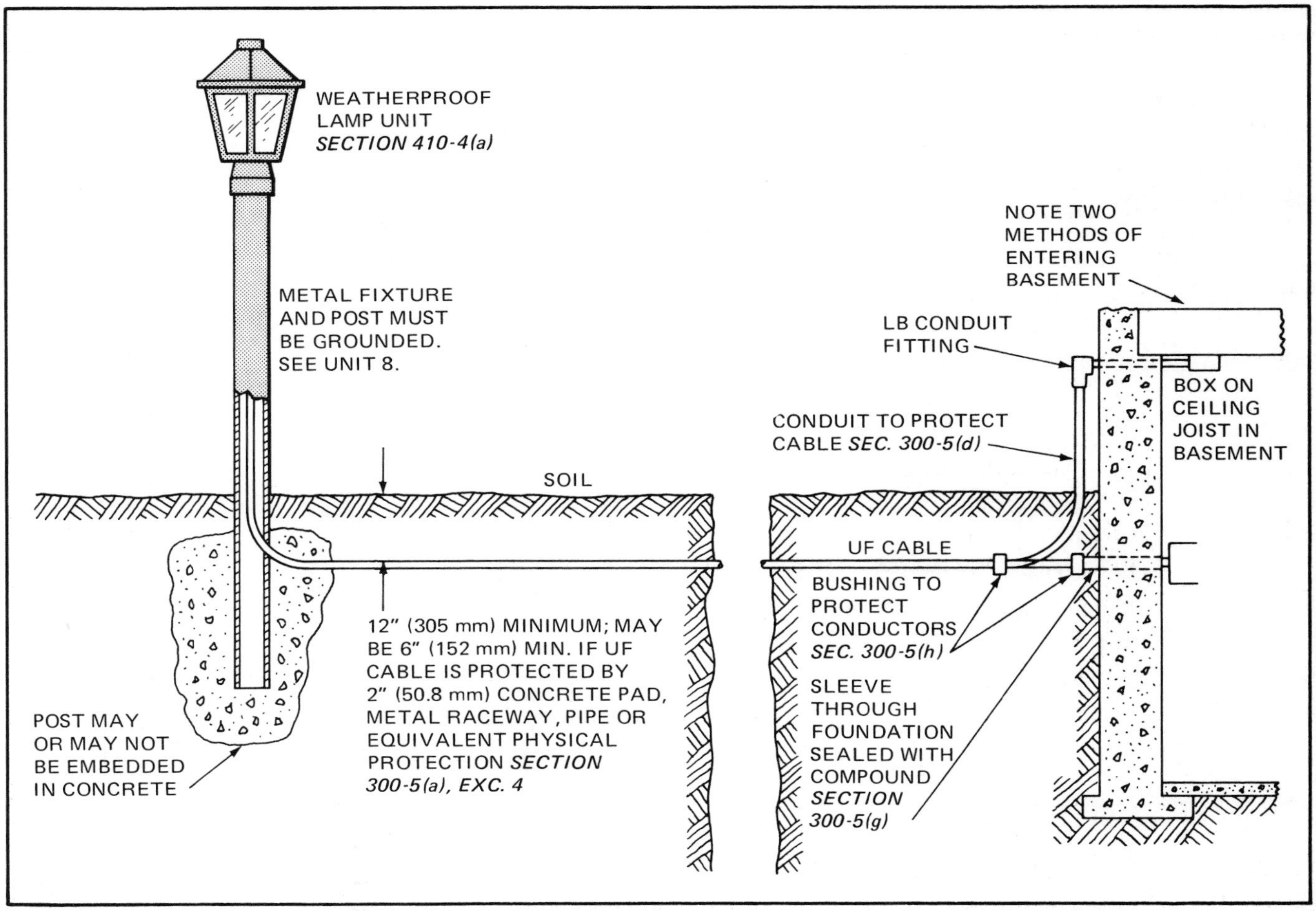

Fig. 9-12 Methods of bringing cable into basement from the post light.

According to the Code, any location exposed to the weather is considered a wet location.

Section 410-4(a) states that outdoor fixtures must be constructed so that water cannot enter or accumulate in lampholders, wiring compartments, or other electrical parts. These fixtures must be marked "Suitable for Wet Locations."

Partially protected areas such as roofed open porches or areas under canopies are defined by the Code as damp locations. Fixtures to be used in these locations must be marked "Suitable for Damp Locations." Many types of fixtures are available. Therefore, it is recommended that the electrician check the UL label on the fixture or refer to the UL *Electrical Construction Materials List* to determine the suitability of the fixture for a wet or damp location.

The post in figure 9-12 may or may not be embedded in concrete, depending on the consistency of the soil, the height of the post, and the size of the lighting fixture. Most electricians prefer to embed the base of the post in concrete to prevent rotting of wood posts and rusting of metal posts.

SWITCH LOCATIONS AND ARRANGEMENTS FOR HALL, FRONT ENTRANCE, AND OUTDOOR LIGHTING

The switch box in the front hall of the residence contains five switches with three pilot lights. If the switches have pilot lights in the handles, a five-gang switch plate can be installed. If separate pilot lights are used, then an eight-gang faceplate must be installed. Alternatively, a separate three-gang faceplate for the pilot lights can only be installed above the switches. If the five-gang faceplate is used, five sectional switch boxes can be ganged together or a five-gang multiple-gang box with a five-gang raised plaster cover can be used. Since the multiple-gang box must accommodate many No. 12 conductors, the electrician must check *Article 370* to determine the box size required.

For added convenience, the front hall fixture may be controlled from any one of the three locations using two three-way switches and one four-way switch. The basic connections for three-way and four-way switches are given in unit 5. However, it is impossible to show every possible connection for these switches. Thus, the electrician must understand the theory and principles of multiple switching and must be able to modify these diagrams to fit the particular installation.

DETERMINING THE CIRCUIT LOAD

The lighting branch circuit for the hall and front entrance consists of:

1 weatherproof GFCI convenience receptacle
1 hall ceiling fixture
1 porch ceiling fixture
3 shrub lights
2 post lights
Total: 8 Outlets

Using the load values in watts given in previous units, the estimated probable connected load on this circuit is:

1 weatherproof convenience receptacle	150 watts
1 hall ceiling fixture	100 watts
1 porch ceiling fixture	150 watts
3 150-watt PAR 38 shrub lamps	450 watts
2 post lights (100 watts each)	200 watts
Total:	1050 watts Estimated Load

This estimated value is well within the 1440-watt limit for a 15-ampere circuit (unit 6).

REVIEW

Note: Refer to the Code or plans where necessary.

1. To what circuit are the hall and front entrance wiring connected? ____________________
2. List each type of switch and receptacle connected to Circuit A29. ____________________

____

3. The cable layout indicates that a three-wire cable runs to Panel A from the hall light. Why? ______

______

______

4. Why is it important that the *hot* conductor in a three-wire circuit be properly connected to opposite phases in a panel? ______

5. In the diagram, Load A is rated at 10 amperes, 120 volts. Load B is rated at 5 amperes, 120 volts.

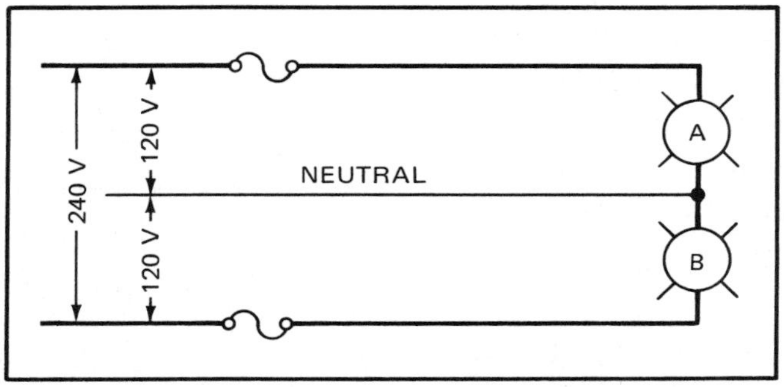

a. When connected to the three-wire circuit as indicated, how much current will flow in the neutral conductor? ______

b. If the neutral should open, to what voltage would each load be subjected, assuming both loads were operating at the time the neutral opened? Show all calculations.

______

______

______

6. Where are the shrub lights to be located? ______

______

7. What type of cable is used to wire the shrub and post lights? ______

8. Why is a spare cable included in the wiring? ______

______

9. What is the minimum depth that must be maintained when UF cable is buried directly in the ground without supplementary protection from injury? ______

10. What is the minimum depth that must be maintained when UF cable is protected with conduit or a 2-inch thick concrete pad? ______

11. What section of the Code prohibits embedding Type UF cable in concrete? ______

12. a. Must the shrub and post lights be grounded? ______
b. What section(s) of the Code determines the answer to (a)? ______

13. How many switches are grouped together just inside the front door? ______

14. How many switches control the hall ceiling fixture? ______

15. From left to right, what do the switches by the front door control? ______

______

16. What is the total number of No. 12 conductors that must be considered when selecting the box to be installed at the five-gang switch location in the front hall? ______________

17. How many feet (meters) of three-wire cable are needed to complete Circuit A29? (Include the feedwire to Panel A and any spare cable.) ______________

18. Who is to select the hall lighting fixture? ______________

19. What is the basic difference between the hall light and the front porch light? ______________

20. What provision is made for heating the hall? ______________

21. a. Where is the thermostat located? ______________
    b. How high is it mounted? ______________

22. Why must all conductors of the same circuit be run through the same metal raceway, or the same knockout of a metal switchbox? ______________

23. Outdoor fixtures must be marked as ______________
    a. suitable for damp locations. b. suitable for dry locations only. c. suitable for wet locations.

24. The following layout is for the lighting circuit for the Front Hall, Entrance, and Front Outside Lighting. Using the cable layout shown in figure 9-1, make a complete wiring diagram of this circuit. Use colored pencils to indicate conductors.

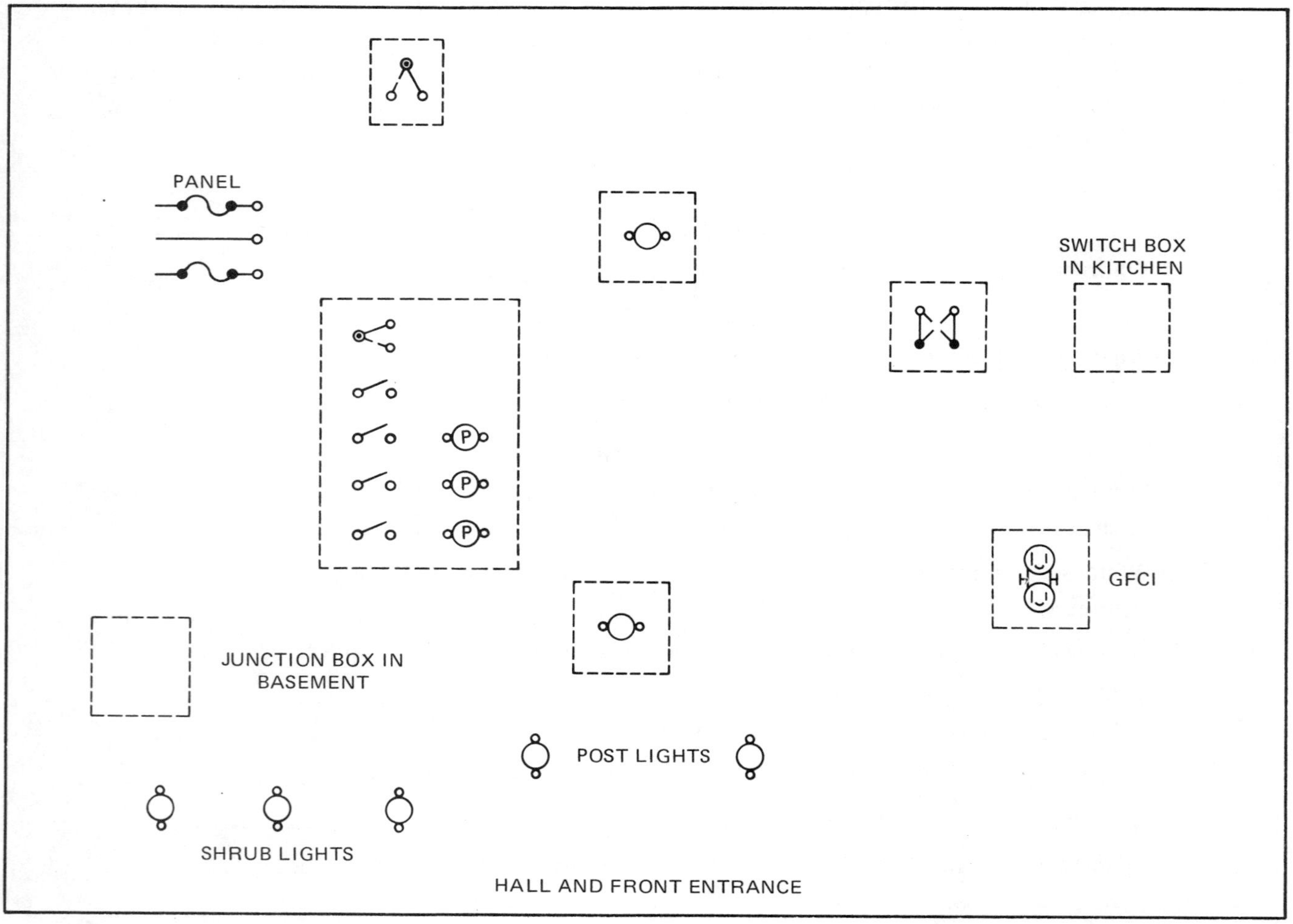

unit 10

Lighting Branch Circuits for Kitchen and Rear Entry and Small Appliance Circuits for Kitchen

OBJECTIVES

After studying this unit, the student will be able to

- list three methods of installing undercabinet lighting fixtures.
- discuss the use of exhaust fans for humidity control.
- install split-circuit (two-circuit) convenience receptacles.
- discuss the general grounding considerations for a residence.

LIGHTING CIRCUIT A31

The outdoor garage lights and the lighting outlets in the kitchen and rear hall are grouped to form a 15-ampere circuit. This circuit is Circuit A31 and is shown in figure 10-1. The feed for this circuit runs to the front hall ceiling outlet box where it is connected to the red and white conductors of the three-wire cable. Unit 9 shows that this three-wire

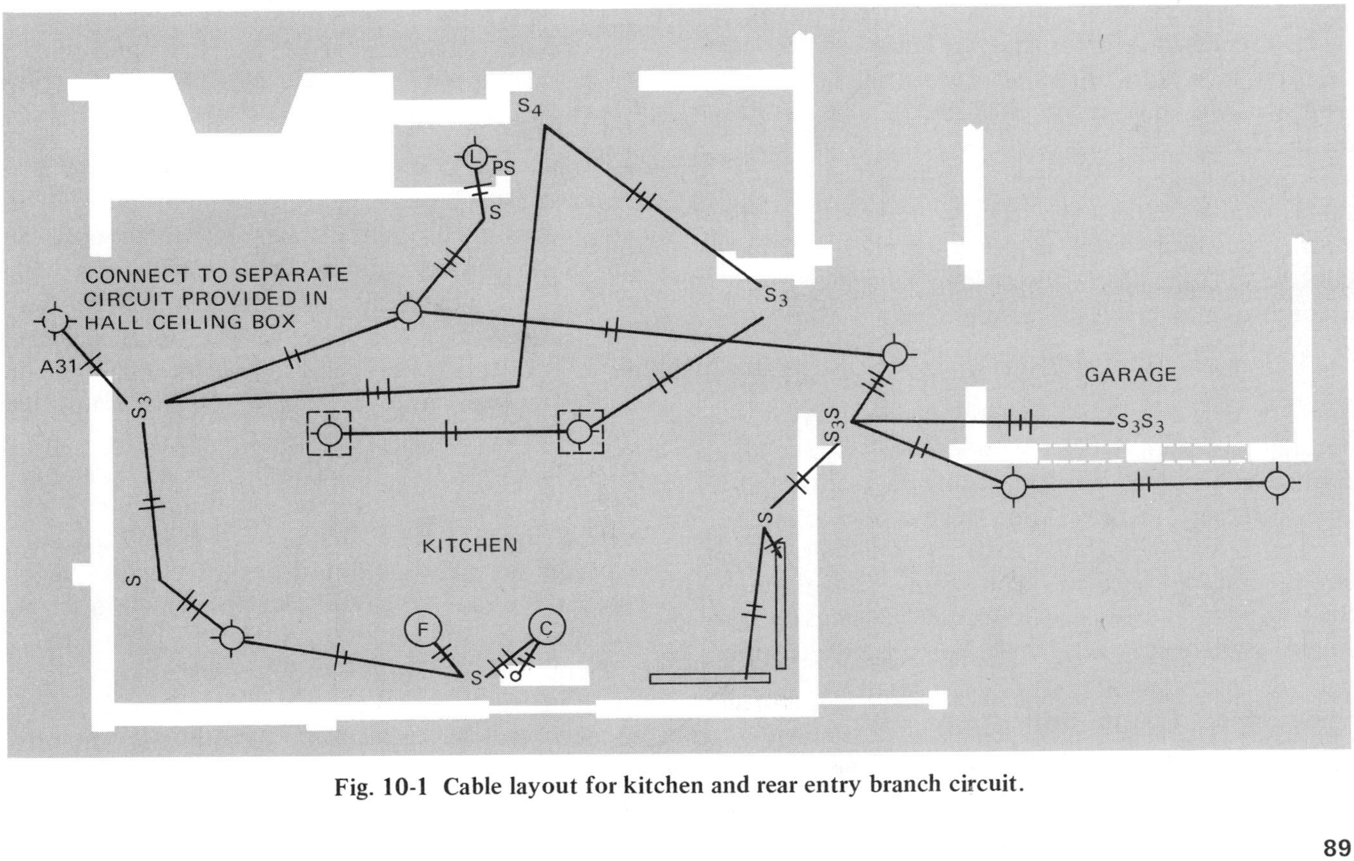

Fig. 10-1 Cable layout for kitchen and rear entry branch circuit.

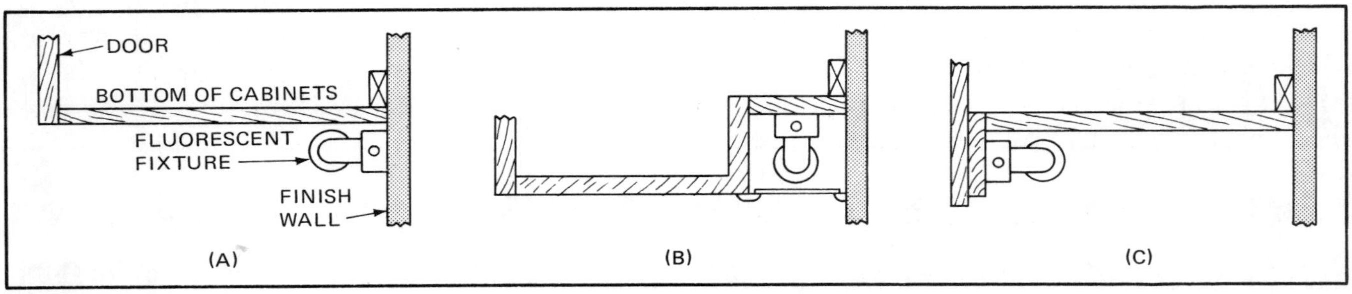

Fig. 10-2 Methods of installing undercabinet lighting fixtures.

cable carries both the front hall and kitchen circuits to Panel A. The red conductor of the cable is connected to Circuit A31 and the black conductor is connected to Circuit A29. The white, neutral conductor of the three-wire cable is connected to the neutral terminal strip and is common to both circuits.

REAR ENTRY LIGHTING

Two side bracket lights are provided on the outside of the garage. These lights are controlled from two three-way switches. One switch is located in the rear hall and the other switch is located in the garage.

KITCHEN LIGHTING FIXTURES

The general lighting for the kitchen is provided by two recessed ceiling fixtures. These fixtures can be turned on from three locations using two three-way switches and one four-way switch. A pull-down ceiling fixture provides light over the dining table. This fixture is controlled by a wall switch near the table. Many fixtures of this type are provided with a three-position switch to permit more exact light control. This switch allows one to select a low, medium, or high level of light when three-way lamps are used.

A goal of good lighting design is to eliminate shadows in work areas. To provide ample light, especially at the kitchen counter work areas, one fixture is installed in the ceiling in front of the large kitchen window, one recessed fluorescent fixture is installed in the soffit over the sink, and two 30-watt fluorescent fixtures are installed under the cabinets in the north corner of the kitchen. Three switches control the work area lighting.

Undercabinet Fixture Installation

Several methods may be used to install the undercabinet lighting fixtures. They can be fastened to the wall just under the upper cabinets, figure 10-2A, installed in a recess provided in the upper cabinets so that they are hidden from view, figure 10-2B, or fastened under and to the front of the upper cabinets, figure 10-2C. The installations shown in figures 10-2B and 10-2C require that the carpenter and the electrician work together closely.

Lamp Type

Good color rendition in the kitchen work areas is achieved by installing warm white deluxe (WWX) 3000-K lamps in the three fluorescent fixtures. These lamps provide lighting similar to that of incandescent lamps. They do not give that "cold" appearance.

Fan Outlet and Clock Outlet

The fan outlet and clock outlet also are connected to branch lighting Circuit A31. Fans can be installed in a kitchen either to exhaust the air to the outside of the building or to filter and recirculate the air. Both types of fans can be used in a residence. The ductless hood fan specified in the plans does not exhaust air to the outside and does not remove humidity from the air.

Homes heated electrically with resistance-type units usually have excess humidity. These heating units neither add nor remove moisture from the air unless humidity control is provided. The proper use of weather stripping, storm doors and storm windows, and a vapor barrier of polyethylene plastic film tends to retain humidity.

Vapor barriers are installed to keep moisture out of the insulation. This protection normally is installed on the warm side of ceilings, walls and floors, under concrete slabs poured directly on the ground, and as a ground cover in crawl spaces. Insulation must be kept dry to maintain its effectiveness. For example, a one percent increase of moisture in insulating material can reduce its efficiency five percent. (Humidity control using a *humidistat* on the attic exhaust fan is covered in unit 21.)

An exhaust fan simultaneously removes moisture from the air and exhausts heated air. The choice of the type of fan to be installed is usually left to the owner. The electrician, builder, and/or architect may make suggestions to guide the owner in making the choice.

For the residence in the plans, a separate wall switch is not required because the speed switch, light, and light switch are integral parts of the fan.

The clock outlet in the face of the soffit may be installed in any sectional switch box or 4-inch square box with a single-gang raised plaster cover. A deep sectional box is recommended because a recessed clock outlet takes up considerable room in the box.

Some decorative clocks have the entire motor recessed so that only the hands and numbers are exposed on the surface of the soffit or wall. Some of these clocks require special outlet boxes (usually furnished with the clock). Other clocks require a standard 4-inch square outlet box. Accurate measurements must be taken to center the clock between the ceiling and the bottom edge of the soffit, figure 10-3. If the clock is available while the electrician is roughing in the wiring, the dimensions of the clock can be checked to help in locating the clock outlet.

SMALL APPLIANCE BRANCH CIRCUITS FOR CONVENIENCE RECEPTACLES

The Code requirements for small appliance circuits in the kitchens of dwellings are covered in *Section 220-3(b)*.

- At least two 20-ampere small appliance circuits shall be installed.
- Either (or both) of the two circuits required in the kitchen is permitted to supply receptacle outlets in other rooms, such as the dining room, breakfast room, or pantry. Figure 10-4 illustrates this portion of the Code.
- These small appliance circuits shall be assigned a load of 1500 watts each when calculating feeders and service-entrance requirements.
- The clock outlet may be connected to an appliance circuit when this outlet is to supply and support the clock only. (In the residence in the plans, the clock outlet is connected to a lighting circuit.)

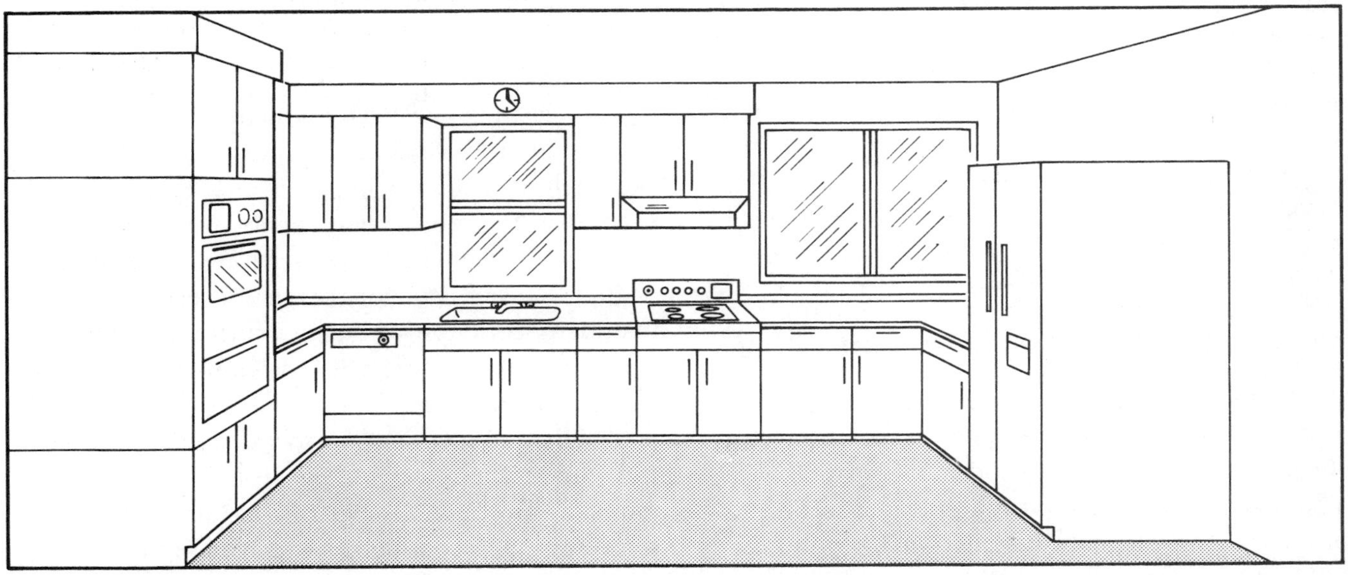

Fig. 10-3 Three-dimensional view of the kitchen.

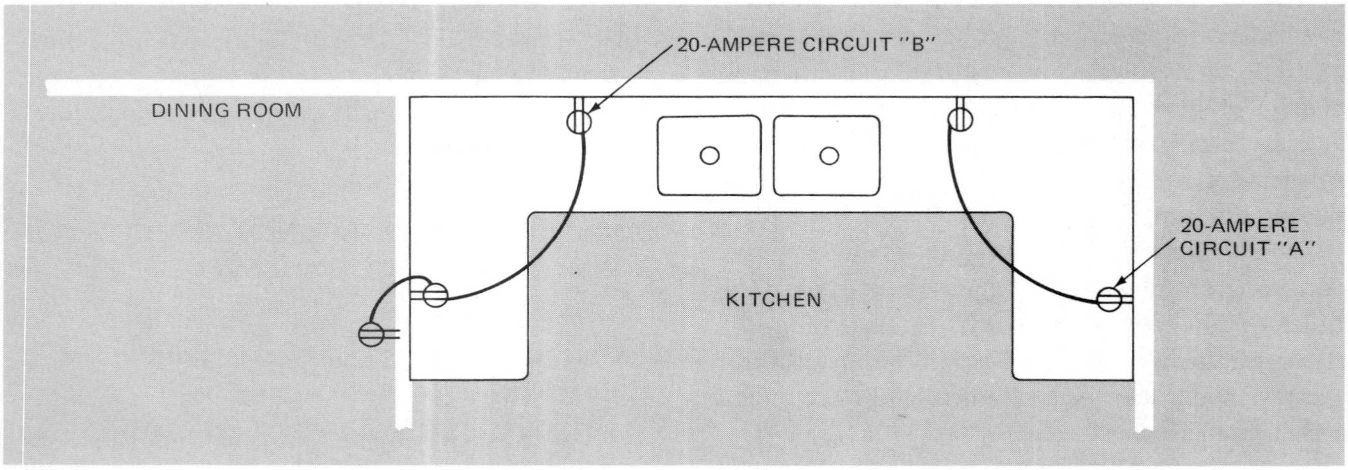

Fig. 10-4 A receptacle outlet in the dining room may be connected to one of the two required small appliance circuits in the kitchen, *Section 220-3(b).*

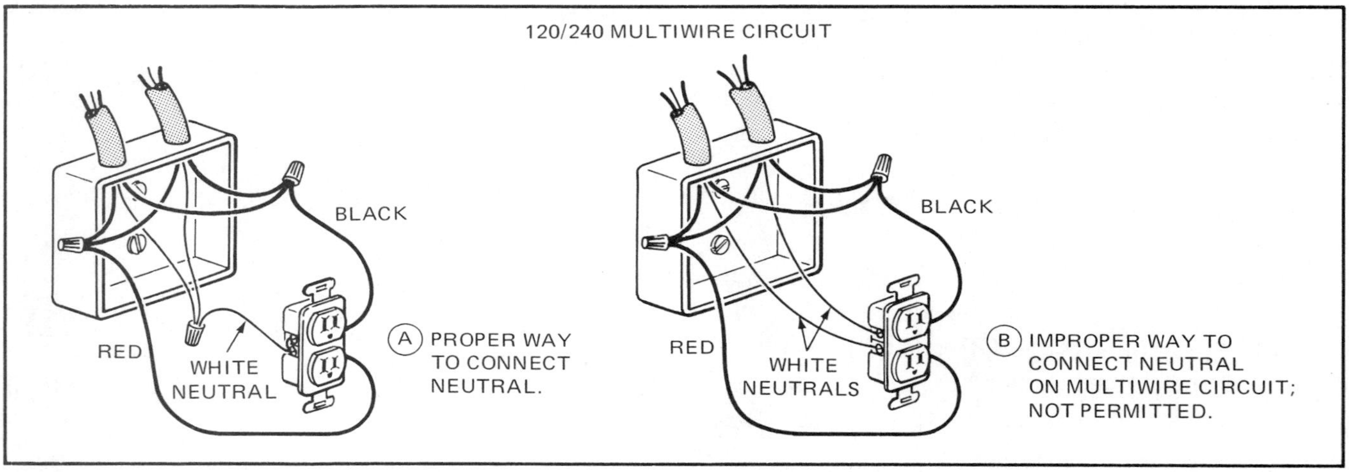

Fig. 10-5 Connecting the neutral in a three-wire curcuit, *Secition 300-13(b).*

The small appliance circuits prevent circuit overloading in the kitchen as a result of the heavy concentration and use of electrical appliances. For example, one type of cord-connected microwave oven is rated 1500 watts at 120 volts. The current required by this appliance alone is:

$$I = \frac{W}{E} = \frac{1500}{120} = 12.5 \text{ amperes}$$

Section 90-1(b) states that the Code requirements are actually the minimum provisions for safety. According to the Code, compliance with these rules will not necessarily result in an efficient, convenient, or adequate installation for good service or future expansion of electrical use.

Exceeding Code requirements, six split-circuit (two-circuit) convenience receptacles are installed in the kitchen of the dwelling. Five receptacles are located over the cabinet work areas and one is located near the kitchen table. A separate circuit is connected to each portion of the receptacles. The neutral is common to both circuits. To complete this wiring, three-wire circuits are installed to feed the receptacles, figure 10-5.

When connecting multiwire circuits, the neutral conductor must *not* be broken at the receptacles. NEC *Section 300-13(b)* states that where more than one conductor is to be spliced, all connections for the identified grounded conductors

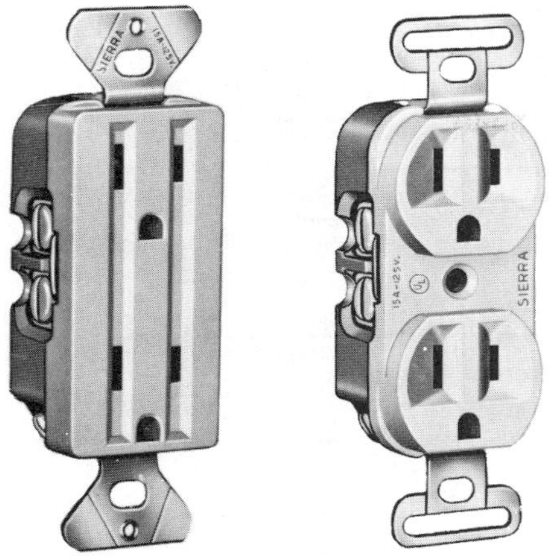

Fig. 10-6 Grounding-type receptacles.

in multiwire circuits must be made up independently of the receptacles, lampholders, and so forth. This requirement is illustrated in figure 10-5. Note that the screw terminals of the receptacle must not be used to splice the neutral conductors. The hazards of an open neutral are discussed in detail in unit 9. Figure 10-6 shows two types of grounding receptacles commonly used for installations of this type.

Receptacles are selected for circuits according to the following guidelines. A single receptacle connected to a circuit must have a rating not less than the rating of the circuit, *Section 210-21(b)*. In a residence, typical examples of this requirement for single receptacles are the laundry outlet (20 amperes), a clothes dryer outlet (30 amperes), or a range outlet (50 amperes).

Circuits rated at 15 amperes supplying two or more receptacles shall not contain receptacles rated at over 15 amperes. For circuits rated at 20 amperes supplying two or more receptacles, the receptacles connected to the circuit may be rated at 15 or 20 amperes.

Figure 10-7 is a suggested cable layout for the kitchen receptacles. Note that three of the six split-circuit receptacles are connected to one three-wire circuit which runs to circuits 20 and 22 in Panel A. The three remaining split-circuit receptacles are connected to a second three-wire circuit which runs to circuits 19 and 21 in Panel A. The two three-wire cables contain No. 12 AWG conductors.

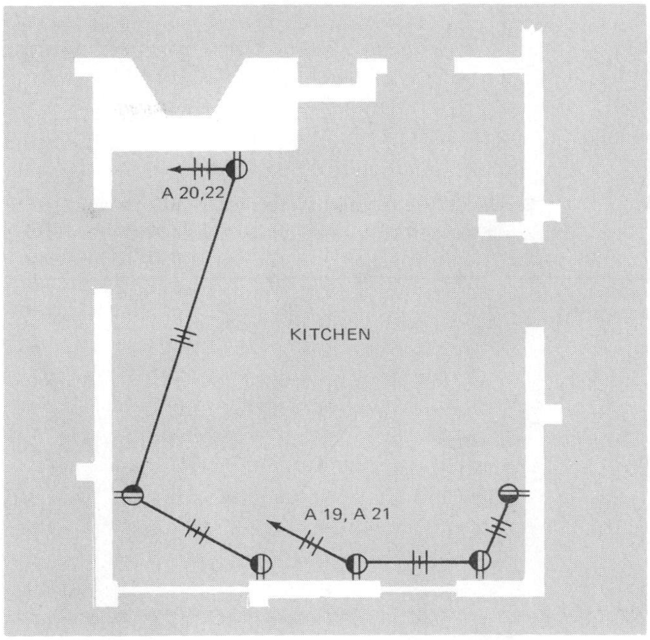

Fig. 10-7 Locations of split-circuit grounding-type convenience receptacles.

The two three-wire circuits are the equivalent of four small appliance branch circuits (circuits 19, 20, 21, and 22 in Panel A).

Much confusion has existed as to whether or not the Code permits the connecting of split-circuit receptacles to a 3-wire, 120/240-volt multiwire circuit, as illustrated in figure 10-8.

The 1981 edition of the Code, *Section 210-6 (c)(1)* clearly permits split-circuit receptacles to be connected to a 3-wire, 120/240-volt multiwire circuit. The section states that the voltage *between conductors* supplying receptacles in dwellings shall not exceed 150 volts. An exception to this is that the voltage may exceed 150 volts between conductors where cord- and plug-connected loads of greater than 1380 watts, or loads of 1/4 horsepower or greater, are connected to such circuits. This sort of loading would be commonly found in kitchen work areas, workshops, and the like.

When multiwire branch circuits supply more than one receptacle on the same yoke (see figure 10-9), a means must be provided to disconnect simultaneously all of the *hot* conductors at the panelboard where the branch circuit originates. By looking at figure 10-8, it can be seen that although 240 volts is present on the wiring device, only 120 volts is connected to each receptacle on the wiring device.

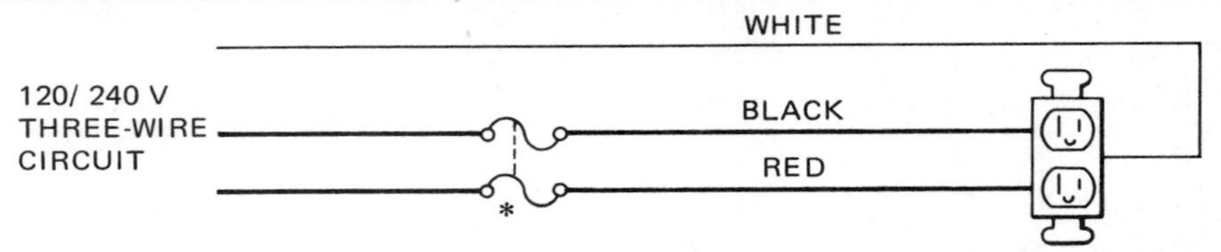

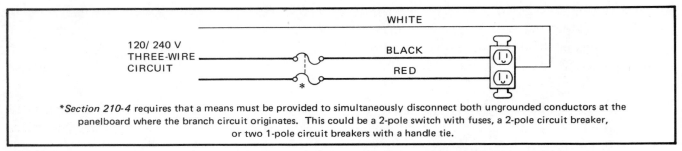

**Section 210-4* requires that a means must be provided to simultaneously disconnect both ungrounded conductors at the panelboard where the branch circuit originates. This could be a 2-pole switch with fuses, a 2-pole circuit breaker, or two 1-pole circuit breakers with a handle tie.

Fig. 10-8 Split-circuit grounding-type convenience receptacle.

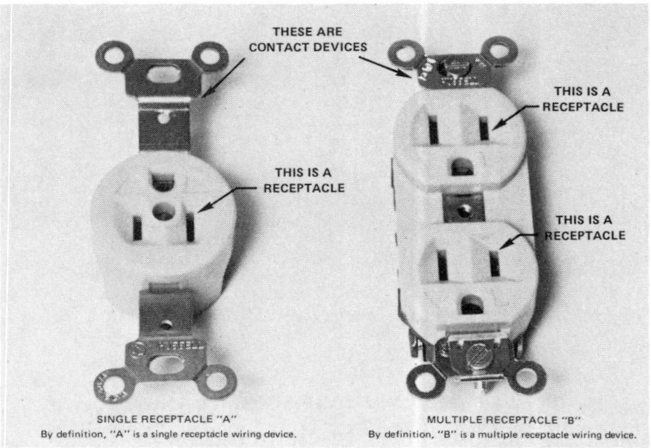

Fig. 10-9 Receptacles. See NEC Definitions, *Article 100.*

RECEPTACLES AND OUTLETS

Article 100 of the NEC gives the definition for a receptacle and an outlet, as illustrated in figures 10-9 and 10-10.

LOCATION OF RECEPTACLES

The number of small appliance circuits in this dwelling exceeds Code requirements. Thus, there is no problem in locating the receptacle outlets to comply with the portions of *Section 210-52* which apply to dwellings. Receptacle outlets are to be placed so that:

- no point along the floor line shall be more than six feet (1.83 m) from a receptacle outlet.
- any wall space two feet (610 mm) wide or greater must have a receptacle outlet.
- counter spaces 12 inches (305 mm) wide or greater must have a receptacle outlet for that space.
- special outlets for refrigerators, disposers, dishwashers, and so forth, are not to be included in locating the required receptacle outlets.

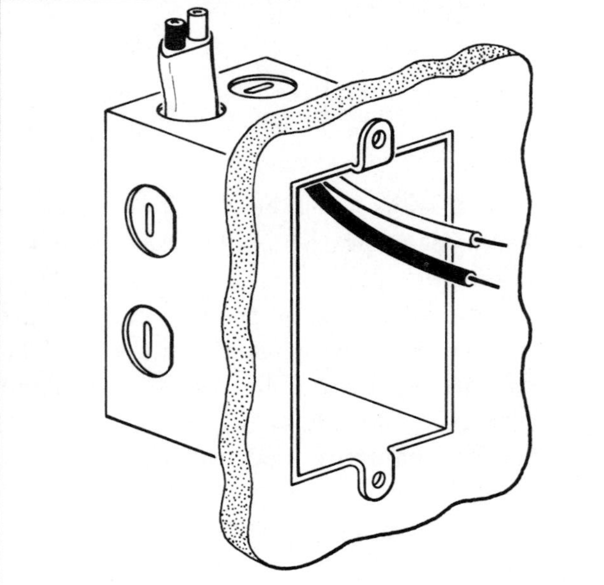

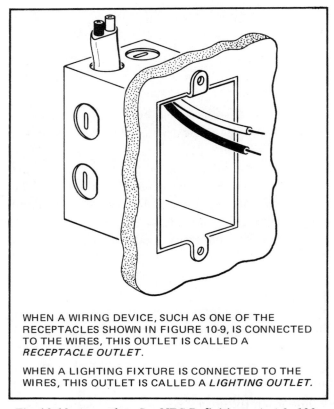

WHEN A WIRING DEVICE, SUCH AS ONE OF THE RECEPTACLES SHOWN IN FIGURE 10-9, IS CONNECTED TO THE WIRES, THIS OUTLET IS CALLED A *RECEPTACLE OUTLET*.

WHEN A LIGHTING FIXTURE IS CONNECTED TO THE WIRES, THIS OUTLET IS CALLED A *LIGHTING OUTLET*.

Fig. 10-10 An outlet. See NEC Definitions, *Article 100.*

DETERMINING THE PROPER BOX SIZE

For the residence, the lighting circuit is shown on one cable layout and the receptacle circuit is shown on another layout. However, the First Floor Electrical Plan shows that the switches and convenience receptacles for five branch circuits (one lighting and four small appliance branch circuits) are to be installed at the same locations. It is obvious that a large number of conductors must enter each switch box at the various locations in the kitchen. To prevent confusion, the electrician must install boxes that will provide sufficient space for the number of conductors entering that particular box. The appropriate box or boxes for an installation of this type can be determined from *Table 370-6(a)*. Instead of ganging the switches and convenience receptacles, separate boxes may be installed for each device. A suitable box for this type of installation is the 4-inch square outlet box with a single-gang plaster cover.

LOAD ESTIMATE FOR THE LIGHTING BRANCH CIRCUIT

The following list gives the total number of outlets connected to lighting branch circuit A31.

1 pull-down fixture
2 kitchen ceiling fixtures (recessed)
1 ceiling fixture over counter
1 20-watt fluorescent fixture over sink
2 30-watt fluorescent fixtures under cabinets
1 closet light
1 rear hall fixture
2 garage bracket fixtures (outdoor)
1 kitchen fan with light
1 clock receptacle
Total 13 Outlets

Thirteen outlets may seem to be too many for one 15-ampere circuit. However, recall that the clock receptacle, the two fluorescent fixtures under the cabinets, and the one closet light add little to the total connected load.

The estimated load for the circuit is:

1 pull-down fixture	200 watts
2 kitchen ceiling fixtures (150 watts each)	300 watts
1 two-lamp fixture over the counter (60 watts each)	120 watts
2 30-watt fluorescent fixtures under the cabinet (multiply by 1.25 for the total wattage: 60 × 1.25 = 75 watts)	75 watts
1 two-lamp, 20-watt fluorescent fixture over the sink (40 × 1.25 = 50 watts)	50 watts
1 closet light	60 watts
1 rear hall fixture	100 watts
2 garage bracket fixtures (100 watts each)	200 watts
1 fan with light	100 watts
1 clock	4 watts
Total:	1209 watts Estimated Load

This load value is well within the recommended 1440-watt limit for a 15-ampere lighting branch circuit.

The circuits for the major fixed appliances (oven, counter-mounted range top, dishwasher, and food waste disposer) are covered in units 19 and 20.

WATTAGE ESTIMATE FOR THE SMALL APPLIANCE BRANCH CIRCUITS

The convenience receptacle load for the four small appliance branch circuits in the kitchen is determined by allowing a load of 1500 watts per circuit. Thus, the total load is 6000 watts (1500 × 4 = 6000 watts), *Section 220-16(a)*.

GENERAL GROUNDING CONSIDERATIONS

The specifications for the residence require that all outlet boxes, switch boxes, and fixtures be grounded. In other words, the entire wiring installation must be a grounded system. The Code also requires that the outside garage lights be grounded, *Section 410-17*. Every electrical component located within reach of a grounded surface must be grounded. Thus, it can be seen that there are few locations where ungrounded boxes or fixtures are permitted. Post lights, weatherproof convenience receptacles, basement wiring, and boxes and fixtures within reach of sinks, ranges or grounded electric units must all be grounded. Hot water heating, hot air heating, and steam heating registers all are grounded surfaces. This means that any electrical equipment, boxes, or fixtures within reach of these surfaces must be grounded. *Section 250-42* requires that any boxes installed near metal or metal lath, tinfoil, or aluminum insulation must also be grounded.

It is confusing to install nonmetallic-sheathed cable with a grounding conductor for certain locations and cable without a grounding conductor for other locations. In general, it is simpler and safer to install either nonmetallic-sheathed cable with a grounding wire throughout the entire installation or to install armored cable to insure good ground continuity. The *grounding* conductor should not be confused with the *grounded* conductor. The grounding conductor is used to ground equipment or the wiring system and the grounded conductor is the white identified (neutral) conductor of a cable, *Article 100.*

REVIEW

Note: Refer to the Code or the plans where necessary.

1. a. What is the ampere rating of the kitchen lighting circuit? ______
   b. What is the number of the circuit? ______
2. Where does the two-wire feed for the kitchen circuit originate? ______
3. How many lighting fixtures are connected to the kitchen lighting circuit? ______
4. From how many points may the recessed ceiling fixtures be controlled? ______
5. What type of fluorescent lamp is recommended for residential installations? ______
6. Where are the switches located that control the outside garage fixtures? ______
7. What is a vapor barrier and what purpose does it serve? ______
8. a. What is the minimum number of 20-ampere small appliance circuits required for kitchens according to the Code? ______
   b. How many are there in this kitchen? ______
9. How many receptacle outlets for small appliances are provided in the kitchen? ______
10. What is meant by two-circuit (split-circuit) convenience receptacles? ______
11. A single receptacle connected to a circuit must have a rating ______ than the ampere rating of the circuit.
12. In kitchen and dining room areas a receptacle outlet shall be installed at each counter space wider than ______ inches (______ millimeters).
13. Tables ______ of the Code supply information about the number of conductors permitted in a box.
14. What is the current consumption if the total load, based on the total estimated wattage, is turned on? Show all calculations. ______
15. A fundamental rule of the Code regarding the grounding of metal boxes, fixtures, etc. is that they *must* be *grounded* when "in reach of ______."
16. Are the outside garage fixtures grounded? ______
17. How many feet (meters) of two-wire cable and three-wire cable are required to complete the kitchen lighting circuit? Include the feed cable to the hall fixture. Two-wire cable ______ feet (______ meters). Three-wire cable ______ feet (______ meters).

18. a. How many circuit conductors enter the box on the southwest wall at the three-way switch location? ______
    b. At the four-way switch location? ______
    c. At the two-gang switch box in the rear hall? ______

19. The Northeast Elevation of the kitchen shows that there are ______ convenience receptacles and ______ switch(es).

20. The following is a layout of the lighting circuit for the Kitchen and Rear Entry. Using the cable layout shown in figure 10-1, make a complete wiring diagram of this circuit. Use colored pencils to indicate conductors.

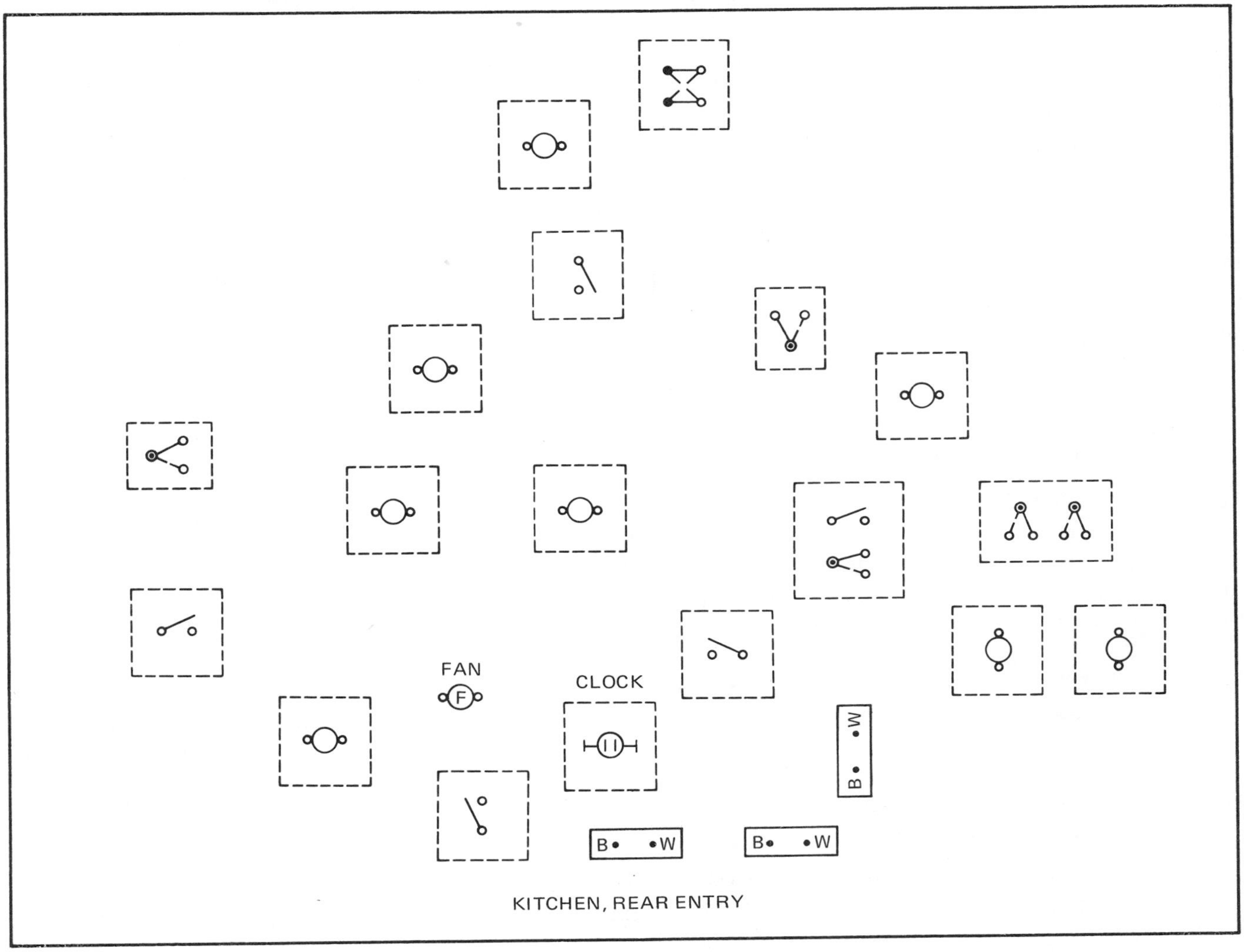

21. How much space is there between the countertop and upper cupboards? ______

22. At what height from the finish floor are the plugs and switches over the kitchen counters to be mounted? ______

23. Give the measurements from the southwest and southeast walls to the center of the pull-down light. From SW wall ______ From SE wall ______

24. a. Where is the thermostat for the kitchen electric baseboard heating unit located?
    ______
    b. How high from the finish floor? ______

25. Where is the speed control for the fan located? ______________________

26. Who is to furnish the range hood? ______________________

27. How many built-in appliances are there in the kitchen? List the appliances. ______________________

____

28. A layout of the Kitchen Convenience Receptacles is as shown. Using the cable layout shown in figure 10-7, make a complete wiring diagram of these four circuits. Use colored pencils to indicate conductors.

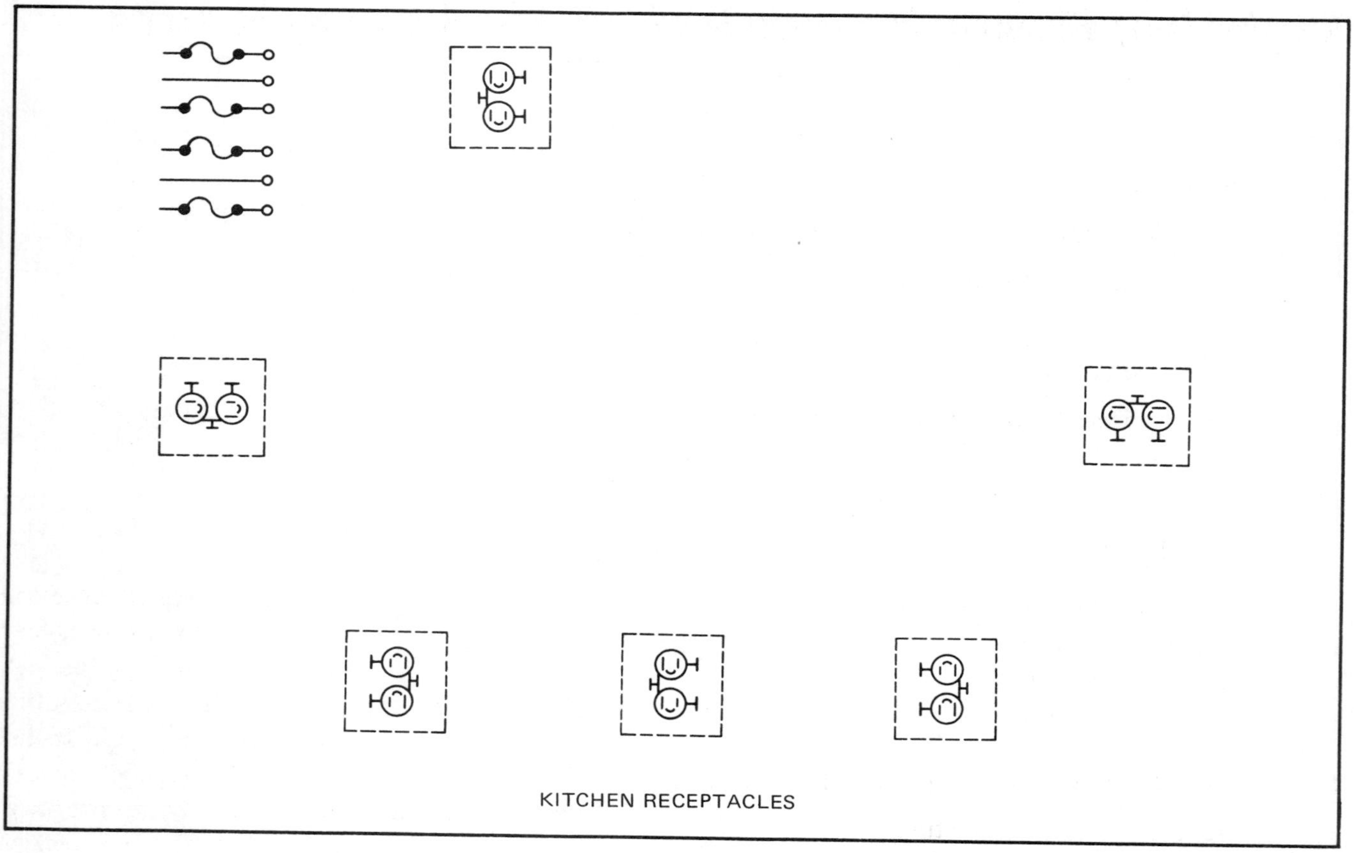

29. Each 20-ampere appliance circuit load demand shall be determined at ______________ .
a. 2400 watts b. 1500 watts c. 1920 watts

30. Is it permitted to connect an outlet supplying a clock receptacle to a 20-ampere small appliance circuit? ______________________

31. The Code requires a minimum of two small appliance circuits in a kitchen. Is it permitted to connect a receptacle in a dining room to one of the kitchen small appliance circuits? What Code section applies to this situation? ______________________

32. According to *Section 210-52*, no point along the floor line shall be more than ________ feet (______ meters) from a receptacle outlet. A receptacle must be installed in any wall space ______ feet (______ meters) wide or greater.

33. The Code states that in multiwire circuits, the screw terminals of a receptacle must *not* be used to "splice" the neutral conductors. Why? ______________________

____

unit 11

Lighting Branch Circuit for the Living Room

OBJECTIVES

After studying this unit, the student will be able to

- install multioutlet assemblies.
- estimate the load requirements of multioutlet assemblies.
- identify various types of dimmer controls.
- connect the proper dimmer controls for incandescent lamp circuits and for fluorescent lamp circuits.
- discuss incandescent lamp load inrush current.
- discuss how internal protection is provided for fluorescent lamp ballasts.

The feed for the living room lighting branch circuit is connected to Circuit 28 in Panel A. The feed is brought from the panel into the two-gang switch box near the front hall.

LIVING ROOM CONVENIENCE RECEPTACLES

There are no ceiling fixtures in the living room. The general lighting for this area is provided by various types of lamps connected to convenience receptacles and valance lighting installed over the living room windows. (Refer to unit 7 for a discussion of the valance lighting which is connected to Circuit A25 in Bedroom No. 1, rather than to living room Circuit A28.)

The seven convenience receptacles are located around the perimeter of the living room. These receptacles are the split-circuit type. That is, the bottom portion of each receptacle is switch-controlled and the top portion of each is hot at all times, figure 11-1. These receptacles are controlled by two three-way switches. One switch is located near the front hall and the other is located next to the door leading to the kitchen, figure 11-2.

The final location of the receptacle outlets in the dwelling may vary somewhat from the layout shown on the plans. Note that some of the wall receptacles in the living room, the bedrooms, the kitchen, and the recreation room may be installed as part of the electric baseboard heaters. (See unit 23, Electric Heating.) The electrician must use imagination in visualizing the final location and method of installing the receptacles because their placement depends upon the electric baseboard heater units. Slight changes from the plan requirements for the

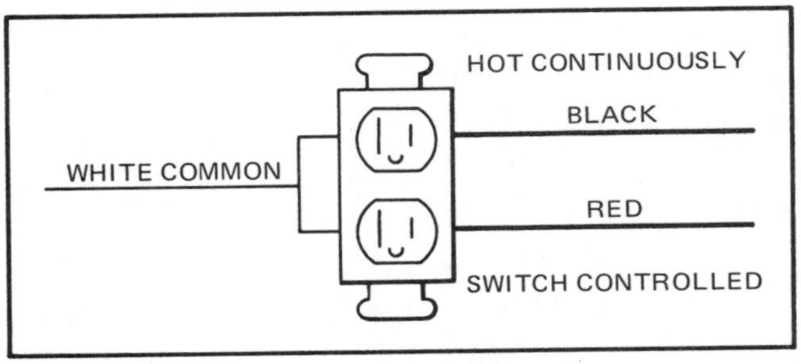

Fig. 11-1 Method of connecting convenience receptacles in the living room.

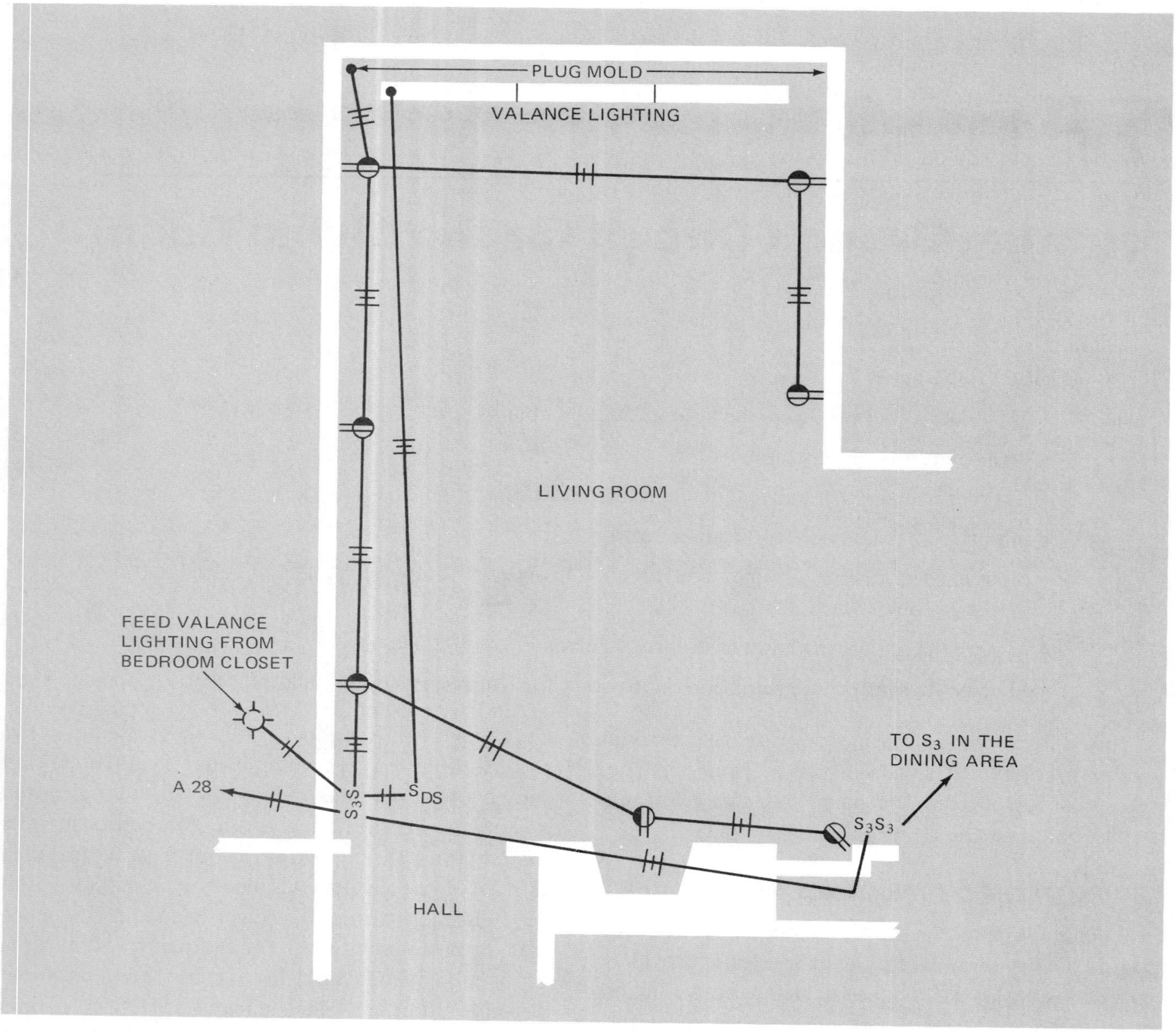

Fig. 11-2 Cable layout for the living room lighting branch circuit.

location of the receptacles or baseboard units, or both, will insure installations that meet Code requirements.

Except for those cases where the receptacle outlets are part of the baseboard heater units, they are to be mounted 12 inches (305 mm) to center from the finish floor. One exception to this statement is the location of the convenience receptacle above the mantel of the fireplace. This receptacle can be used for a clock or a small lamp.

Installation of Mantel Outlet

Two dimensions must be known before the convenience receptacle can be installed over the mantel. These measurements are the height of the mantel, 4′0″ (1.22 m) as given on the Living Room Fireplace Detail, sheet 7 of the plans, and the height of the receptacle above the mantel, 52 inches (1.32 m) from the finish floor to the center of the receptacle, as given on the First Floor Electrical Plan, sheet 8 of the plans.

As shown in figure 11-3, the width of the single-gang wall plate is approximately 2 3/4 inches (69.85 mm). If this receptacle is installed in a horizontal position above the mantel at a height of 52 inches (1.32 m) to center from the finish floor, then the distance from the bottom edge of the receptacle to the top edge of the mantel is approximately 2 5/8 inches (66.67 mm). If the owner desires, the receptacle can be installed with less than 2 5/8-inch (66.67 mm) clearance to the top of the mantel. The smaller clearance means that the receptacle is less conspicuous.

The electrician must make accurate measurements before roughing in the box for the mantel

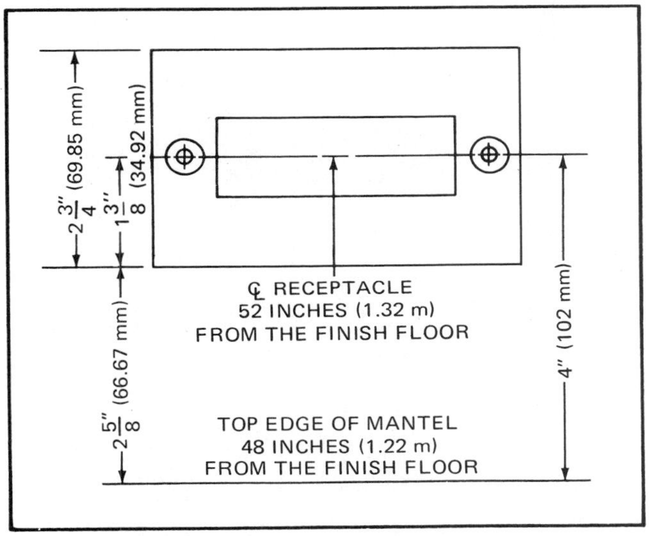

Fig. 11-3 Location of receptacle over mantle.

receptacle. Both the electrician and the carpenter must refer to the fireplace details to insure that the finished job has a neat appearance.

MULTIOUTLET ASSEMBLY (*ARTICLE 353*)

Section F-F on sheet 4 of the plans shows that there is not enough space in the wall under the living room window to install switch boxes for conventional receptacles. The entire space is obstructed by the wood sill which blocks the two 2" x 4" sole plates.

Convenience receptacles can be installed in areas where the space is limited by using a *multioutlet assembly*, figure 11-4. Multioutlet assemblies can be used wherever convenience receptacles are to be mounted close together. These assemblies can also be adapted to kitchen work areas where many electrical appliances are used.

The residence plans specify multioutlet assemblies for the living room and the workshop. The installation of multioutlet assemblies must conform to the requirements of *Article 353*.

Table 220-2(b) gives the general lighting loads in watts per square foot (0.093 m^2) for types of occupancies. The footnote to this table states that the multioutlet assembly load need not be calculated in amperes. The reason is that the multioutlet assembly load in a residential occupancy is included in the watts per square foot (0.093 m^2) rating for general lighting.

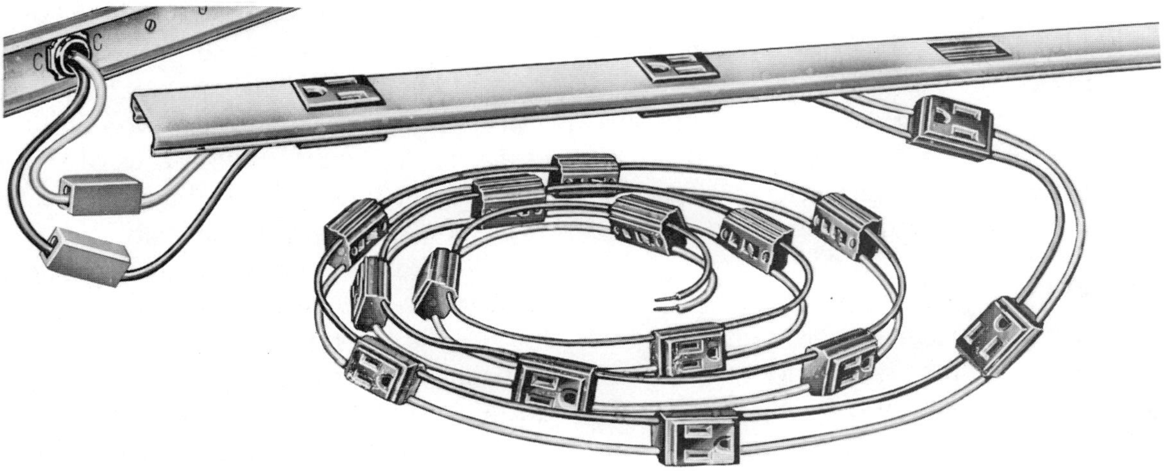

Fig. 11-4 A multioutlet assembly.

However, many electricians still determine the load by adding 1 1/2 amperes for each five feet (1.52 m) of multioutlet assembly. This estimate was a requirement of previous editions of the Code. In this dwelling, therefore, the estimated load in Circuit A28 is determined by using 1 1/2 amperes for each five feet (1.52 m) of multioutlet assembly. The resultant load is more than adequate for the multioutlet receptacles used in the residence.

Many types of convenience receptacles are available for various sizes of multioutlet assemblies. For example, duplex, single-circuit grounding, split-circuit grounding, and duplex split-circuit receptacles can be used with multioutlet assemblies.

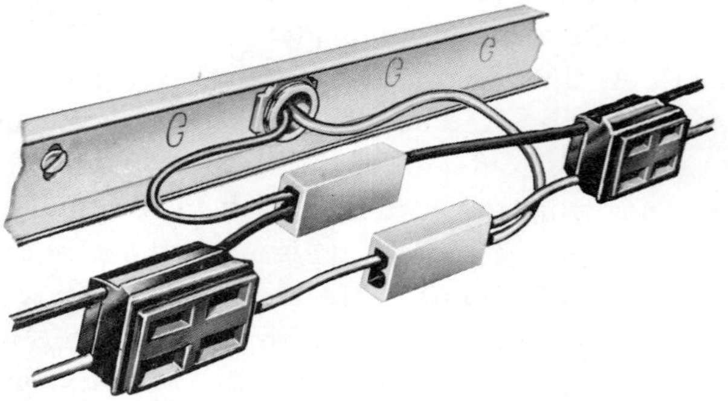

Fig. 11-5 Prewired multioutlet assembly.

Wiring of the Multioutlet Assembly

The electrician first attaches the metal base of the multioutlet assembly to the wall or baseboard. The receptacles may be wired on the job. If the receptacles are factory prewired, they are snapped into the base, figure 11-5. Covers are then cut to the lengths required to fill in the spaces between the receptacles. Manufacturers of these assemblies can supply precut covers to be used when evenly spaced receptacles are installed. These receptacles are usually installed 12 inches (305 mm) or 18 inches (457 mm) apart.

Receptacles can also be installed by snapping them into the cover and then attaching the cover to the metal base.

In general, multioutlet installations are not difficult because of the large number of fittings and accessories available. Connectors, couplings, ground clamps, blank end fittings, elbows for turning corners, and end entrance fittings can all be used to simplify the installation.

Prewired multioutlet assemblies normally are wired with No. 12 AWG conductors and have either 15- or 20-ampere receptacles.

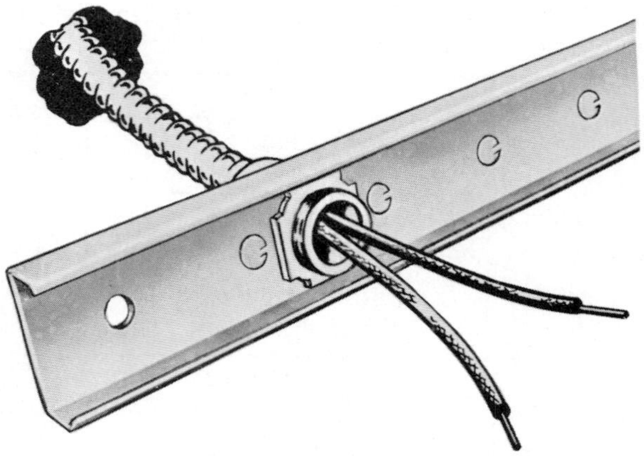

Fig. 11-6 Feed for multioutlet assembly.

Installation of the Multioutlet Assembly in the Residence

According to the residence plans and specifications, alternate receptacles in the multioutlet assembly are switch controlled by the three-way switches that control the other wall convenience receptacles in the living room. In the multioutlet assembly under the living room window, the spacing between the receptacles is 18 inches (457 mm) on center. The total number of receptacles is nine. Thus, the total distance between the center of the first receptacle and the center of the last receptacle is exactly twelve feet (3.66 m). Eight covers are required to fill the spaces between the receptacles, one less than the number of receptacles. Once the nine receptacles and eight covers are snapped into place, the electrician must cut two additional covers to fill in the spaces on both ends of the assembly between the last receptacles and the baseboards on the southeast and northwest walls.

Knockouts are provided in the back of the assembly for the feed, figure 11-6. However, the blocking under the living room window makes it difficult to use these knockouts except at the extreme ends of the assembly (to the right and left of the window blocking). Because of the obstruction, the feed for the multioutlet assembly is brought into a special end fitting approved for this purpose.

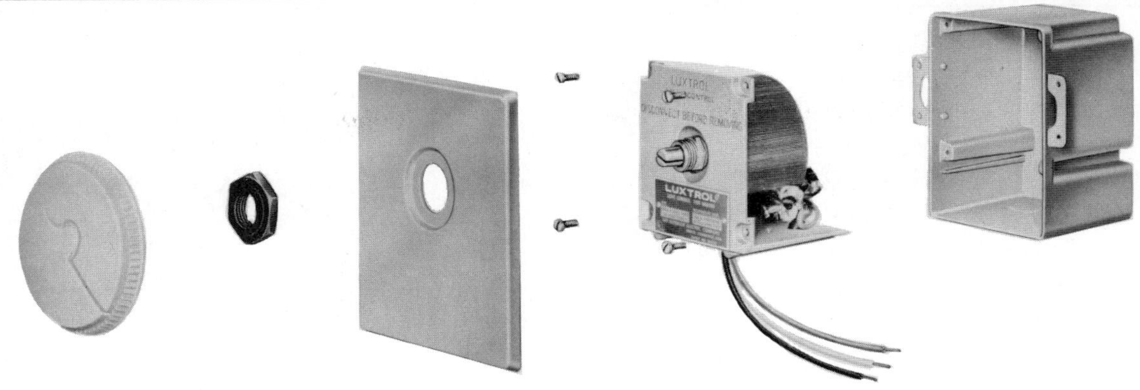

Fig. 11-7 Dimmer control, autotransformer type.

VALANCE LIGHTING

It was shown previously that the living room valance lighting is fed by Circuit A25 in Bedroom No. 1. The valance lighting consists of three 40-watt, rapid-start fluorescent fixtures. These fixtures are installed above the living room window and are controlled by a dimmer unit. Only fixtures with special ballasts designed for use with dimmers can be used. The dimmer control is mounted above the single-pole and three-way switches on the southeast wall of the living room. WWX (deluxe warm white) fluorescent lamps or 3000-K fluorescent lamps are used in these fixtures.

DIMMER CONTROLS

One type of dimmer control has a continuously adjustable autotransformer which changes the lighting intensity by varying the voltage to the lamps, figure 11-7. When the knob of the transformer is rotated, a brush contact moves over a bared portion of the transformer winding to change the lighting intensity from complete darkness to full brightness. This type of control uses only the current required to produce the lighting desired.

Dimmer controls are used on 120-volt, 60-hertz, single-phase, alternating-current systems. The operation of such controls depends on transformer action. Thus, they will not operate on direct-current lines.

The electrician can prevent overloading of the dimmer control autotransformer by checking the total load in watts to be controlled by the dimmer against the rated capacity of the dimmer. The maximum wattage that can be connected safely is shown on the nameplate of the dimmer. In general, overcurrent protection is built into dimmer controls to prevent burnout due to an overload.

Instructions furnished with most dimmers state that they are for control of lighting only, and are not to be used to control receptacle outlets. The reasons for this are:

- serious overloading of the dimmer will result, since it may not be possible to determine or limit what other loads might be plugged into the receptacle outlets.
- reduced voltage supplying a television or radio, stereo components, home computer, or such motors as vacuum cleaners and food processors, can result in costly damage to the appliance.

Dimmer Controls for Incandescent Lighting

The wiring diagram in figure 11-8 shows the circuit used to control incandescent lamps. Note the internal switch installed in the OFF position that permits the entire load to be cut off.

Electronic Solid-state Dimmer Controls

Solid-state electronic dimmers of the type sold for home use are used to control incandescent loads only, figure 11-9. These dimmers are more compact than the autotransformer dimmers. For residential installations, the dimmer commonly used is a 600-watt unit which fits into a standard single-gang wall box. Both single-pole and three-way dimmers are available. The control knob is pushed

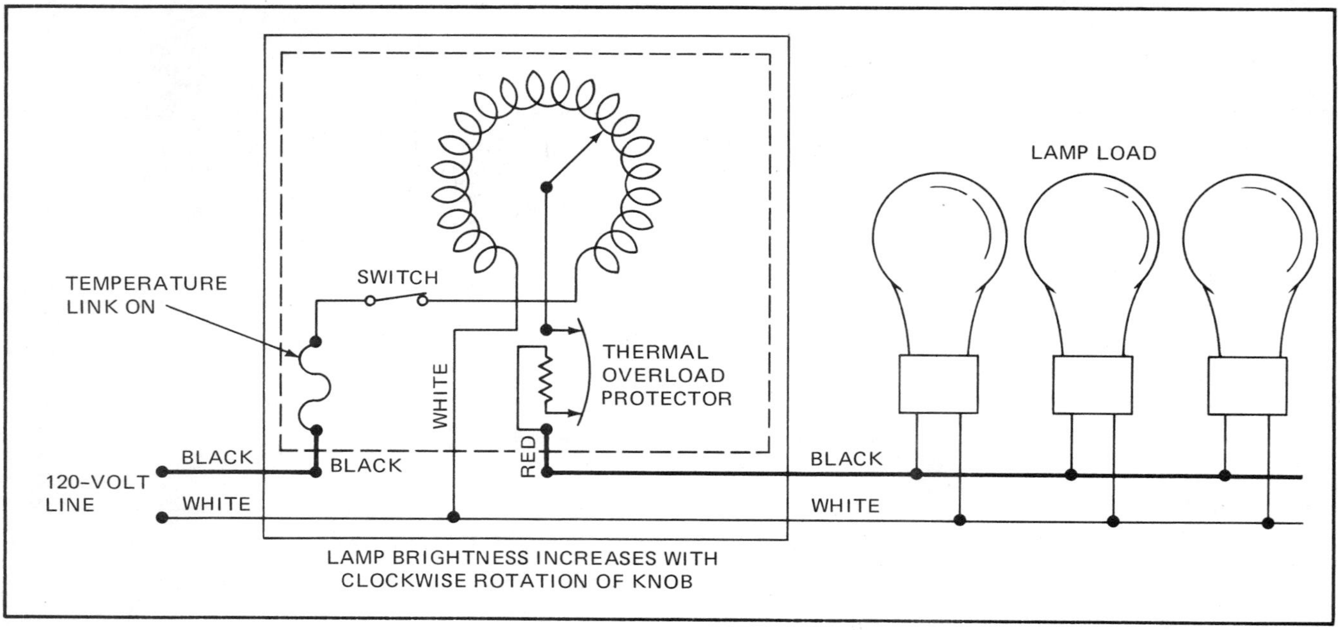

Fig. 11-8 Dimmer control (autotransformer) wiring diagram for incandescent lamps.

to turn the lights on or off and is rotated to dim the lights. Compact, 1000-watt dimmer units are available for use in two-gang wall boxes. Figure 11-10 shows how single-pole and three-way solid-state dimmer controls are used in circuits.

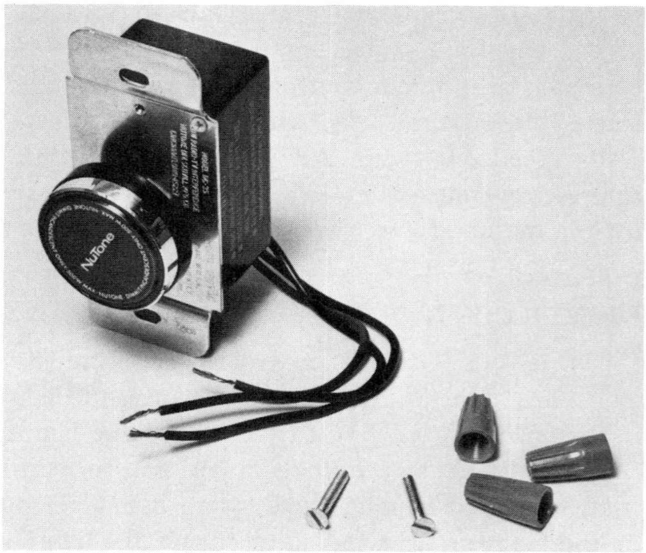

Fig. 11-9 Solid-state dimmer control.

Incandescent Lamp Load Inrush Currents

An unusual action occurs when a circuit is energized to supply a tungsten filament lamp load. A tungsten filament lamp has a very low resistance when it is cold. When the lamp is connected to the proper voltage, the resistance of the filament increases very quickly. This increase occurs within 1/240th of a second or one-quarter cycle after the circuit is energized. During this period, there is an inrush current that is 10 to 20 times greater than the normal operating current of the lamp.

In unit 5 it is stated that T-rated snap switches are available. If a T-rated switch is used in a circuit supplying tungsten filament lamps, then the momentary high surge of current has little effect on the branch circuit under ordinary conditions. However, tungsten filament lamps may be connected through an autotransformer-type dimmer control. Assume that the dimmer is turned from the off to high (bright) position or to a preset low (dim) setting and the circuit is turned on with a separate switch. The resulting inrush current lasts slightly longer due to the slower heating of the lamp filament.

This prolonged surge of current may cause certain types of circuit breakers to trip. These breakers have a highly sensitive magnet-tripping mechanism. When this problem occurs, the sensitive breaker must be replaced with one that has high magnetic-tripping characteristics. Most manufacturers whose breakers react to the inrush

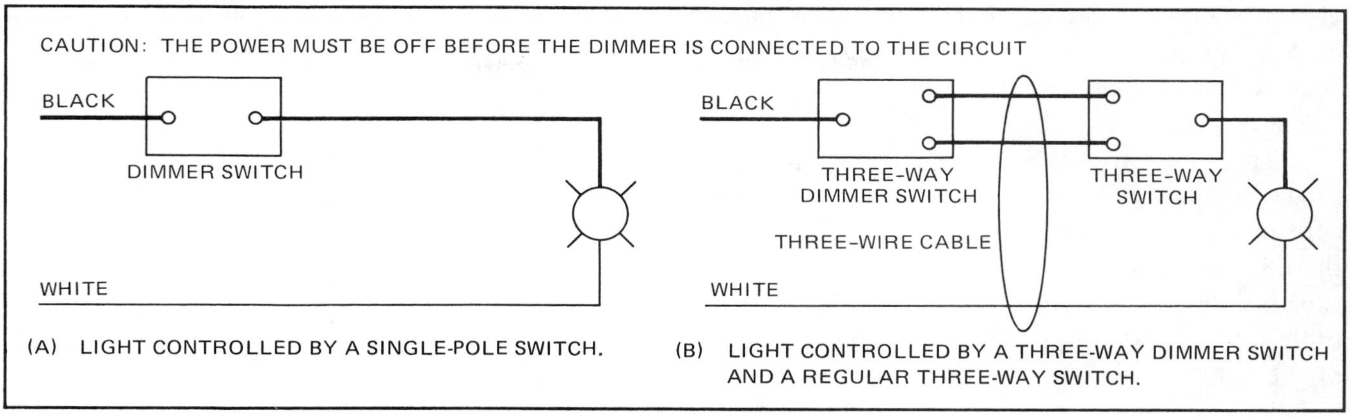

Fig. 11-10 Use of single-pole and three-way solid-state dimmer controls in circuits.

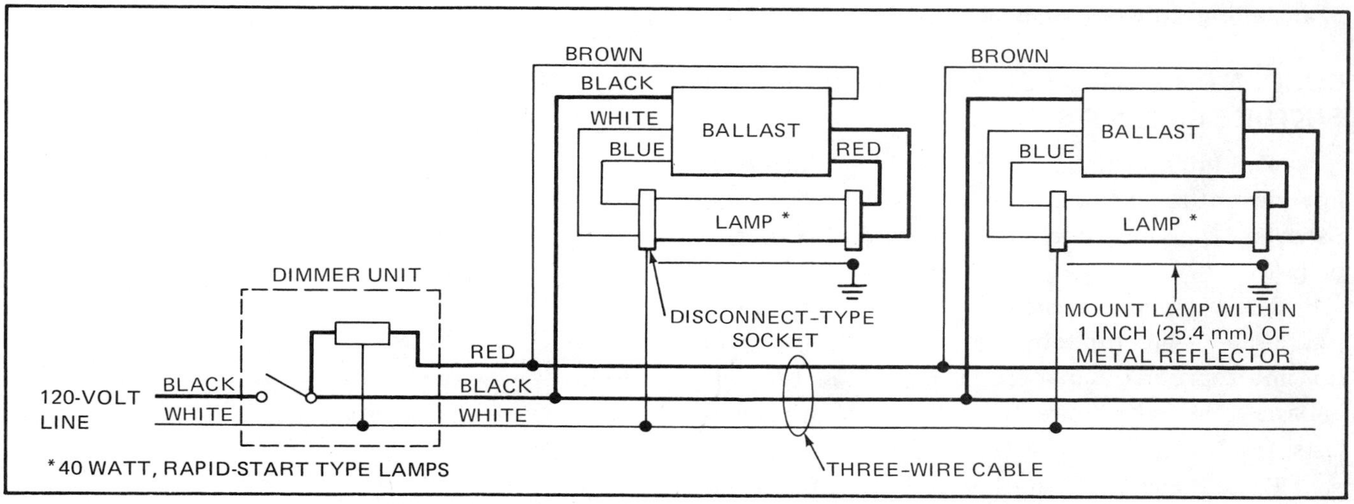

Fig. 11-11 Dimmer control wiring diagram for fluorescent lamps.

current surge of tungsten filament lamp loads can supply the high magnetic-tripping circuit breaker. Branch-circuit fuses are not affected by the inrush current.

An external wall switch is not required, but one may be installed ahead of the dimmer. This switch turns on the lighting load to a preset value. Extra switches are required for multiple on-off switch control. The actual dimming action can be controlled at the dimmer only.

USE OF DIMMER CONTROLS WITH FLUORESCENT FIXTURES

With slight changes in the wiring, dimmer controls can be used to control rapid-start fluorescent lamps (type F40T12RS). Only special ballasts designed for dimmers can be used. These special ballasts energize the cathodes of the fluorescent lamp to a voltage that can maintain the proper cathode operating temperature. The dimmer unit varies the current in the arc.

For fluorescent lamp installations, the electrician must install three conductors between the fluorescent dimming ballast and the dimmer control. The connections for a fluorescent lamp dimmer control are shown in figure 11-11. The lamp reflector, the ballast case, and the white grounded wire of the system must be solidly grounded to insure proper operation of the lamps. If a nonmetallic reflector is used, a grounded metal strip must be mounted parallel to the lamp and within 1 inch (2.54 mm) of the lamp. This strip is about 1-inch (2.54 mm) wide and 45 inches (1.14 m) long. The dimming lead of the ballast must return to the controlled lead of the dimmer. Because incandescent lamps and fluorescent lamps have different characteristics, these lamps cannot be controlled simultaneously with a single dimmer control.

The residence plans show that a separate single-pole switch is provided ahead of the dimmer control. A two-wire cable must be installed between the bedroom closet ceiling outlet box and the two-gang switch box in the living room. This two-wire cable feeds the single-pole switch connected ahead of the dimmer. A three-wire cable (with ground) is installed between the dimmer and the valance fixtures.

Figure 11-12 illustrates one permissible method of installing fluorescent valance lighting.

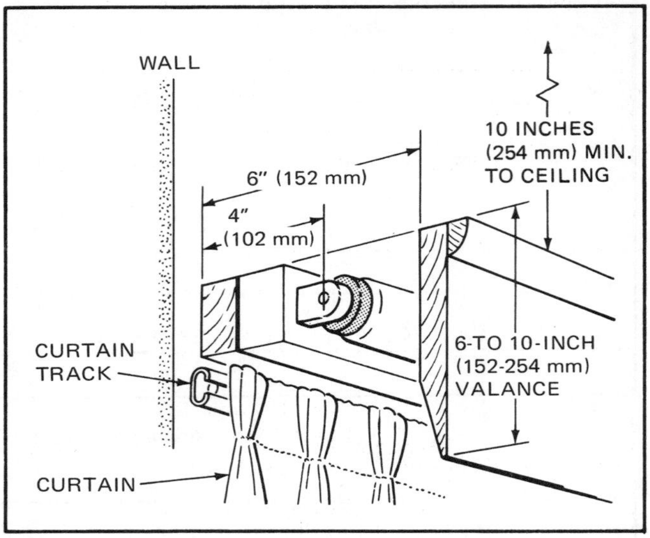

Fig. 11-12 Fluorescent valance with draperies.

CURRENT RATINGS FOR FLUORESCENT FIXTURE BALLASTS

The table in figure 11-13 shows the change in line current for different types of 40-watt dimming ballasts depending upon the power factor of the ballast.

Note: The designations A, B, C, and D in figure 11-13 are for reference only. These letters do not represent a particular model or type of ballast. The ratings are typical for dimming ballasts used with type F40T12 fluorescent lamps.

Before installing the lighting, the electrician must determine how many dimming ballasts can be connected to the circuit. If Type A ballasts are used, it appears that 30 ballasts can be used, $1440 \div 48 = 30$ ballasts. (Recall that a 15-ampere, 120-volt circuit at 80 percent load can handle 1440 watts: $15 \times 120 \times 0.80 = 1440$.)

The line current taken by these ballasts is $30 \times 0.85 = 25.5$ amperes. This value exceeds the capacity of a 15-ampere circuit. Thus, it is very important that the electrician check the current rating of all fluorescent ballasts (both dimming and regular) before the installation is started. The conductors and overcurrent devices must be rated to carry the connected load.

Ballast	Line Current	Watts Input	Line Power Factor
A	0.850	48	0.48
B	0.450	56	0.90
C	0.500	44	0.60
D	0.550	63	0.90

Fig. 11-13 Variation of line current for dimming ballasts (40 watt).

BALLAST PROTECTION

Section 410-73(e) states that fluorescent ballasts must have integral thermal protection within the ballast, except for simple reactor-type ballasts.

Ballasts provided with built-in thermal protection are listed by the Underwriters' Laboratories as Class P ballasts. Under normal conditions, the Class P ballast has a case temperature not exceeding 90°C. The built-in ballast protection must open when the case temperature reaches 110°C.

An exception to *Section 410-73(e)* states that fluorescent fixtures used indoors do not require integral ballast protection when simple reactance ballasts are used and the lamps are rated at 20 watts or less. The simple reactance ballast has only two lead wires and is connected in series with the fluorescent lamp, figure 11-14.

Additional backup protection can be provided by an in-line fuseholder connected in series with the black line lead and the black ballast lead, figure 11-15. The in-line fuseholder acts as a disconnect for the ballast. Thus, when a ballast is to be replaced, the fuse is removed to disconnect the ballast, but

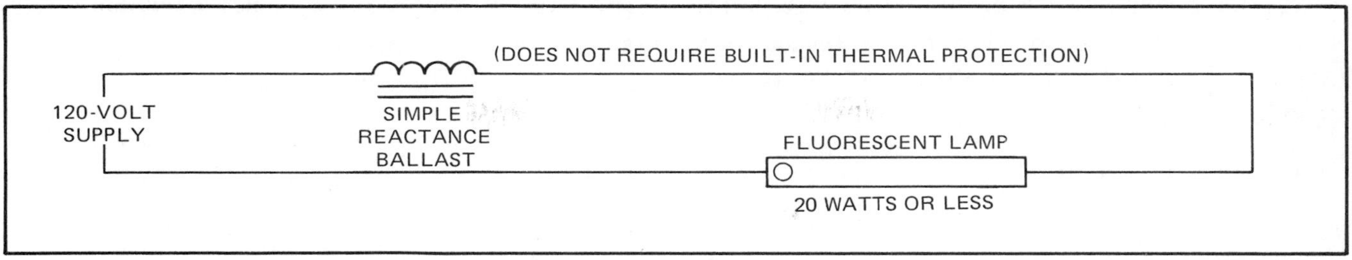

Fig. 11-14 Wiring diagram for a simple reactance ballast.

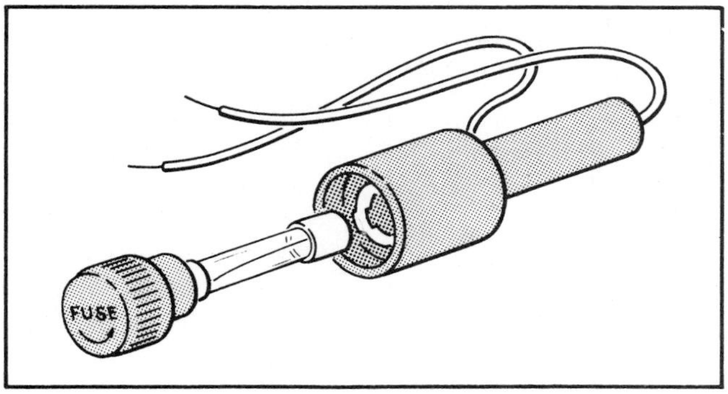

Fig. 11-15 In-line fuseholder for ballast protection.

the entire circuit is not shut off. In general, the ballast manufacturer will recommend the correct fuse size to be used with the ballast.

WATTAGE RATING FOR LIVING ROOM CIRCUIT

The living room convenience receptacles are connected to Circuit A28 in Panel A. The following receptacles are part of this circuit:

Seven convenience receptacles (150 watts each)	1050 watts
Multioutlet assembly, 12 feet (3.66 m) (1 1/2 amperes are allowed per five feet (1.52 m) of assembly, or fraction thereof; thus 12 feet (3.66 m) is equivalent to 4 1/2 amperes, or 4 1/2 X 120 = 540 watts)	540 watts
Total:	1590 watts

REVIEW

Note: Refer to the Code or the plans where necessary.

1. To what circuit is the valance lighting connected? ______
2. How many convenience receptacles are connected to the living room circuit? ______
3. a. How many wires enter the two-gang switch box below the dimmer control? ______
   b. What type and size of box may be installed at this location? ______
4. Are the convenience receptacles in the living room all mounted 12 inches (305 mm) to center from the finish floor? List other heights, if any. ______
5. How high is the convenience receptacle located over the mantel? Give the dimension from the center to the finish floor. ______

6. Where space prohibits the installation of standard convenience receptacles, or where many receptacles must be spaced together, an electrician can install a(an) ____________ . The installation must conform to *Article* ____________ of the Code.

7. What type of fluorescent lamp is installed for the valance lighting? (Underline one)
   a. instant start b. rapid start c. trigger start

8. How many wires must be run between an incandescent lamp and its dimmer control?
   ____

9. Complete the wiring diagram for the dimmer and lamp.

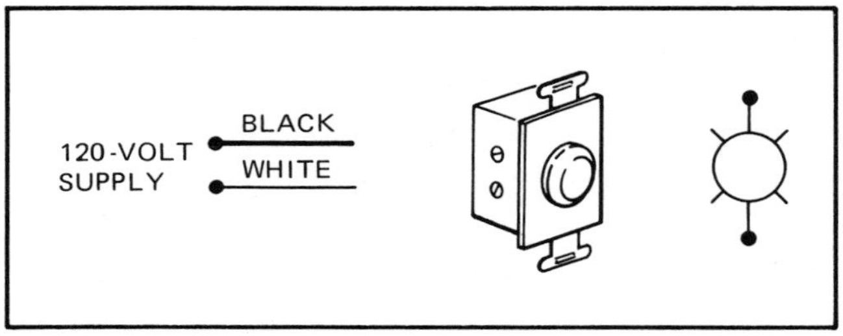

10. What is meant by incandescent lamp load inrush current? What type of switch is required? ____
    ____
    ____

11. Is is possible to dim standard fluorescent ballasts? ____________

12. a. How many wires must be run between a dimming-type fluorescent ballast and the dimmer control? ____________
    b. Is a switch needed in addition to the dimmer control in fluorescent installations?
    ____

13. Explain why fluorescent lamps having the same wattage draw different current values.
    ____
    ____

14. Fluorescent ballast protection is provided by ____________ ballasts. Additional backup short-circuit protection can be provided by connecting a(an) ____________ with the proper size fuse.

15. Sectional view ____________ on the plans reveals the space available for the installation of convenience receptacles under the living room windows.

16. List the type and number of the outlets under the living room window on the southwest wall. Explain the switching arrangement of these outlets. ________

17. What is the total wattage of the baseboard electric heating units controlled by the living room thermostat? ________

18. What is the total current consumption of these electric heating units? Show all calculations. ________

19. At what height is the thermostat mounted? ________

20. The thermostat is (underline one)
   a. single pole, double throw.
   b. single pole, single throw.
   c. double pole, double throw.
   d. double pole, single throw.

21. What is the purpose of $_{L3}$? ________

22. How many television outlets are provided in the living room? ________

23. a. Where is the telephone outlet located in the living room/dining room area? ________

   b. How high is the telephone outlet mounted? ________

24. When laying out the receptacle outlets around the living room, what important factor must be considered because of the baseboard electric heaters? ________

25. A layout of the outlets, switches, dimmer and valance lighting for the Living Room is shown in the following diagram. Using the cable layout of figure 11-2, make a complete wiring diagram of this circuit. Use colored pencils to indicate conductors.

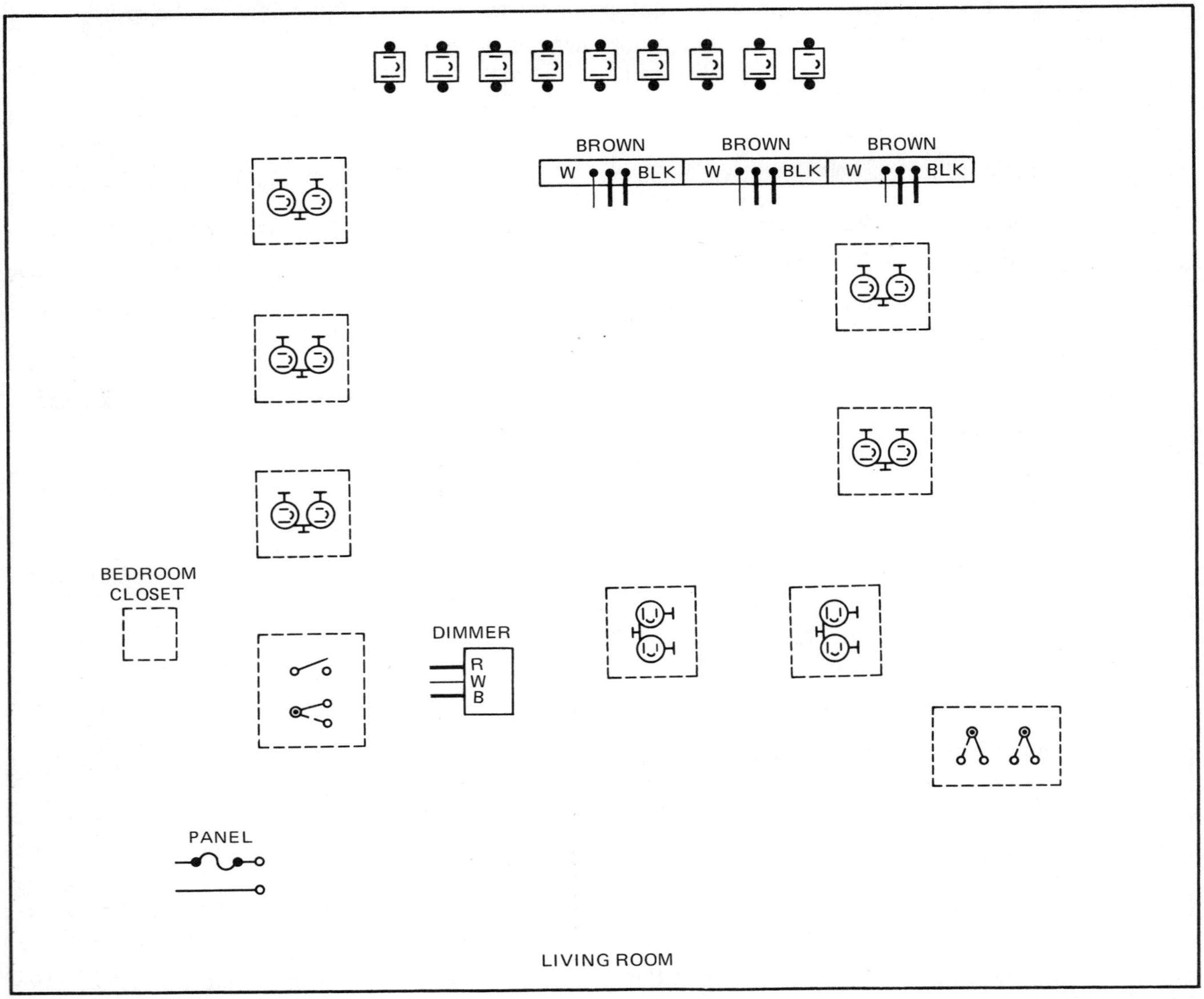
BROWN
BROWN
BROWN
W
BLK
W
BLK
W
BLK
BEDROOM CLOSET
DIMMER
R
W
B
PANEL
LIVING ROOM

unit 12

Lighting Branch Circuit for the Dining Area, Porch and Cornice, Garage Storage Area, and Attic

OBJECTIVES

After studying this unit, the student will be able to

- install a three-way switch with a built-in dimmer control.
- discuss the Code requirements for the installation of outlets and receptacles in damp or wet locations.
- install attic wiring according to Code requirements.

The dining area, porch, garage storage area, attic lighting fixtures, and cornice floodlights form lighting branch Circuit A32. This is a 15-ampere circuit from Panel A.

Figure 12-1 is a suggested cable layout showing one possible combination of outlets and receptacles for this circuit. A three-wire branch circuit runs from Panel A to the ceiling outlet box in the dining room. At this point, the three-wire circuit divides so that the black conductor feeds the lighting outlets in the areas listed and the red conductor is spliced into the black conductor of the two-wire cable supplying garage circuit A30. The white *neutral* conductor is common to both Circuits A32 and A30. Diagrams in unit 9 show the proper method of installing and connecting a three-wire branch circuit.

DINING AREA

The general lighting in the dining area is provided by a pulldown, track-mounted fixture, figure 12-2. This fixture can be raised or lowered and moved closer to or farther from the wall along the track. The length of the track governs how far the fixture can be moved from the wall.

The pulldown fixture is controlled by an electronic solid-state dimmer to provide subdued or full intensity lighting in the dining area. The dimmer is an integral part of the three-way switch as shown in figure 12-1.

In unit 11, it is stated that at least two wires must be installed between the dimmer and the incandescent load it is to control. The cable layout in figure 12-1 shows that a three-wire cable is installed between the ceiling outlet box and the three-gang switch box. This three-wire cable is used because both the circuit conductors (black and white wires) and the switch loop (red wire) must be run between these boxes, figure 12-3. (The switches for the porch fixture and the cornice lighting are not shown in this diagram).

The three convenience receptacles in the dining area, figure 12-4, are not included in lighting Circuit A32 to comply with NEC *Section 220-3(b)*. These convenience receptacles are fed by Circuit A18, a separate 20-ampere small appliance circuit.

Regardless of the type of installation, it is important that the electrician review all Code requirements on the use of switch boxes and outlet boxes before the boxes are selected.

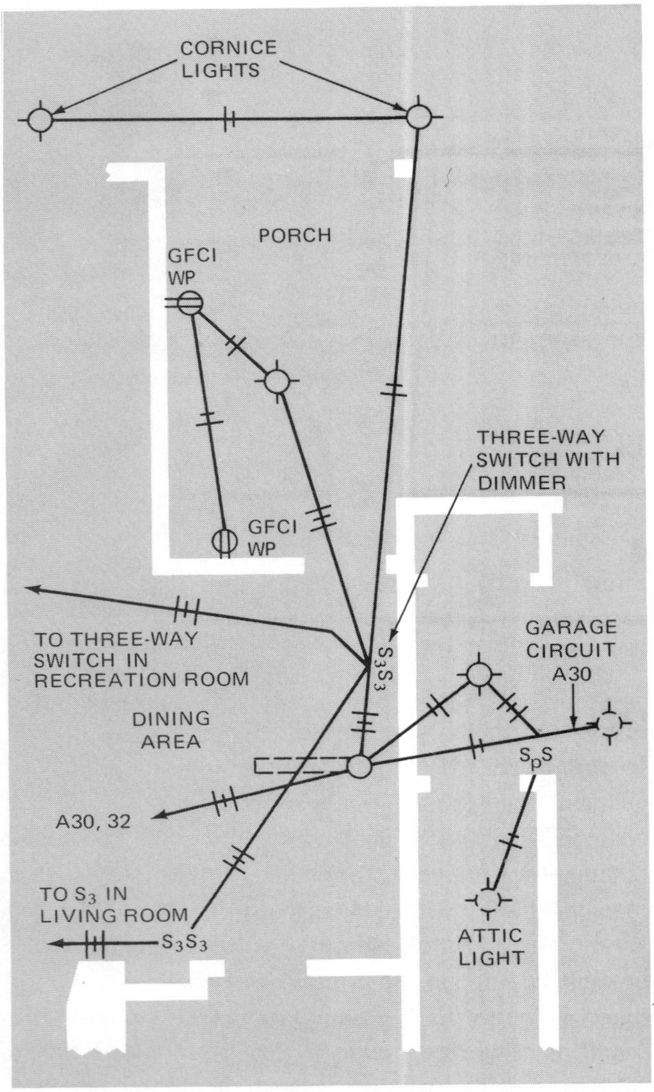

Fig. 12-1 Cable layout for the dining room, porch, cornice, garage storage area, and attic.

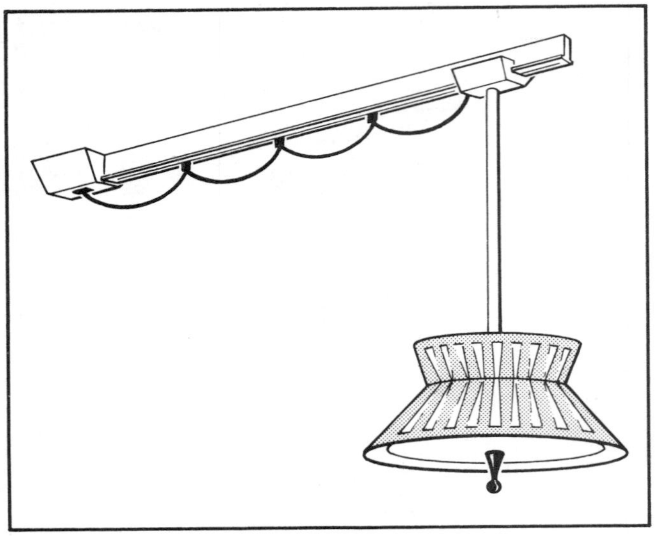

Fig. 12-2 Pulldown lighting fixture, track mounted.

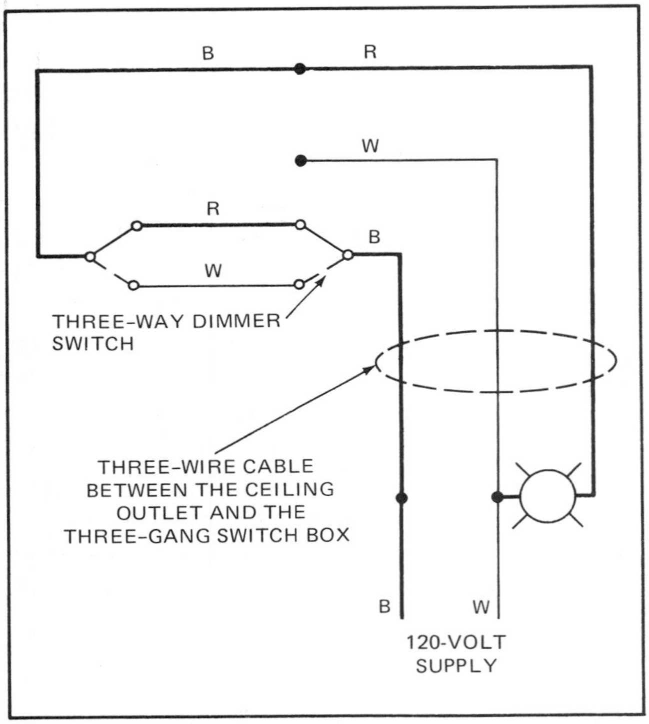

Fig. 12-3 Wiring diagram for dining room dimmer fixture.

PORCH AREA

Section 210-7 states that only grounding-type receptacles can be installed on 15- and 20-ampere circuits. The receptacle outlets on the porch must have ground-fault circuit protection (unit 6). In addition, *Section 410-57* requires weatherproof receptacles for wet or damp installations. Therefore, all receptacles installed on the outside of the residence, including the porch area, must be weatherproof.

According to *Section 410-4*, fixtures located in wet or damp locations must be constructed so that water will not enter their wiring compartments, lampholders, or other electrical parts. These fixtures are marked "Suitable for Wet Locations."

Receptacles located outdoors in damp locations such as porches and under canopies, or in wet locations directly exposed to the weather, must be weatherproof when the self-closing cover of the receptacle is closed, *Section 410-57*. Such a receptacle is weatherproof when something is not plugged into it.

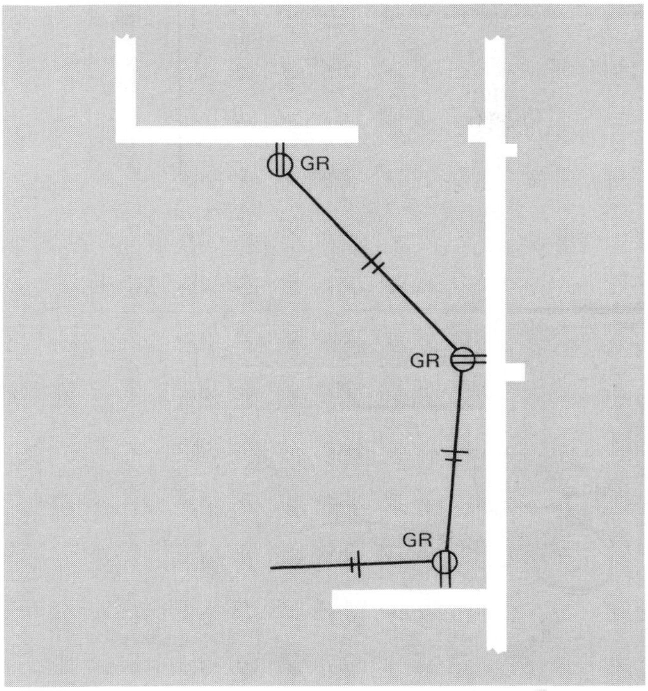

Fig. 12-4 Dining area convenience receptacles.

Switches installed in wet locations or outside a building must be protected in a weatherproof enclosure according to *Section 380-4*. Figure 12-5 shows two weatherproof electrical devices commonly used.

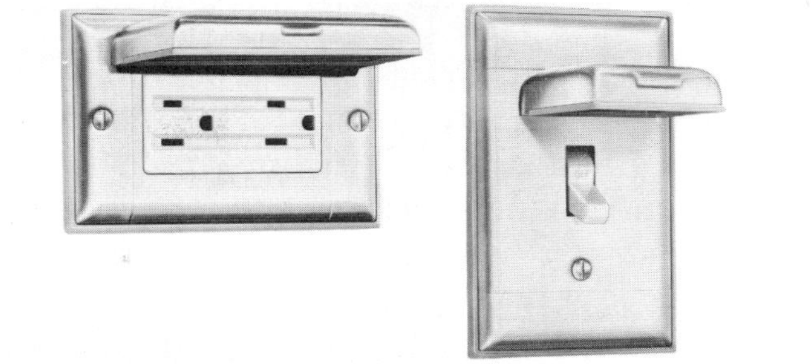

Fig. 12-5 Weatherproof devices commonly used.

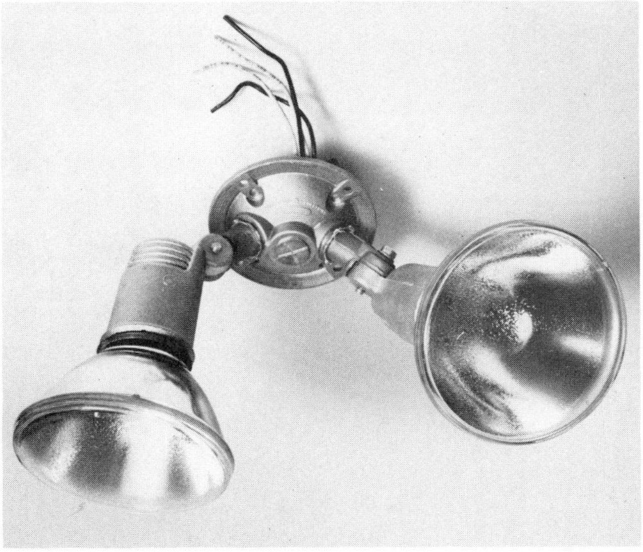

Fig. 12-6 Outdoor fixture with lamps.

REAR YARD

The rear yard lighting consists of four 150-watt PAR38 reflector flood lamps. Two of these lamps are installed at each of two fixtures located under the cornices of the dwelling. A typical outdoor fixture is shown in figure 12-6. The flood lamps are controlled by two three-way switches. One switch is in the dining room and the other is in the recreation room in the basement.

The electrician must check the plans carefully to determine if switch control from different rooms (which may be on different floors) is specified for particular fixtures. For example, the residence in the plans has multiple switch control in two rooms on different floors. If the electrician does not notice this face and completes roughing in the first floor wiring prior to roughing in the basement wiring, the carpenter or lather may begin to close up the walls on the first floor before the oversight is discovered. This results in considerable rework for both the electrician and the carpenter. Such rework can be very expensive and can cause the loss of the electrical contractor's profit margin on the job.

GARAGE STORAGE AREA AND ATTIC

The general lighting for the garage storage area and the attic is part of Circuit A32. The cable layout in figure 12-1 shows that the attic light switch has a pilot light to indicate when the attic light is on. A switch with a pilot light in the handle or a separate pilot light may be installed. Wiring diagrams for switch and pilot light combinations are given in unit 9. Electricians commonly install porcelain lampholders in garage and attic areas. These lampholders have an integral receptacle and can be mounted on almost all types of outlet boxes.

INSTALLATION OF CABLE IN ATTIC AREA

According to the plans, the distance from the top edge of the 2" x 10" ceiling joists to the bottom edge of the ridgeboard is approximately 4 1/2 feet (1.37 m). A steep stairway in the garage storage area provides access to this attic. A scuttle hole is located at the head of the stairs.

The wiring in the attic is to be done in cable, and must meet the requirements of *Section 336-9* (for nonmetallic-sheathed cable). This section refers the reader directly to *Section 333-12* which describes how the cable is to be protected, figure 12-7.

In accessible attics, figure 12-7A, cables must be protected by guard strips when:

- they are run across the top of floor joists (1).
- they are run across the face of studs (2) or rafters (3) within seven feet (2.13 m) of the floor or floor joists.

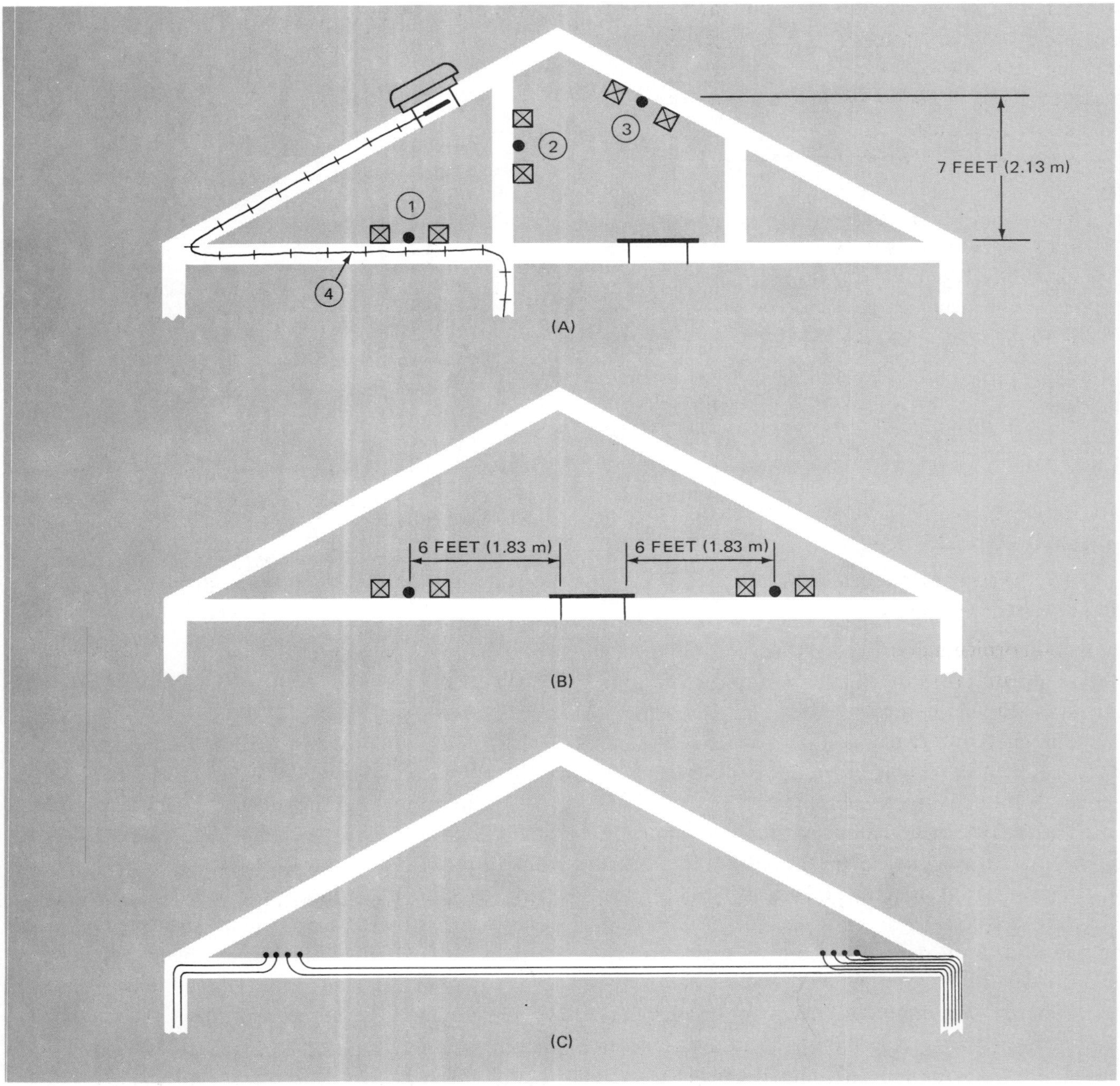

Fig. 12-7 Protection of cable in attic.

Guard strips are *not* required if the cable is run along the sides of rafters, studs, or floor joists (4).

In attics not accessible by permanent stairs or ladders, figure 12-7B, guard strips are required only within six feet (1.83 m) of the nearest edge of the scuttle hole or entrance.

Figure 12-7C illustrates a cable installation that most electrical inspectors consider to be safe. Because the cables are installed close to the point where the ceiling joists and the roof rafters meet, they are protected from physical damage. It would be very difficult for a person to crawl into this space or for cartons to be stored in an area with such a low clearance.

Although the attic space is not large, the owner may desire to install flooring in part of the attic to obtain additional storage space. Because of the large number of cables required to complete the circuits, it would interfere with the flooring to install guard strips wherever the cables run across the tops of the joists, figure 12-8. However, the cables can be run through holes bored in the joists and along the sides of the joists and rafters. In this way, the cables do not interfere with the flooring.

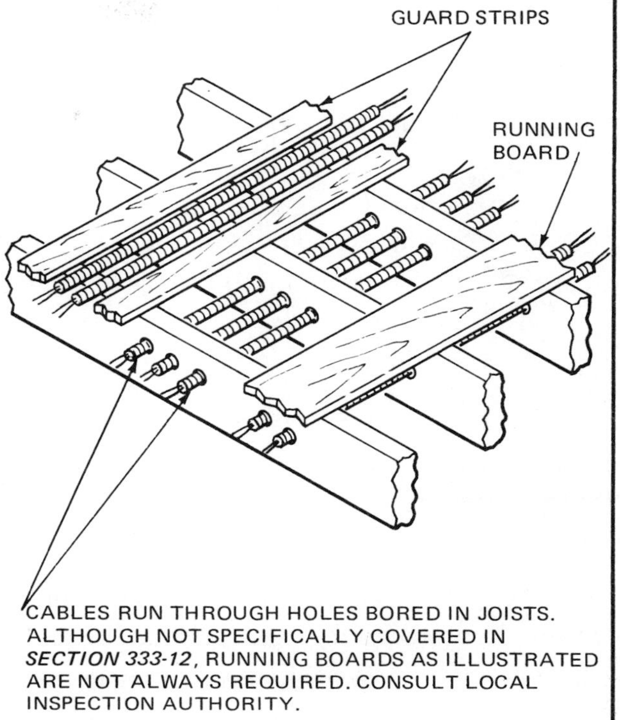
GUARD STRIPS
RUNNING BOARD
CABLES RUN THROUGH HOLES BORED IN JOISTS. ALTHOUGH NOT SPECIFICALLY COVERED IN *SECTION 333-12*, RUNNING BOARDS AS ILLUSTRATED ARE NOT ALWAYS REQUIRED. CONSULT LOCAL INSPECTION AUTHORITY.

Fig. 12-8 Methods of protecting cable installations.

WATTAGE ESTIMATE FOR LIGHTING BRANCH CIRCUIT A32

The fixtures and convenience receptacles that are part of Circuit A32 are:

4 cornice lights
1 porch fixture
1 dining area fixture
1 garage storage fixture
1 attic fixture
2 convenience receptacles on porch
Total: 10 Outlets

Using the values recommended in previous units, the estimated connected load in watts on Circuit A32 is:

4 150-watt PAR38 lamps	600 watts
1 porch ceiling fixture	150 watts
1 dining area fixture (dimmer control)	200 watts
1 garage storage fixture	100 watts
1 attic fixture	100 watts
2 convenience receptacles on porch (150 watts each)	300 watts
Total:	1450 watts Estimated

This value is acceptable because it is evident that all of the outlets will not be in use at the same time.

REVIEW

Note: Refer to the Code or the plans where necessary.

1. a. What types of circuits supply the dining area? ______
   b. What are the numbers of these circuits? ______
   c. What are the ampere ratings of these circuits? ______

2. a. List the number of switches and convenience receptacles connected to Circuit A32. ____________________
   ____
   b. How many lighting outlets are connected to Circuit A32? ____________________
3. a. How many circuit wires enter the dining area ceiling box? ____________________
   b. What size wires are used? ____________________
4. a. Are the dining area convenience receptacles connected to the dining area lighting circuit? ____________________
   b. What section of the Code refers to this situation? ____________________
5. What type of box and hanger can be installed for the dining area ceiling outlet? ____________________
   ____
   ____
6. What type of switches and receptacles must be installed in damp or wet locations? ____________________
7. What type of convenience outlets are found on the porch? ____________________
8. What section of the Code requires that receptacles on the porch have ground-fault protection? ____________________
9. From how many locations may the cornice lights be controlled? ____________________
10. What type of lamps are used in the cornice fixtures? ____________________
11. The suggested cable layout shows the northeast porch convenience receptacle fed from the southeast porch convenience receptacle. Is it possible to feed this outlet from the porch ceiling outlet? Explain why or why not. ____________________
    ____
12. What special item is provided at the attic light switch control? ____________________
13. When installing cables in an attic along the top of the floor joists, ____________________ or ____________________ must be installed to protect the cables.
14. If branch circuits A30 and A32 are both loaded to their full estimated load, what is the current in the neutral conductor of the three-wire feed? Show all calculations.
    ____
    ____
    ____
15. What is the width of a two-gang switch plate? ____________________
16. a. What is the clearance between the wood casings of the bookcase and the doorway leading to the kitchen? ____________________
    b. Will a two-gang switch plate fit at this location? ____________________
17. a. At what height are the switches in the dining area mounted? ____________________
    b. At what height are the convenience receptacles mounted? ____________________
18. Where should the three-gang switch box on the northwest dining room wall be mounted? Give the measurement from the southwest wall. ____________________
    ____

19. Approximately how high above the terrace are the south cornice lights mounted?

____

20. How many feet (meters) of three-wire cable are required for the run between the switches controlling the cornice lighting? Indicate where the cable is run. ______________

____

____

21. If an attic is accessible through a scuttle hole only, guard strips are installed to protect cables run across the top of the joists only within (6 feet) (1.83 m), (7 feet) (2.13 m), (12 feet) (3.66 m), of the scuttle hole. (Underline one answer.)

22. The following is a layout of the lighting circuit for the dining room, porch, garage, attic, and cornice. Using the cable layout of figure 12-1, make a complete wiring diagram of this circuit. Use colored pencils to indicate conductors.

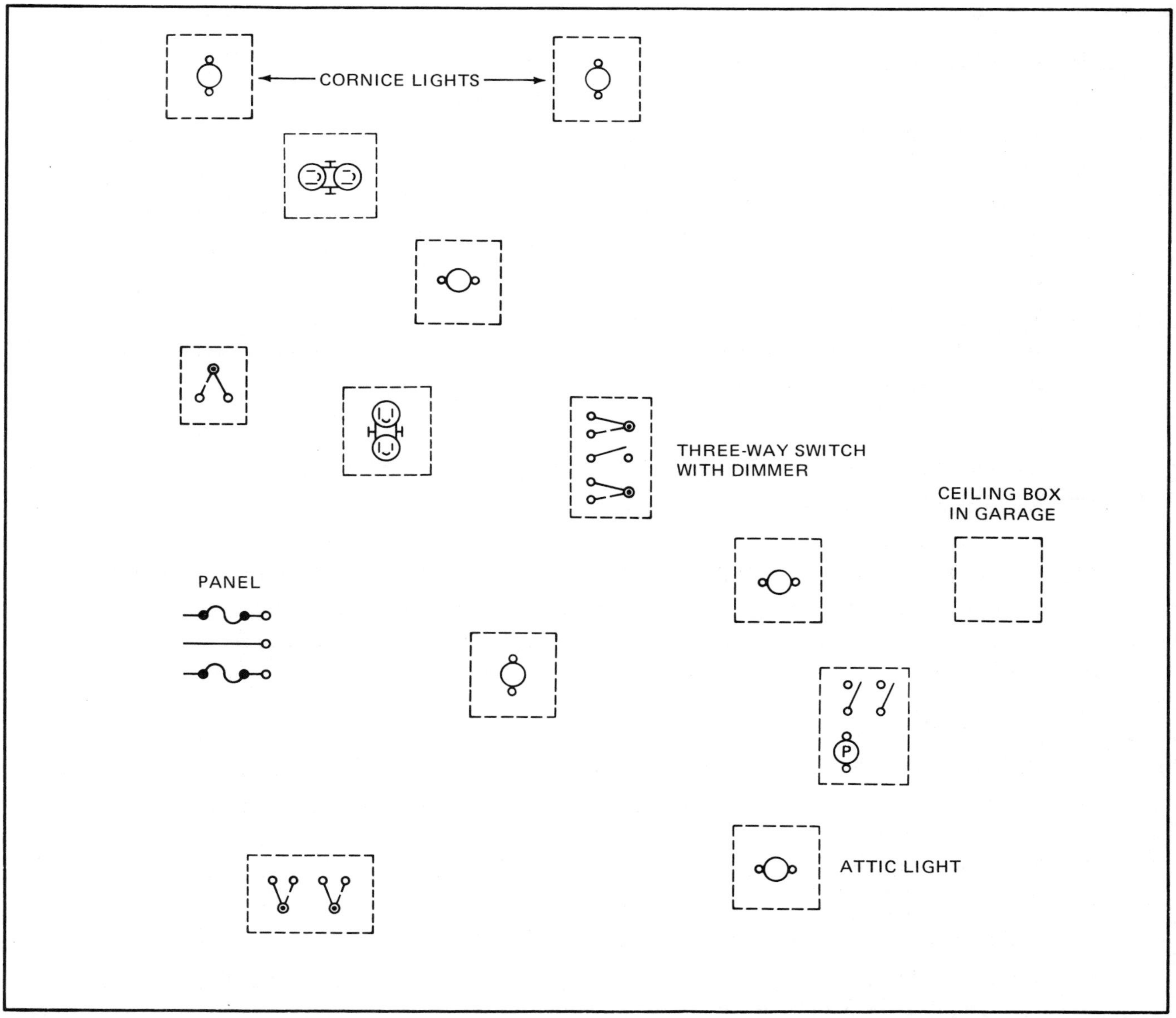

unit 13

Lighting Branch Circuit for the Garage

OBJECTIVES

After studying this unit, the student will be able to

- select the proper wiring method (cable or conduit) based on the installation requirements.
- determine the most economical layout for an installation.

Lighting branch circuit A30, figure 13-1, consists of the garage lights and convenience receptacles, the exterior weatherproof convenience receptacles, the bracket fixture, and the post light. A two-wire cable runs from the first ceiling fixture in the garage to the dining area ceiling outlet box. Here the cable is connected to the red and white conductors of the three-wire cable originating in Panel A.

FIXTURES

To provide interior garage lighting, it is recommended that one 100-watt lamp be mounted on the ceiling above each side of an automobile. For a one-car garage, two fixtures are required. Three fixtures are required for a two-car garage. Lighting fixtures arranged in this manner eliminate shadows between automobiles. Such shadows are a hazard because they hide from view objects which may cause a person to trip or fall.

The three fixtures in the garage can be turned on and off from three locations. A three-way switch is located at the rear door leading to the rear porch, a four-way switch is located at the door leading to the kitchen hall, and a three-way switch is located between the overhead garage doors.

The cable layout in figure 13-1 is only one way in which the garage can be wired. The recommended cable runs result in an economical installation. Often it is less expensive to install a few extra feet of cable if this means that the number of connections is reduced or a smaller switch or outlet box can be used.

The electrician must consider both material and labor costs when planning the most economical cable layout. Remember that the shortest cable runs do not always form the most economical wiring installation.

RECEPTACLE OUTLETS

All receptacle outlets connected to Circuit A30 must be of the grounding type, *Section 210-7*. In addition, the receptacle outlets located outdoors must be of the weatherproof type, *Section 410-57*.

All of the receptacles located outdoors also must have ground-fault circuit interrupter protection, *Section 210-8(a)(3)*.

In the garage area, all of the receptacle outlets must have ground-fault circuit interrupter protection. One exception to this rule is for outlets that are not readily accessible, such as the receptacles provided for the overhead door openers. Another exception is for the receptacle outlets installed for appliances occupying dedicated space and which are cord-connected.

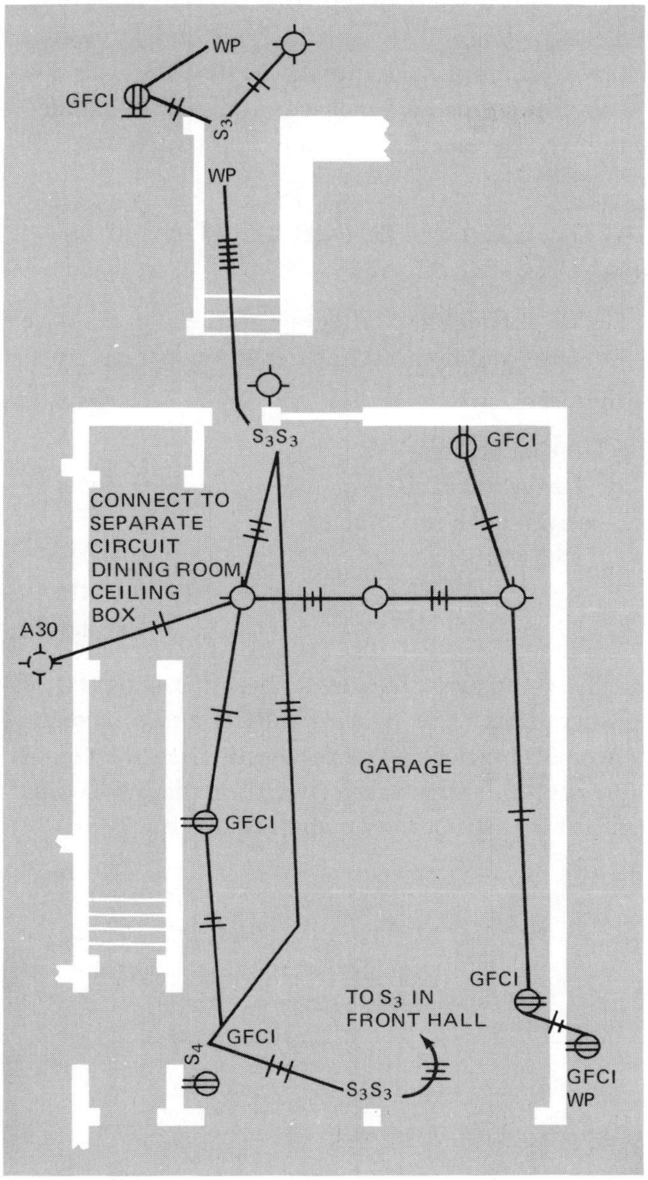

Fig. 13-1 Cable layout for the garage lighting branch circuit.

Among the possible installations to provide ground-fault circuit interrupter protection for Circuit A30 are:

- install a GFCI device on Circuit A30 in the main panel.
- install GFCI devices at each receptacle location. (This would be prohibitive from a cost standpoint.)
- install one GFCI feedthrough receptacle outlet in the garage as shown in figure 13-2.

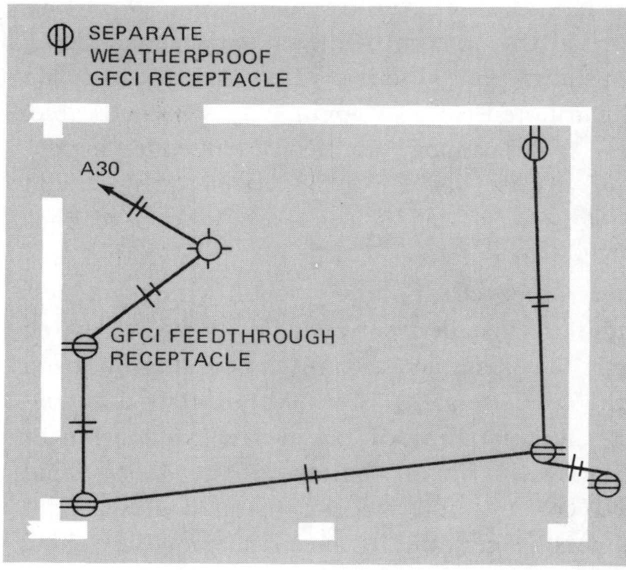

Fig. 13-2 Another possible circuit layout for the garage area receptacle outlets requiring GFCI protection, *Section 210-8.*

- install several GFCI devices at certain locations on Circuit A30, using judgment as to economy and ease of installation.

For purposes of this text, ground-fault circuit interrupter protection is provided for all of the receptacles located in the garage, as indicated in figure 13-1. Note that a separate, weatherproof GFCI has been installed for the receptacle outlet outdoors at the bottom of the steps near the 3-way switch.

Caution should be used when installing extremely long runs of a circuit beyond the GFCI. Extremely long cable runs beyond the GFCI have been known to result in nuisance tripping of the GFCI, due to leakage currents. The manufacturer's instructions furnished with the GFCI may provide circuit length limitations.

See unit 6 for a detailed discussion on ground-fault circuit interrupters.

WIRING OF CIRCUIT

The Northwest Elevation, sheet 5 of the plans, refers to Section E-E. This section is located on sheet 4, Northeast Elevation. Section E-E shows that 3/16-inch (4.76 mm) flexboard is used on the garage ceiling and walls. As a result, the installation is done in cable since all of the wiring in the garage can be concealed.

The weatherproof receptacle and three-way switch at the bottom of the concrete steps leading from the garage to the rear yard cannot be installed using armored cable or nonmetallic-sheathed cable. These cables cannot be buried or embedded in concrete. Even Type UF cable, *Article 339*, should not be used because of the severe wear conditions expected in the gravel fill in the unexcavated area under the porch. Following good wiring practice, conduit is installed between the two-gang switch box in the garage and the switch box at the bottom of the concrete steps. Conduit is also used to connect the weatherproof outlet and the post light to the switch box at the foot of the steps. Rigid metal conduit, intermediate metal conduit, rigid nonmetallic conduit, or electrical metallic tubing may be used for these installations, *Articles 345, 346, 347,* and *348.*

Type TW, RW, THW or any other moisture-resistant conductors approved for general use may be pulled into the conduit, *Table 310-7*. The conduit leading from the garage to the foot of the steps contains the circuit conductors (black and white), a switch return conductor (red), and two travelers (blue or other color except black, white and red). *Table 3A, Chapter 9* of the Code indicates that 1/2-inch conduit is required for five No. 12 TW conductors. One-half inch conduit is also installed to the weatherproof outlet and post lamp.

WATTAGE ESTIMATE FOR LIGHTING BRANCH CIRCUIT A30

The estimated wattages for the convenience receptacles and lighting fixtures in the garage are as follows:

5 lighting outlets (100 watts each)		500 watts
6 convenience receptacles (150 watts each)		900 watts
	Total:	1400 watts Estimated

In summary, the garage circuit has a total of eleven outlets (five lighting outlets and six convenience receptacles). The total estimated wattage is 1400 watts. This value is within the 1440-watt limit for a 15-ampere branch lighting circuit.

REVIEW

Note: Refer to the Code or plans where necessary.

1. What circuit supplies the garage? ____________
2. How many GFCI-protected receptacle outlets are connected to the garage circuit?
____
3. How many weatherproof convenience receptacles are connected to the garage circuit?
____
4. a. How many cables enter the southeast ceiling box in the garage? ____________
b. How many circuit conductors enter this box? ____________
c. How many grounding conductors enter this box? ____________
5. a. How many lights are recommended for a one-car garage? ____________ For a two-car garage? ____________ For a three-car garage? ____________
b. Where are these lights to be located? ____________
____
6. From how many points in the garage may the ceiling lights be controlled? ____________
7. What section(s) of the Code requires that the convenience outlets on the exterior of the garage be weatherproof? ____________

8. GFCI breakers or GFCI receptacles cost approximately $30.00 each. How would *you* arrange the wiring for the garage circuit to make an economical installation that complies with the Code? ______

______

______

9. a. What size conduit is required between the two three-way switches controlling the post light and the rear garage light? ______
   b. What type or types of wire can be installed in this conduit? ______

   ______

10. The total estimated wattage of the garage circuit is equal to how many amperes? Show all calculations. ______

______

______

______

11. Following the suggested cable layout, how many feet (meters) of two-wire and three-wire cable are required to complete the garage circuit (A30) up to the dining area ceiling fixture?
    a. Two-wire cable ______ feet (______ meters)
    b. Three-wire cable ______ feet (______ meters)

12. How high are the weatherproof convenience receptacles to be mounted? ______

13. How high are the switches and the convenience receptacles in the garage to be mounted?

______

14. The garage floor is: (underline one)
    a. level with the kitchen floor.
    b. 7 1/2 inches (190 mm) lower than the kitchen floor.
    c. 15 inches (381 mm) lower than the kitchen floor.

15. In what type of material is the three-way switch at the bottom of the rear steps mounted?

______

16. The outdoor convenience receptacle on the northwest side of the garage is mounted in: (underline one)
    a. concrete.
    b. stone.
    c. vertical wood siding.
    d. horizontal wood siding.

17. The following is a layout of the garage branch lighting circuit. Using the cable and conduit layout shown in figure 13-1, make a complete wiring diagram of this circuit. Use colored pencils to indicate conductors.

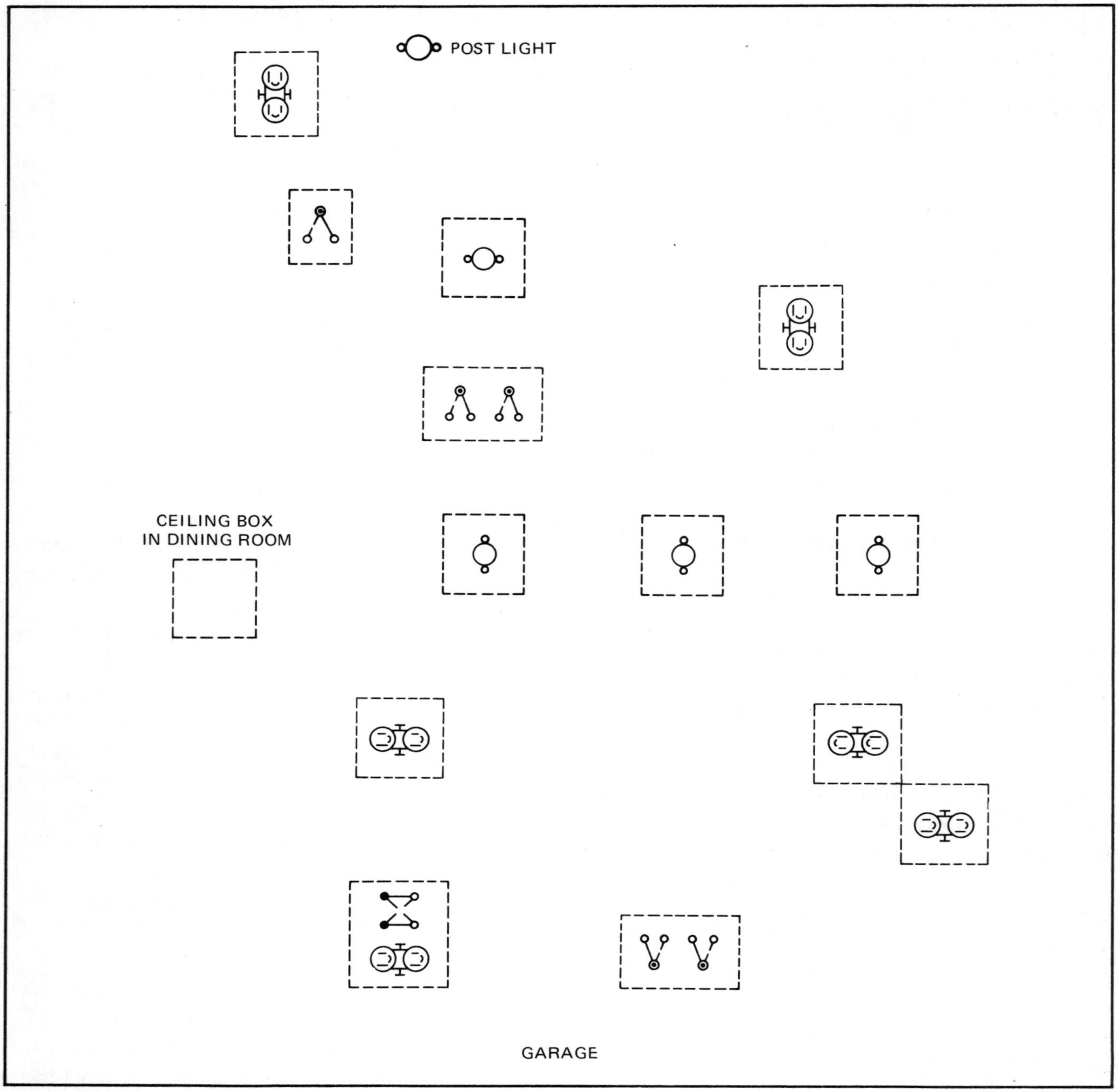

unit 14

Lighting and Receptacle Branch Circuits for the Terrace, Recreation Room, and Utility Room

OBJECTIVES

After studying this unit, the student will be able to

- select the proper box sizes for areas where the construction makes it impossible to install deep boxes.
- describe how conduit is installed in concrete floors and on cinder block walls.
- list the requirements for installing conduit in an exposed location.
- explain why the convenience receptacles in the utility room are connected to 20-ampere circuits.

TERRACE AND RECREATION ROOM

The circuits supplying the terrace, recreation room, and the utility room are shown in the cable and conduit layout of figure 14-1. A three-wire circuit runs from the ceiling outlet box in the passage over the lavatory ceiling to Panel A. At Panel A, the red conductor is connected to Circuit A34, the black conductor is connected to Circuit A36, and the white neutral conductor is connected to the neutral terminal strip. At the passage ceiling outlet box, the three-wire circuit divides. The red conductor of the three-wire cable is connected to the black conductor of the two-wire cable supplying the convenience receptacles in the recreation room. The black conductor of the three-wire cable supplies the terrace and recreation room lighting. The white neutral conductor is common to both circuits. Therefore, the recreation room lighting is fed from Circuit A36 and the convenience receptacles are fed from Circuit A34.

Installation of Boxes for Outlets and Convenience Receptacles

It is very important that the Code restrictions on the number of conductors in a box are not violated. Before an installation, the electrician must check the number of conductors terminating in or passing through a given box.

The plans show that the outer walls of the basement are concrete and the inner wall partitions are cinder blocks (except for certain partitions in the lavatory). Outlet boxes and switches may be fastened to masonry walls with lead anchors, shields, concrete nails, or power-actuated studs. In addition, boxes and switches may be glued to such walls with an adhesive approved for the installation.

The walls in the recreation room and passage are finished with knotty pine paneling attached to 2″ x 2″ furring strips. Deep switch boxes cannot be installed because of the thickness of the furring strips. Shallow switch boxes do not provide enough room for the conductors entering the boxes. Thus, 4 inch square by 1 1/2 inch-deep outlet boxes are installed with suitable single-gang or two-gang raised covers. Typical outlet boxes and raised covers are shown in figure 14-2. The depth of the raised cover depends upon the thickness of the paneling. This measurement can be checked with the carpenter.

The three-gang box shown in figure 14-2B is to be installed for the switches next to the door

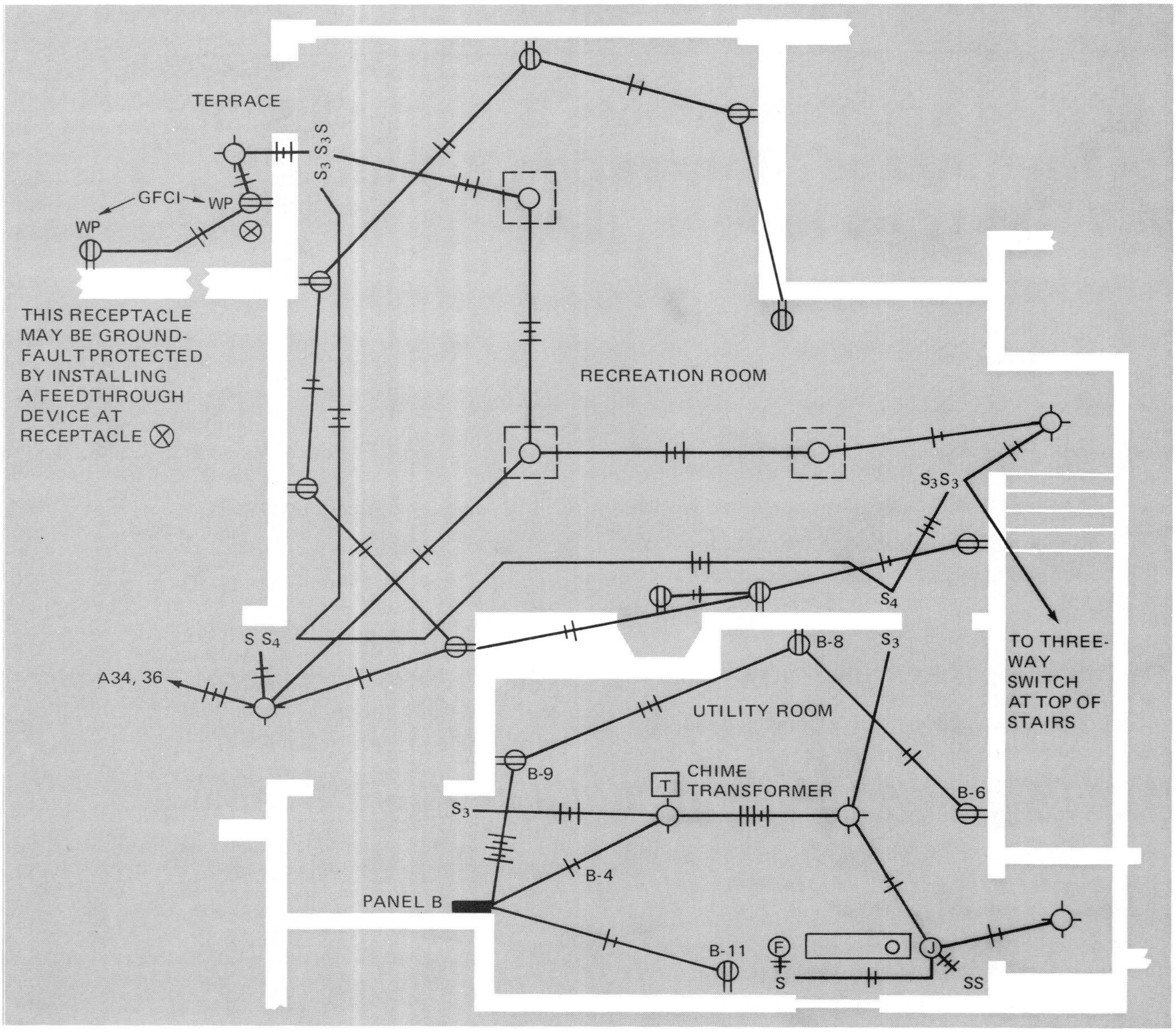

Fig. 14-1 Terrace, recreation room, and utility room cable and conduit layout.

leading from the recreation room to the terrace. One-half inch conduit from this box connects the terrace bracket fixture and the terrace convenience receptacles. Grounding-type receptacles must be used on the terrace. In addition, the receptacles must be weatherproof (unit 12) and have ground-fault protection (unit 6).

Recreation Room Wiring

The walls of the recreation room and passage are finished with knotty pine paneling attached to furring strips. Detail F-F and the basement fireplace detail show that the ceiling of these areas is finished with 1/2-inch acoustical tile mounted on 3/8-inch (9.5 mm) Sheetrock®. There is enough room above the ceiling and behind the knotty pine paneling to conceal all of the required wiring. The one exception in the recreation room is the short run of conduit connecting the box above the mantel to the box to the left of the fireplace (as viewed when facing the fireplace). The conduit cannot be connected inside the fireplace wall since the entire fireplace is masonry construction.

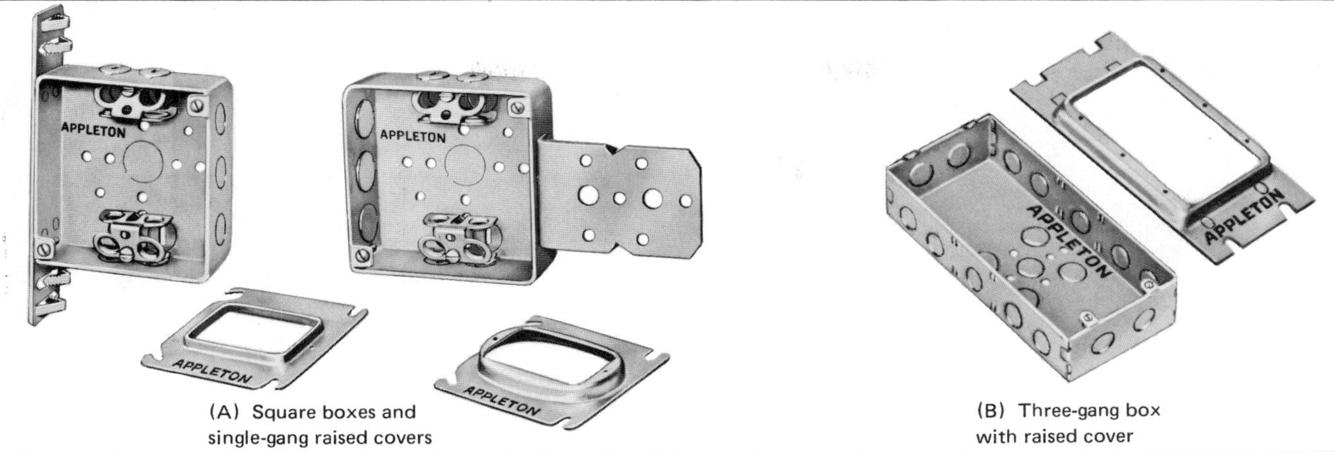

(A) Square boxes and single-gang raised covers

(B) Three-gang box with raised cover

Fig. 14-2 Typical outlet boxes and raised covers.

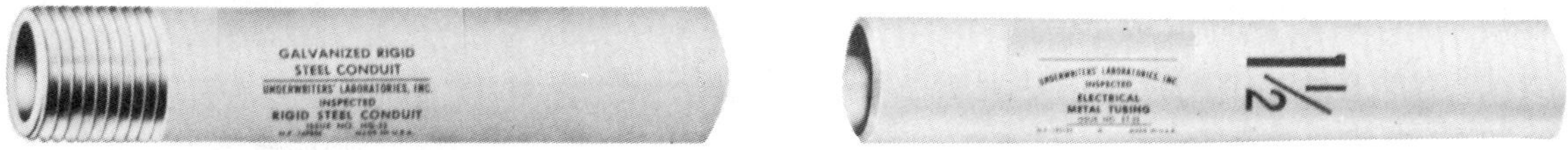

(A) Rigid steel conduit and coupling

(B) Electrical metallic tubing

Fig. 14-3 Types of electrical conduit.

The spacing of receptacle outlets in the recreation room may require the use of the receptacle outlets as an integral part of the electric baseboard heating units. Units 11 and 23 contain more information on this topic.

UTILITY ROOM

The utility room walls and ceiling are not finished. Thus, some of the wiring in these areas is not concealed. To insure that the wiring conforms to good wiring standards, rigid conduit or electrical metallic tubing must be used, figure 14-3. Most electricians and electrical contractors prefer to use electrical metallic tubing or EMT (also known as *thinwall* conduit or *steel tubing*) because it is easier to install.

Installation of Conduit

Conduit is not visible on the southeast, southwest, or northwest walls of the utility room. The specifications state that the conduits for these walls are to be roughed in by the electrician into the concrete floor. The conduits are then stubbed up into the partitions of the walls. After the basement concrete floor sets, the electrician must work closely with the mason. As the mason builds up the wall partitions with cinder blocks, the electrician adds conduit to the conduit stubs until the switch and receptacle height is reached (48 inches on center). Boxes and covers of the proper sizes are then attached to the conduits. The electrician pulls in the required number of conductors, makes all connections, and attaches the devices and wall plates to complete the installation. (*Note:* See *Table 1, Chapter 9* of the Code for the percent fill of conductors in conduit.) The conduits are run concealed within the hollow spaces of the cinder blocks from the top of the switch boxes (where the three-way switches are installed) to the ceiling. At the ceiling level, the conduits are brought out to run across the bottom of the ceiling joists.

Conduit may be fastened to masonry surfaces using EMT straps, figure 14-4A. Cable may be fastened to wood surfaces using staples, figure 14-4B or C.

The conduits on the outer northeast wall and on the ceiling of the utility room are exposed to view. The electrician must install the exposed wiring in a neat and skillful manner while complying with the following practices:

- the conduit runs must be straight,

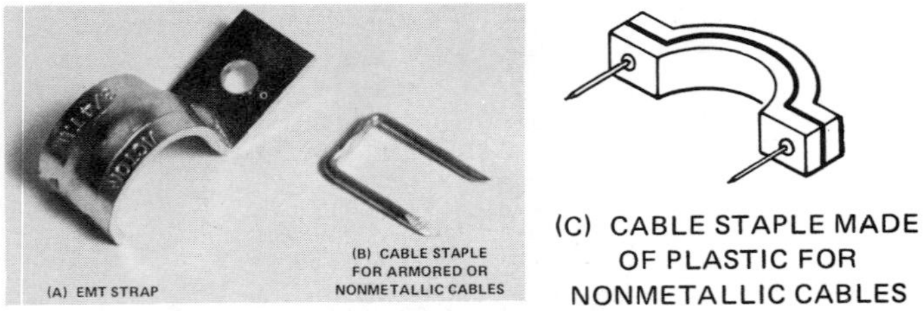

Fig. 14-4 Devices used for attaching conduit and cable to masonry surfaces (A), and wood surfaces (B) or (C).

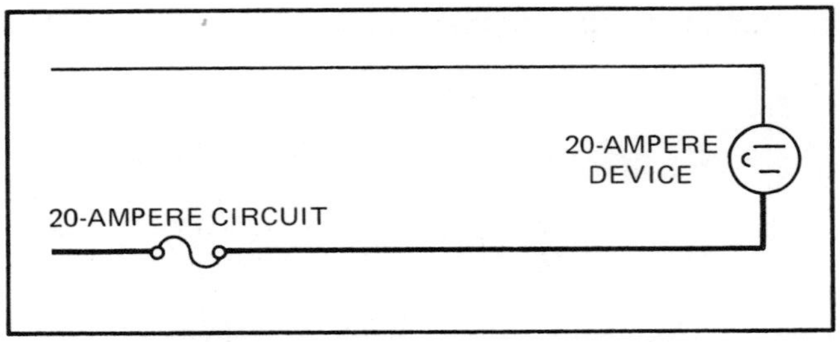

Fig. 14-5 Single receptacle must have same rating as branch circuit, *Section 210-21(b) (1).*

- bends and offsets must be true, and
- vertical runs down the surfaces of the walls must be plumb.

All conduits and boxes on the utility room ceiling are fastened to the underside of the wood joists. The outlet box in the pump room is mounted directly on the underside of the concrete ceiling slab.

Convenience Receptacle Circuits in the Utility Room

Four grounding-type convenience receptacles are installed in the utility room as required by *Section 210-7.* The appliances normally connected or plugged into these receptacles draw a large amount of current. Thus, each receptacle is connected to a separate 20-ampere circuit in Panel B.

When a single receptacle is installed on an individual branch circuit, the receptacle rating must be not less than the rating of the branch circuit, figure 14-5. Each receptacle in the utility room must be rated at 20 amperes for use on a separate 20-ampere circuit, *Section 210-21(b)(1).*

Figure 14-1 shows that Circuit B6 supplies the receptacle on the northwest wall, Circuit B8 supplies the freezer outlet, Circuit B9 supplies the receptacle on the southeast wall, and Circuit B11 supplies the washer outlet on the northeast wall. Circuits B6 and B8 are three-wire circuits. The four appliance circuits are connected to 20-ampere overcurrent devices in Panel B.

NEC *Section 220-3(c)* states that the laundry outlet circuit (B11 in the dwelling) shall have no other outlets.

Most large freezers have an alarm or pilot light to indicate when the power to the freezer is turned off or disconnected. It is recommended that a clock or nightlight be connected to the freezer outlet as an indicator to warn that the power is off.

Except for the dryer and washer outlets, all switches and convenience receptacles in the utility room are mounted 48 inches (1.22 m) to center from the finish floor. The washer and dryer outlets are mounted 30 inches (762 mm) to center from the finish floor and are hidden from view behind the appliances. The dryer outlet is covered in unit 18. According to *Section 210-50(c)*, outlets for special appliances must be mounted within six feet (1.83 m) of the planned location of the appliance.

Lighting Circuit

The lighting in the utility room and pump room is connected to Circuit B4, a 15-ampere lighting branch circuit. Figure 14-1 shows that the chime transformer is also connected to this lighting circuit. The transformer is mounted on one of the ceiling outlet boxes. It is easy to connect the transformer to the lighting circuit from this location. However, the transformer may be mounted on Panel B.

A fluorescent fixture containing two 40-watt lamps is installed above the laundry equipment.

Exhaust Fan

An exhaust fan is mounted in the utility room to remove the excess moisture resulting from the use of the laundry equipment. Exhaust fans may be installed in walls or ceilings, figure 14-6. The wall-mounted fan can be adjusted to fit the thickness of the wall. If a ceiling-mounted fan is used, sheet metal duct must be installed between the fan unit and the outside of the house. The fan unit terminates in a metal hood or grille on the exterior of the house. The fan has a shutter which opens as

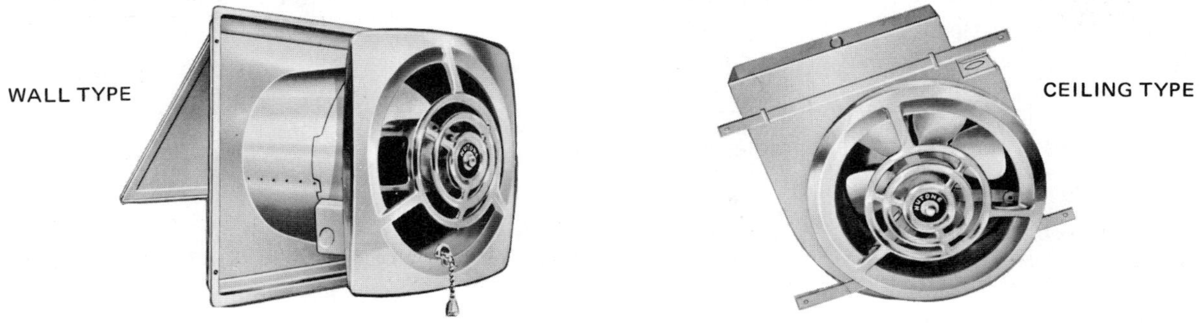

Fig. 14-6 Exhaust fans.

the fan starts up and closes as the fan stops. The fan may have an integral pull-chain switch for starting and stopping or it may be used with a separate wall switch. In either case, single-speed or multi-speed control is available. The fan in use has a very small power demand.

To provide better humidity control, both ceiling-mounted and wall-mounted fans may be controlled with a *humidistat.* This device starts the fan when the humidity reaches a certain value. When the humidity drops to a preset level, the humidistat stops the fan.

The exhaust fan in the dwelling in the plans is connected to Circuit B4. The fan is mounted between the ceiling joists in the utility room and is vented to the outside. The speed control unit is connected in series with the hot wire leading to the fan motor. Generally, speed control is achieved by varying the voltage to the motor.

SUMMARY OF CIRCUITS

The following circuits are provided in the recreation room and utility room. The estimated load in watts for each circuit is shown.

Circuit A34:
9 convenience receptacles (150 watts each) — Total: 1350 watts

Circuit A36:

3 recreation room ceiling fixtures (150 watts each)	450 watts
1 fixture at the foot of the stairs	100 watts
1 terrace fixture	100 watts
1 passage fixture	100 watts
2 grounding-type convenience receptacles on terrace (150 watts each)	300 watts
Total:	1050 watts

Circuit B4:

2 ceiling fixtures (100 watts each)	200 watts
1 two-lamp, 40-watt fluorescent fixture (80 × 1.25)	100 watts
1 pump room fixture	100 watts
1 chime transformer	10 watts
1 exhaust fan	100 watts
Total:	510 watts

Circuit B6:
Convenience receptacle
(Separate 20-ampere circuit)

Circuit B8:
Convenience receptacle
(Separate 20-ampere circuit)

Circuit B9:
Convenience receptacle
(Separate 20-ampere circuit)

Circuit B11:
Convenience receptacle
(Separate 20-ampere circuit)

REVIEW

Note: Refer to the Code or the plans where necessary.

1. a. What circuit supplies the recreation room lighting? ______________
   b. What circuit supplies the recreation room convenience outlets? ______________

2. a. Is it permitted to connect loads other than the clothes washer to a laundry receptacle outlet? ______
   b. Where are ground-fault circuit interrupter receptacles located in this circuit?
   ______
   c. How many GFCI receptacles are included? ______
3. How many three-way switches are used in the:
   a. recreation room? ______ b. utility room? ______ c. passage? ______
4. How many circuit wires enter the following outlet boxes?
   a. Passage ceiling outlet box ______
   b. Junction box next to utility room fluorescent fixture ______
   c. Passage switch box ______
   d. Ceiling outlet box at foot of stairs ______
   e. Southeast utility room ceiling outlet box ______
5. How many two-gang switch plates are required for the area covered in this unit?
   ______
6. What type of boxes are installed for the convenience receptacles in the recreation room? ______
7. What special-purpose outlets are provided in the utility room? Sketch the symbols used for these outlets. ______
   ______
8. What electric panel supplies the utility room lighting circuit? ______
9. How is EMT fastened to masonry? ______
10. When a single receptacle is installed on an individual branch circuit, the receptacle must have a rating ______ than the rating of the branch circuit.
11. a. How many lighting fixtures are connected to the recreation room lighting circuit?
    ______
    b. What is the total estimated wattage of these fixtures? ______
12. What type of fixtures are installed in the recreation room? ______
13. What space is required between a recessed fixture and the wood joists when no support is provided? Give the Code section reference. ______
14. a. What is the wattage of the electric baseboard heating units in the recreation room?
    ______
    b. What is this load in amperes? ______
15. a. What size conductors supply Panel B? ______
    b. What size conduit? ______
    c. Is Panel B flush or surface mounted? ______
16. The conduits feeding convenience outlets B8 and B9 run from Panel B, under the concrete floor, to the cinder block wall where they are stubbed up into the partitions at the proper locations. Give the measurements from the northeast and northwest concrete basement walls to the places where these conduits are stubbed up. Take these measurements from sheet 2 of the plans.

	From Northeast Wall	From Northwest Wall
B8	______ feet ______ inches	______ feet ______ inches
B9	______ feet ______ inches	______ feet ______ inches

17. Approximately how much 1/2-inch EMT is used for the lighting and convenience receptacle circuits in the recreation room, passage, utility room, pump room, and terrace?

____

18. The following is a layout of the lighting circuits for the terrace and recreation room area. Using the cable layout in figure 14-1, make a complete wiring diagram. Use colored pencils to indicate conductors.

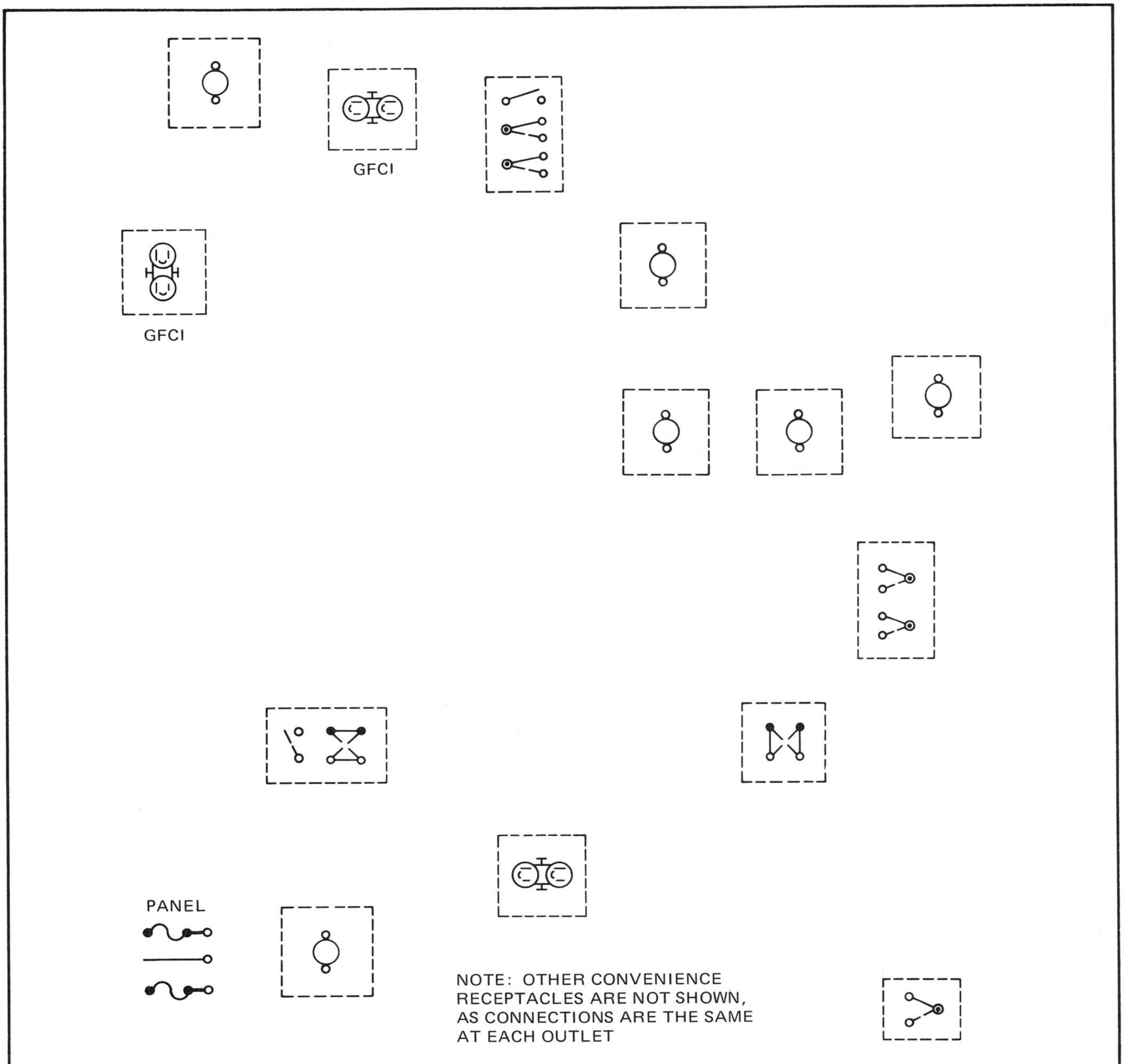
GFCI
GFCI
PANEL
NOTE: OTHER CONVENIENCE RECEPTACLES ARE NOT SHOWN, AS CONNECTIONS ARE THE SAME AT EACH OUTLET

TERRACE AND RECREATION ROOM

19. The following is a layout of the lighting circuit for the utility room and pump room areas. Using the conduit layout shown in figure 14-1, make a complete wiring diagram of this circuit. Use colored pencils to indicate conductors.

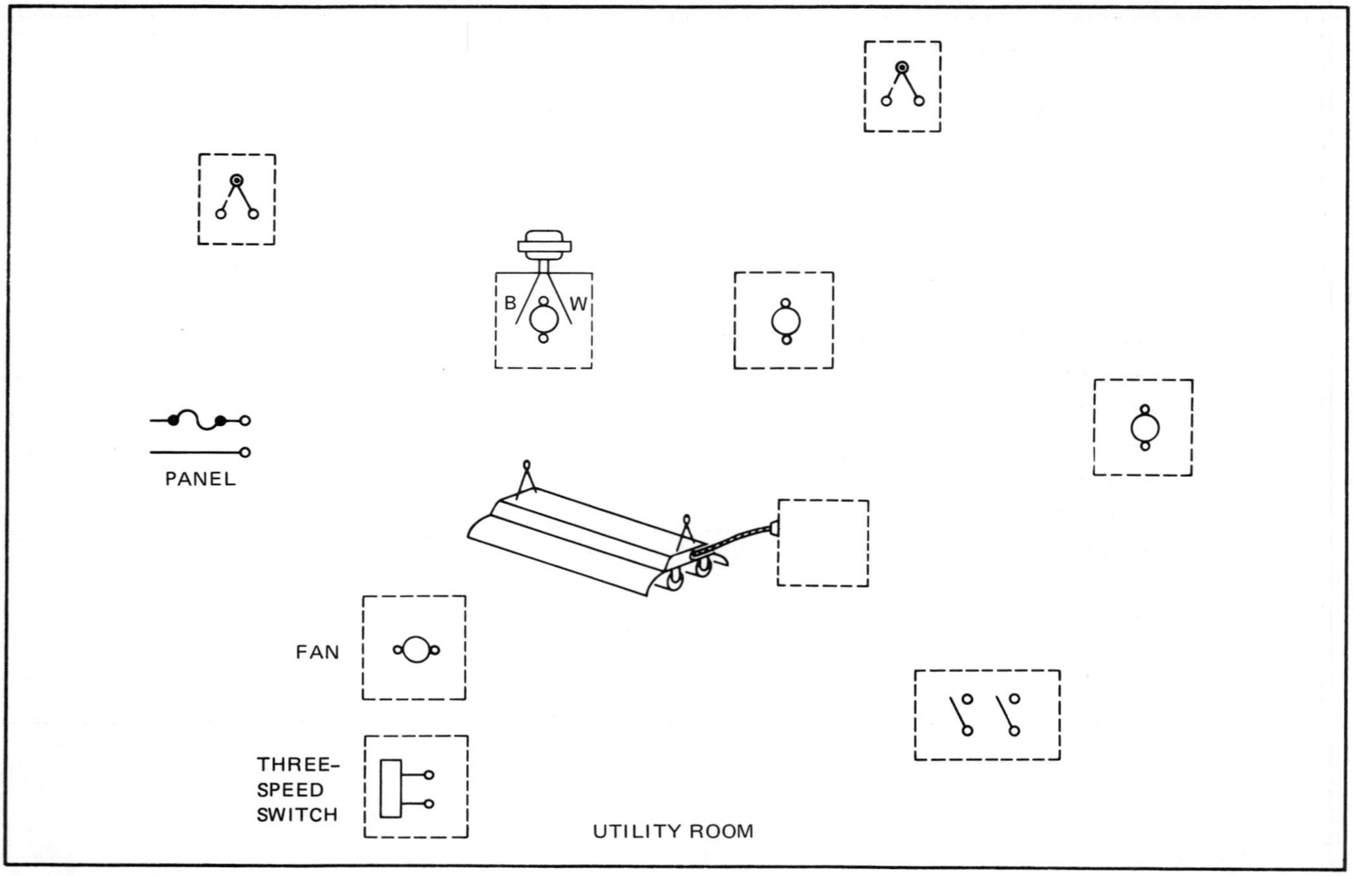

20. The following is a layout of the convenience receptacles for the utility room. Using the conduit layout in figure 14-1, make a complete wiring diagram of these circuits. Use colored pencils to indicate conductors.

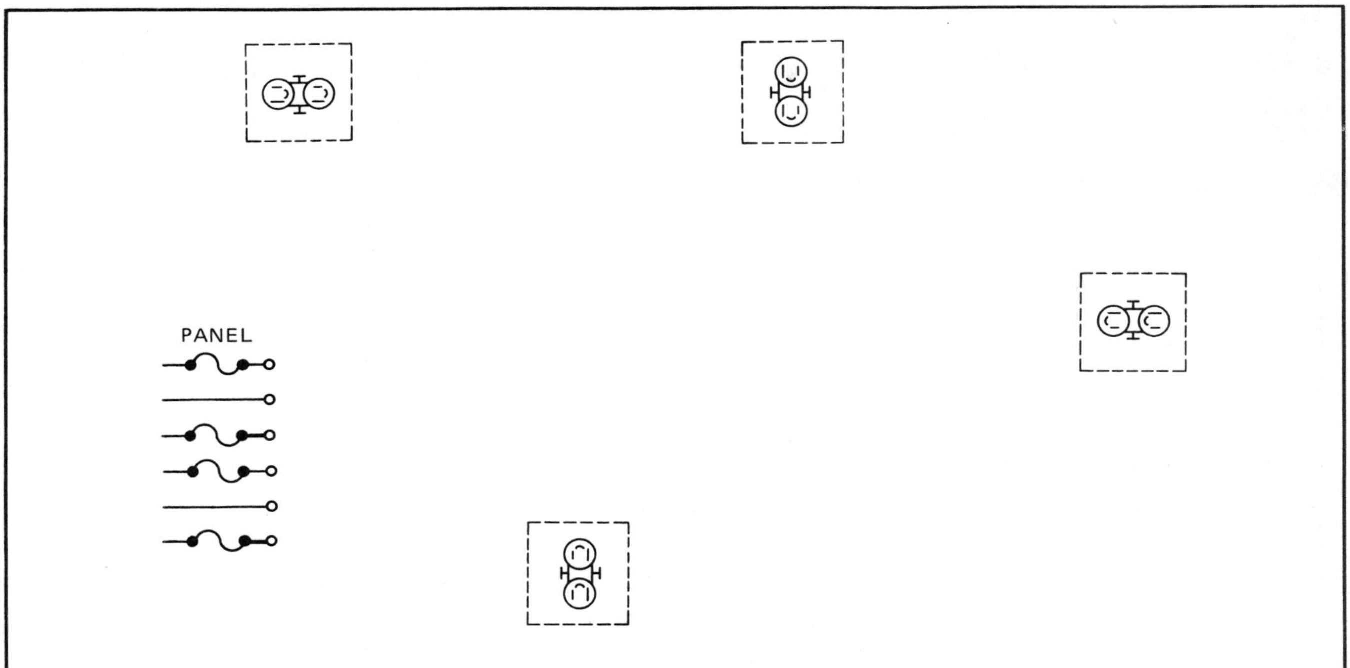

UTILITY ROOM RECEPTACLES

unit 15

Lighting and Convenience Receptacle Branch Circuits for the Lavatory, Workshop, and Storage Room

OBJECTIVES

After studying this unit, the student will be able to

- select the proper material for installing exposed wiring in an unfinished basement (cable or conduit).
- select the proper outlet box for surface mounting in a basement area.
- calculate the allowable fill for specific conductors in conduit.
- select the conduit size based on the allowable fill value.
- apply the proper ampacity derating factors to conductors for certain conditions.

The basement floor plan shows that the lavatory ceiling and walls are enclosed or finished. (The finishing methods used are as described for the recreation room and passage walls and ceilings.) Thus, cable can be installed in the lavatory by concealing it behind the wall paneling and above the finished ceiling. The plans also show that the ceiling joists in the lavatory, workshop, and storage room run parallel to the outside southeast wall. As a result, it is convenient to install 1/2-inch conduit (EMT) to carry lighting branch Circuit A35 between the southeast ceiling fixture in the workshop, the lavatory ceiling fixture, and the storage room ceiling light, figure 15-1.

CODE REQUIREMENTS FOR THE INSTALLATION OF CABLE AND CONDUIT

The Code permits cable to be installed in the lavatory, workshop, and storage room areas. To comply with the specifications, however, conduit (EMT) is installed in the basement wherever the wiring is exposed to view. Cable is installed where the wiring is concealed. In summary, conduit (EMT) must be installed in the workshop, storage room, utility room, and pump room. The lavatory, recreation room, and passage area are wired with cable.

All switch boxes, outlet boxes, and fixtures installed with cable must be grounded. Either one of two grounding methods can be used: the grounding conductor in grounding-type nonmetallic-sheathed cable or the armor and grounding conductor in armored cable. Specific Code requirements must be followed when installing cable or conduit.

Installation of Cable in Basement Areas

Unit 4 covers the Code sections dealing with the installation of nonmetallic-sheathed cable and armored cable. Cable assemblies smaller than two No. 6 conductors or three No. 8 conductors must be run through holes bored in joists or on running boards when installed in unfinished basements, *Section 336-8*.

If armored cable is installed in an unfinished basement, then it must follow the building surface or running boards to which it is fastened, *Section 333-11*. An exception to this requirement is armored cable run on the undersides of floor joists in basements where supported at each joist and so located as not to be subject to physical damage, (*Section 333-11, Exception No. 2*). *Note:* The local inspection authority usually interprets the meaning of "subject to physical damage." Similar methods are used to protect cables in exposed areas in basements and cables in attics (see unit 12).

Installation of Conduit in Basement Areas

Some states do not permit either nonmetallic-sheathed cable or armored cable to be installed in any basement where the walls of the basement are in direct contact with the earth. In this case, Type UF cable or conduit with Type TW conductors or other approved conductors having moisture-resistant insulation may be installed.

A local electrical code may specify that conduit must be used in unfinished basements. However, cable is generally permitted if not more than 18 inches (457 mm) of the cable are exposed to view. In other words, short lengths of cable may be exposed. For example, a cable may be dropped into the basement from a switch at the head of the basement stairs. Transitions between cable and conduit also must be made for certain conditions. The electrician must check the national and local codes before selecting conduit or cable for the installation to prevent costly wiring errors.

Since local codes vary, the electrician should also check with the local inspection authority that enforces the code. Electrical utility companies can supply additional information on local regulations. The electrician should never assume that the National Electrical Code is the recognized standard everywhere. Any city or state may pass electrical installation and licensing laws. In many cases, these laws are more stringent than the National Electrical Code.

LAVATORY

The medicine chest in the lavatory is prewired by the manufacturer. Two 20-watt fluorescent fixtures are attached to the sides of the medicine chest and an approved junction box is provided. (*Note:* Prewired units must be listed by Underwriters' Laboratories.) The medicine chest is connected by running the cable from the switch controlling the side lights into the junction box provided. The conductors of the prewired unit and the cable from the wall switch are spliced together in the junction box. The grounding conductor of the cable must be connected to the medicine chest so that the chest itself is grounded.

WORKSHOP AND STORAGE ROOM

Lighting Circuit

A two-lamp, 40-watt fluorescent fixture is mounted above the workbench to reduce shadows over the work area. The electrical plans show that this fixture is controlled by a single-pole wall switch. A junction box is mounted immediately above or adjacent to the fluorescent fixture so that the connections can be made readily. (A similar fixture is installed over the laundry tubs in the utility room using the same method.)

Armored cable, flexible metal conduit, or flexible cord (Type S, SJ, or equivalent) can be

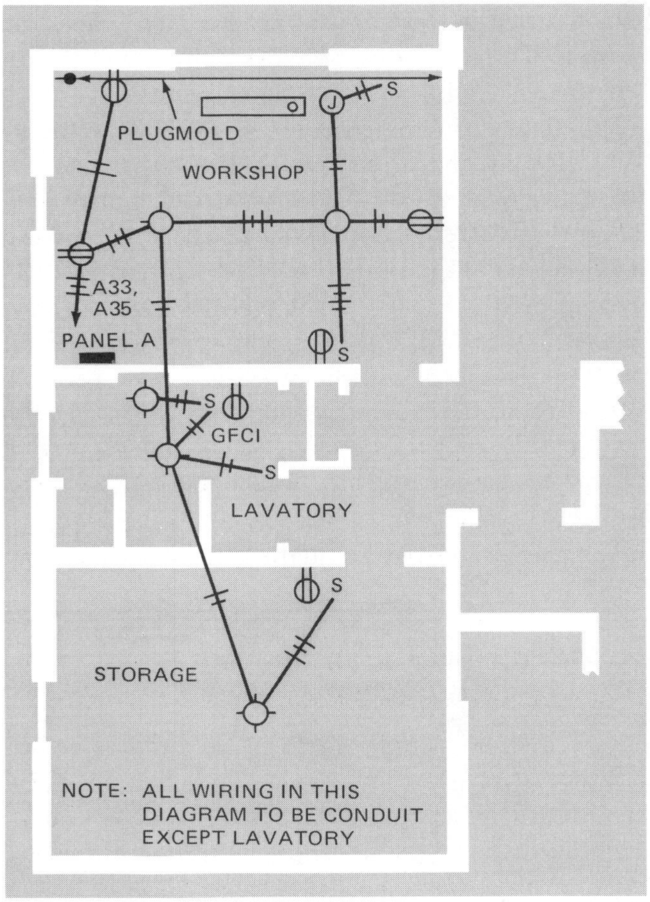

Fig. 15-1 Conduit and cable layout for the workshop, lavatory, and storage room.

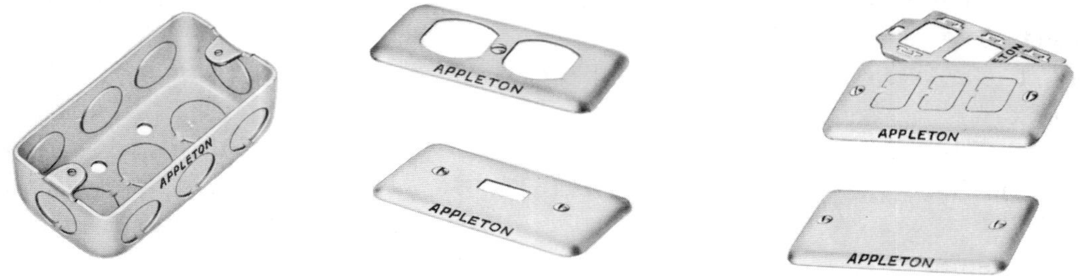

Fig. 15-2 Handy box and covers.

used to connect the fixture to the junction box. The fluorescent fixture must be grounded. Thus, any flexible cord used must contain a third conductor for grounding only.

See unit 8 for a discussion on types of fluorescent lamps to install.

Convenience Receptacle Circuit

A multioutlet assembly is installed above the workbench and three other convenience receptacles are installed in the workshop. These receptacles are connected to Circuit A33, a 20-ampere, 120-volt circuit in Panel A. According to *Section 210-7*, all of these devices must be grounding-type receptacles.

For safety reasons, as well as from an adequate wiring viewpoint, it is recommended that any workshop receptacles be connected to a separate circuit. In the event of a malfunction in any power tool commonly used in the home workshop – such as a saw, planer, lathe, or drill – then only receptacle Circuit A33 is affected. The lighting in the workshop is unchanged by a power outage on the receptacle circuit.

Circuits A33 (workshop receptacles) and A35 (lighting) are wired as a three-wire circuit using No. 12 AWG cable. Since the workshop wiring is installed in conduit (EMT), the electrician can connect the receptacle circuit to the red conductor and the lighting circuit to the black conductor. The white neutral conductor is common to both circuits.

Outlet Boxes for Use in Exposed Installations

For surface wiring, such as in the workshop and storage room, various types and sizes of outlet boxes may be used. For example, a *handy box*, figure 15-2, can be used for the fluorescent fixture over the workbench and the convenience receptacle on the northwest wall of the workshop. The remaining workshop and storage area boxes can be 4-inch square outlet boxes with raised covers, figure 15-3. The box size and type is determined by the maximum number of conductors contained in the box, *Article 370.*

Switch and outlet boxes are available with knockouts of the following sizes: 1/2 inch, 3/4 inch, and 1 inch. The knockout size selected depends upon the size of the conduits entering the box. The conduit size is determined by the maximum percentage of fill of the total cross-sectional area of the conduit by the conductors pulled into the conduit.

See page 54 for a discussion of the number of wires permitted in boxes.

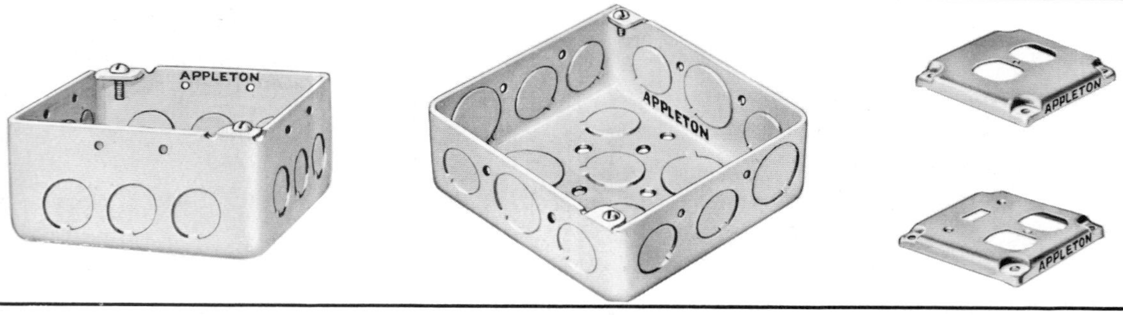

Fig. 15-3 Four-inch square boxes and raised covers.

CONDUIT FILL CALCULATIONS

The electrician must be able to select the proper conduit size to meet various requirements. If the conductors to be installed in a conduit are the same size, then *Tables 3A, 3B,* and *3C* of *Chapter 9* of the Code can be used to determine the conduit size.

For example, three No. 10 TW conductors require 1/2-inch conduit, five No. 12 RHW conductors require 3/4-inch conduit, and seven No. 14 THW conductors require 3/4-inch conduit.

When the conductors to be used have different sizes, the cross-sectional areas of all conductors must be added. This sum must not exceed the allowable percentage fill of the interior cross-sectional area of the conduit as given in *Table 4* of the Code.

For example, if three No. 6 TW conductors and two No. 8 TW conductors are to be installed in one conduit, the proper conduit size is determined as follows:

1. Find the cross-sectional area of the conductors (using *Table 5, Chapter 9* and the dimensions in *Column 5*).

Three No. 6 TW conductors ($0.0819\ in^2$ each)	$0.2457\ in^2$
Two No. 8 TW conductors ($0.0471\ in^2$ each)	$0.0942\ in^2$
Total Area	$0.3399\ in^2$

2. Check *Table 1, Chapter 9* to determine the percentage of fill permitted. When three or more conductors are installed in a conduit, the percentage of fill is 40 percent.
3. Check *Column 5* (40% fill) of *Table 4* to find the conduit size. A 3/4-inch conduit holds up to 0.21 square inch of conductor fill and 1-inch conduit holds up to 0.34 square inch of conductor fill. Therefore, 1-inch conduit is the minimum size that can be used to carry the conductor area determined in step 1.

CONDUIT BODIES [*SECTION 370-6(c)*]

Conduit bodies are used with conduit installations to provide an easy means to turn corners, terminate conduits, and mount switches and receptacles. They are also used to provide access to conductors, to provide space for splicing (when permitted) and to provide a means for pulling conductors. (See figure 15-4.)

A conduit body

- must have a cross-sectional area not less than twice the cross-sectional area of the largest conduit to which it is attached, figure 15-5A. This is a requirement only when the conduit body contains No. 6 AWG conductors or smaller.
- may contain conductors in the amounts as permitted in *Table 1, Chapter 9* of the Code for the size conduit attached to the conduit body.
- with provisions for less than three conduit entries, must *not* contain splices, taps, or devices *unless* the conduit is marked with its cubic-inch area, figure 15-5B.
- the conductor fill volume must be properly calculated according to *Table 370-6(b)* which lists the free space that must be provided for *each* conductor within the conduit body. If the conduit body is to contain splices, taps, or devices, it must be supported rigidly and securely.

DERATING FACTORS

If a raceway is to contain more than three conductors, the student must check *Note 8* to *Tables 310-16* through *310-19*. *Note 8* states that

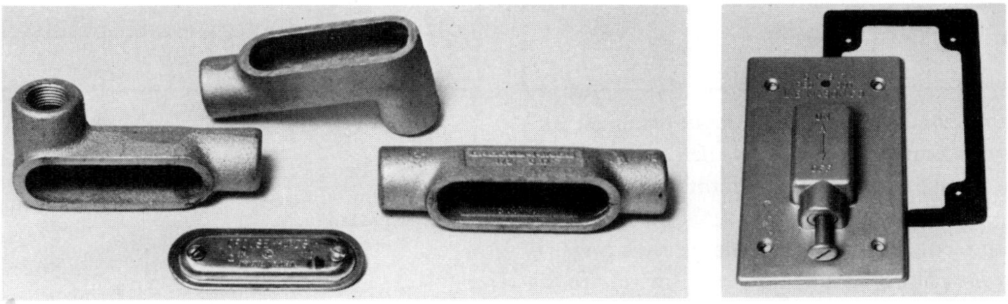

Fig. 15-4 A variety of common conduit bodies and covers.

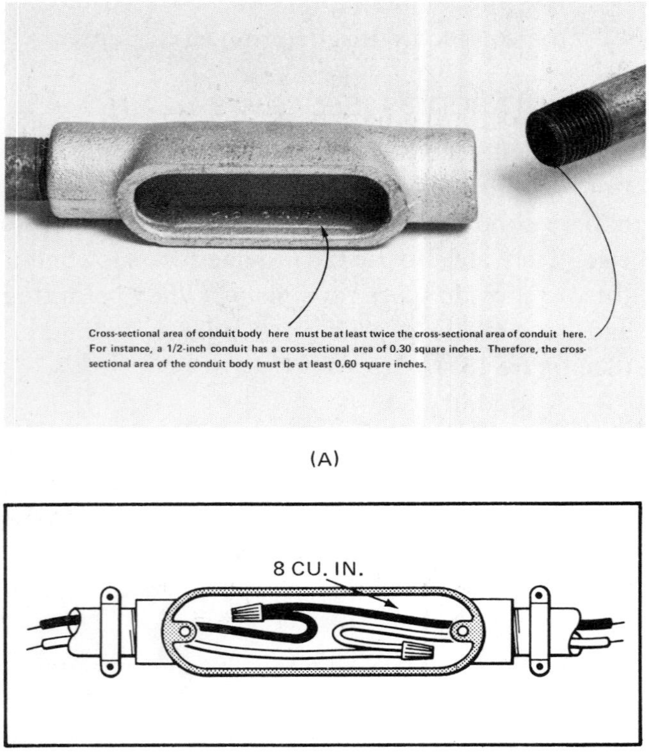

Cross-sectional area of conduit body here must be at least twice the cross-sectional area of conduit here.
For instance, a 1/2-inch conduit has a cross-sectional area of 0.30 square inches. Therefore, the cross-sectional area of the conduit body must be at least 0.60 square inches.
(A)
8 CU. IN.
(B)

Fig. 15-5 Requirements for conduit boxes.

Number of Conductors	Percent of Values in Tables 310-16 and 310-18
4 to 6	80
7 to 24	70
25 to 42	60
43 and above	50

Fig. 15-6 Ampacity reduction factors.

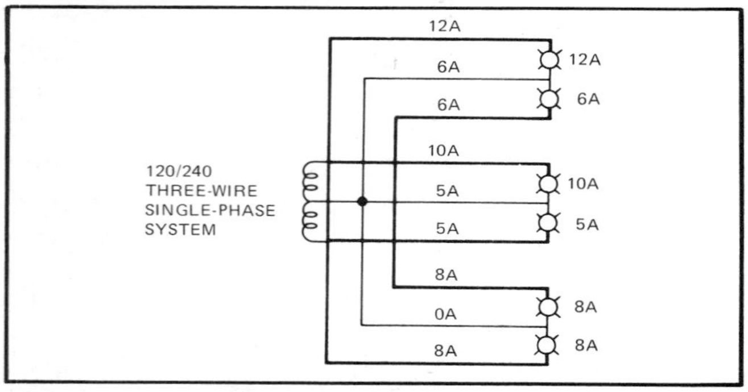

Fig. 15-7 A neutral conductor carrying the unbalanced currents from the other conductors need *not* be counted when determining the ampacity of the conductor, according to *Note 8* to *Tables 310-6* through *310-10* (see figure 15-6 also). The Code would recognize this example as *two current-carrying conductors.* Thus, in the three multiwire circuits shown, the actual total is nine conductors, but only six current-carrying conductors are considered when the derating factor of figure 15-6 is applied.

for the conductors listed in the tables, the maximum allowable ampacity must be reduced when more than three conductors are installed in a raceway. The reduction factors based on the number of conductors are listed in figure 15-6 and unit 4.

According to *Note 10*, a neutral conductor carrying only unbalanced currents shall not be counted in determining current-carrying capacities as provided for in *Note 8*. See figure 15-7.

When the derating factors in figure 15-6 are used, the 80 percent loading factor given in *Section 210-22(c)*, and the derating factor given in *Sections 220-2(a)* and *220-10(b)* do not apply. Also see *Note 8* to *Table 310-16, Exception No. 2.*

When the conductors are to be installed at temperatures above 86°F (30°C) additional correction factors must be applied (see *Note 13* to *Tables 310-16* through *310-19*). For example, a No. 3 RH conductor has an ampacity of 100 amperes (*Table 310-16*). If this conductor is to be installed at a temperature of 95°F (35°C), the necessary correction factor is 0.88 (for temperatures between 86°F and 104°F). Thus, the maximum allowable load current is 100 × 0.88 = 88 amperes.

MAXIMUM LOAD CURRENT

An important *footnote* below NEC *Table 310-16* must be observed when installing No. 14, No. 12, and No. 10 AWG conductors. Notice in the *Table* many obelisk symbols (†) (also called daggers) referring to these sizes of conductors. As can be seen, the maximum permitted loading and sizing of the overcurrent protection is less than the ampacities listed in the *Table*, as shown in figure 15-8.

Example:

To illustrate what has been learned so far about derating and maximum permitted load currents, suppose that eight (8) THW copper current-carrying conductors are to be installed in the same

Wire Size	Maximum Load Current and Overcurrent Protection	
	Copper	Aluminum or Copper-clad Aluminum
No. 14 AWG	15 amperes	(No. 14 AWG aluminum *not* permitted)
No. 12 AWG	20 amperes	15 amperes
No. 10 AWG	30 amperes	25 amperes

Fig. 15-8 *Footnote* to NEC *Table 310-16.*

conduit. The connected load will be 18 amperes. The size wire to use may be determined as follows.

- No. 14 THW copper wire
  a. The maximum permitted load current from figure 15-8 (*footnote* to *Table 310-16*) is 15 amperes.
  b. The ampacity of No. 14 THW copper wire from *Table 310-16* is 20 amperes.
  c. Apply the derating factor (8 current-carrying wires in the raceway) from figure 15-9 (*Note 8* to *Table 310-16*).

     20 X 0.70 = 14 amperes maximum permitted load

  Thus, No. 14 THW wire is *too small* to supply the 18-ampere load.
- No. 12 THW copper wire
  a. The maximum permitted load current from figure 15-8 (*footnote* to *Table 310-16*) is 20 amperes.
  b. The ampacity of No. 12 THW copper wire from *Table 310-16* is 25 amperes.
  c. Apply the derating factor (8 current-carrying wires in the raceway) from figure 15-9 (*Note 8* to *Table 310-16*).

     25 X 0.70 = 17.5 amperes maximum permitted load

  Thus, No. 12 THW is *too small* to supply the 18-ampere load.
- No. 10 THW copper wire
  a. The maximum permitted load current from figure 15-8 (*footnote* to *Table 310-16*) is 30 amperes.
  b. The ampacity of No. 10 THW copper wire from *Table 310-16* is 35 amperes.
  c. Apply the derating factor (8 current-carrying wires in the raceway) from figure 15-9 (*Note 8* to *Table 310-16*).

     35 X 0.70 = 24.5 amperes maximum permitted load

  Thus, No. 10 THW copper wire is adequate to supply the 18-ampere load.

Derating Factors
(*Note 8, Tables 310-16* through *310-19*)

Number of Current-carrying Conductors in Raceway	Percent of Ampacity Values in *Tables 310-16* and *310-18*
4 thru 6	80 percent
7 thru 24	70 percent
25 thru 42	60 percent
43 and above	50 percent

Fig. 15-9 ***Note 8*** **to NEC** ***Table 310-16.***

Double Derating

Double derating is not necessary. *Exception No. 2* to *Note 8* of *Table 310-16* clearly states that if a conductor has been derated because more than 3 current-carrying wires are installed in the same raceway, then it is *not* necessary to derate again because of "continuous loading," as required by *Sections 210-22(c), 220-2(a),* and *220-10(b).*

Continuous loading occurs when the circuit's maximum permitted load continues for periods of three hours or more.

Correction Factors

When conductors are installed in locations where the temperature is higher than 30°C (86°F), the correction factors noted below *Table 310-16* must be applied.

For example, consider that four current-carrying No. 3 RHW copper conductors are to be installed in one raceway in an ambient temperature of 35°C. The maximum permitted load current for these conductors is determined as follows.

- The ampacity of No. 3 RHW copper conductors from *Table 310-16* = 100 amperes.
- Apply the correction factor for 35°C.

  100 X 0.88 = 88 amperes
- Apply the derating factor for four current-carrying wires in one raceway.

  88 X 0.80 = 70.4 amperes

Therefore, 70.4 amperes is the maximum permitted load current for the example given.

WATTAGE ESTIMATE

The estimated load in watts for Branch Circuit A35 (supplying the lighting in the workshop, lavatory, and storage area) is determined as follows:

Workshop	
2 ceiling fixtures (100 watts each)	200 watts
1 two-lamp, 40-watt fluorescent fixture (80 X 1.25)	100 watts
Storage	
1 ceiling fixture	100 watts
1 convenience receptacle (grounding)	150 watts
Lavatory	
1 ceiling fixture	150 watts
1 convenience receptacle (grounding)	150 watts
1 medicine chest with two 20-watt fluorescent lamps (40 X 1.25)	50 watts
8 Outlets Total	900 watts

REVIEW

Note: Refer to the Code or the plans where necessary.

1. What type of convenience receptacles are used in the workshop, lavatory, and storage room? ______________________________

2. What form of wiring is used in the workshop? ______________________________

3. How many No. 12 AWG conductors enter the workshop ceiling outlet box? Four conduits are connected to this box. ______________________________

4. What size box may be installed at the location given in question 3? Refer to *Table 370-6(a).* ______________________________

5. What devices are connected to Circuit A33? ______________________________

6. What size conductors are used for wiring Circuit A33? ______________________________

7. List the proper conduit size for the following, assuming that it is new work and that TW insulated conductors are used.
   a. 3 No. 14 AWG ______________
   b. 4 No. 14 AWG ______________
   c. 5 No. 14 AWG ______________
   d. 6 No. 14 AWG ______________
   e. 3 No. 12 AWG ______________
   f. 4 No. 12 AWG ______________
   g. 5 No. 12 AWG ______________
   h. 6 No. 12 AWG ______________

8. According to the Code, what size conduit is required for each of the following combinations of conductors? Assume that these are new installations. Show all calculations.
   a. Three No. 14 AWG TW, Four No. 12 AWG TW______________
   b. Two No. 12 AWG TW, Three No. 8 AWG TW______________
   c. Three No. 3 RHW, One No. 4 Bare ______________
   d. Three No. 3/0 RHW, Two No. 8 TW ______________

9. According to the Code, what is the maximum allowable load current of a No. 10 AWG, type TW copper conductor when installed in an area having an average temperature of 90°F (32.2°C)? ______________________________

10. a. What type of lighting fixtures are installed on the workshop and storage room ceilings? ______________________________
    b. Who furnishes these fixtures? ______________________________

11. When more than three current-carrying conductors are installed in one raceway, the allowable ampacities must be reduced according to ______________ to *Tables 310-16* through *310-19.*

12. How many convenience outlets are required to complete the multioutlet assembly?
    ____

13. On what type of surface is the wall box for the storage room mounted? (Underline one.)
    a. concrete b. concrete block c. cinder block

14. Following the suggested wiring layout for the workshop, lavatory, and storage room, determine how many feet (meters) of each of the following are required.
    a. 1/2-inch EMT ________ feet (meters) c. Two-wire cable ________ feet (meters)
    b. 3/4-inch EMT ________ feet (meters) d. Three-wire cable ________ feet (meters)

15. Approximately how many feet (meters) of No. 12 TW wire are required for the workshop, lavatory, and storage room? Do not include cable or prewired plugmold. Allow 10 percent extra for splicing and waste and follow the suggested wiring layout. ____________

16. Using the cable layout shown in figure 15-1, make a complete wiring diagram of the following circuit. Use colored pencils to indicate conductors.

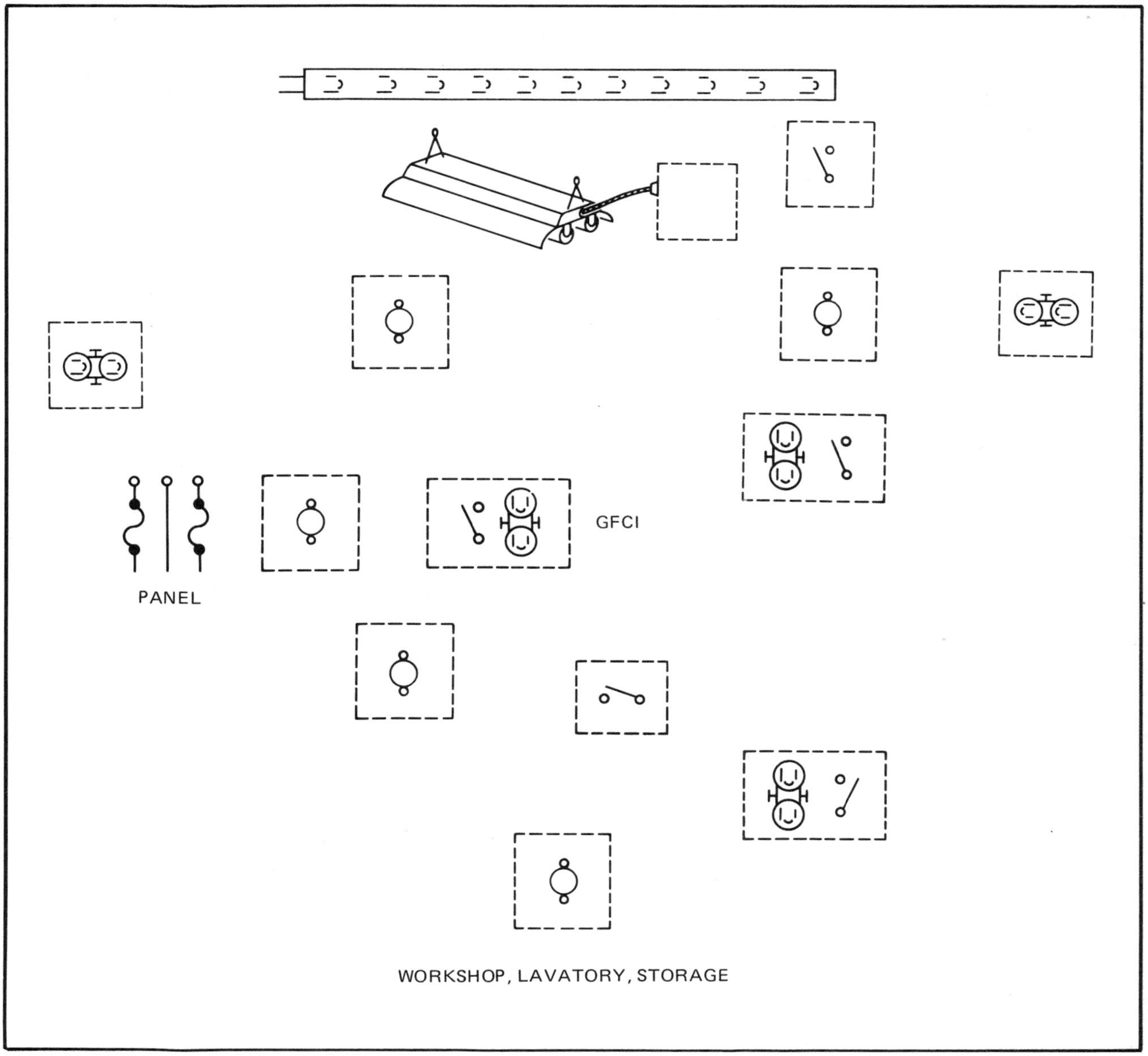

17. A conduit body must have a cross-sectional area not less than (two) (three) (four) times the cross-sectional area of the (largest) (smallest) conduit to which it is attached. (Underline the answers.)

18. A conduit body may contain splices if marked with
   a. the UL logo.
   b. its cubic-inch area.
   c. the size of conduit entries.

unit 16

Special-Purpose Outlets for Portable Heating Units

OBJECTIVES

After studying this unit, the student will be able to

- identify special-purpose outlet designations.
- calculate the load to be connected to a special outlet.
- make the proper grounding connections when using armored cable and nonmetallic-sheathed cable.
- make the proper connections to insure grounding continuity between a grounded outlet box and the grounding circuit of the receptacle.

The National Electrical Code installation requirements for lighting branch circuits and small appliance circuits are covered in previous units. Units 16–21 are concerned with special-purpose outlets. Modern dwellings use special-purpose outlets for which the loads must be calculated. The electrician must be able to select and install these outlets properly.

Circuits for special-purpose outlets are separate and distinct from one another and from the lighting and small appliance circuits. Each special outlet circuit has a specific function and must have an individual branch-circuit protective device.

Special outlets for portable electric heating units are described in this unit. Figure 16-1, showing the floor plan of the basement of the residence, indicates the special-purpose outlets only. The symbol for a special-purpose outlet is a triangle within a circle plus a subscript letter. A number may be added to the subscript letter to define the outlet even more.

GROUNDING-TYPE RECEPTACLE FOR HEATERS

Figure 16-2 shows that grounding-type convenience receptacles are used in the workshop and the utility room. A 1750-watt, 120-volt ac portable electric heater will be plugged into each of these receptacles. The heaters can be mounted on the wall, the ceiling, or the floor.

The rating of each heater in amperes is:

$$\text{amperes} = \frac{\text{watts}}{\text{volts}}$$

$$= \frac{1750}{120} = 14.58 \text{ amperes}$$

The manufacturer recommends that each heater be connected to a separate 120-volt, 20-ampere circuit using No. 12 TW conductors. Therefore, a 20-ampere, 125-volt receptacle is installed for each heater in the workshop and the utility room. *Section 210-23(a)* states that the rating of any one cord- and plug-connected appliance shall not exceed 80 percent of the branch-circuit rating.

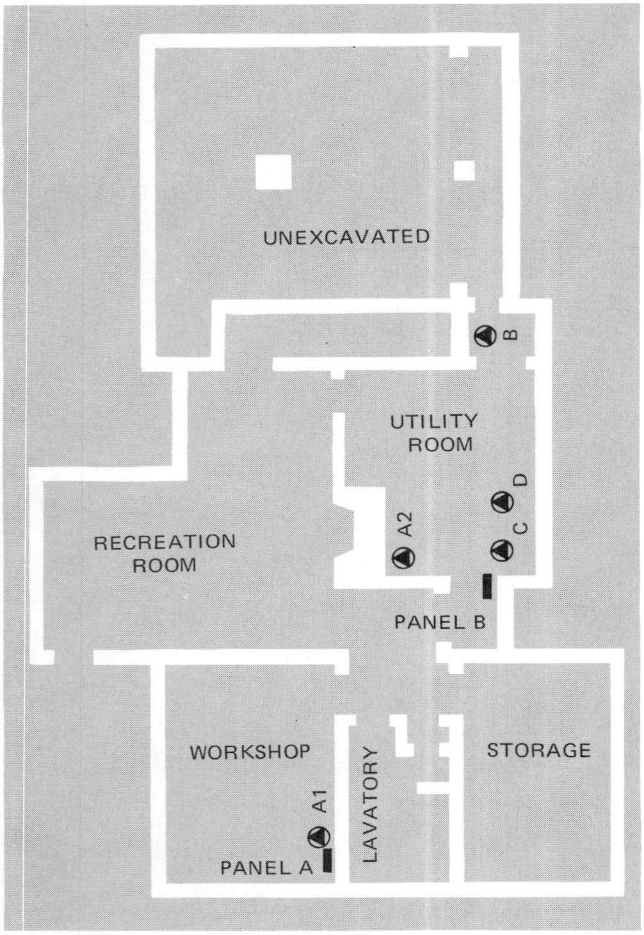

Fig. 16-1 Special-purpose outlets, basement.

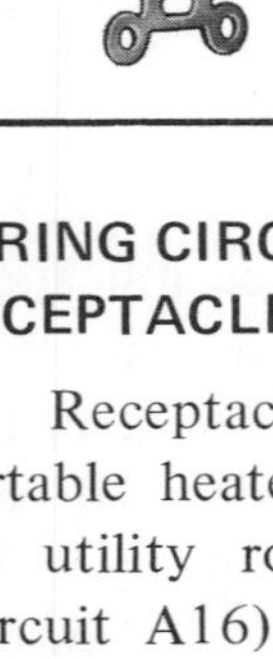

Fig. 16-2 Grounding-type convenience receptacle.

WIRING CIRCUITS FOR GROUNDING-TYPE RECEPTACLES A1 A2

Receptacle A1 (Circuit A14) supplies the portable heater in the workshop. The heater in the utility room is supplied by receptacle A2 (Circuit A16). A short length of 1/2-inch conduit (EMT) runs from the top of Panel A to the 4-inch square outlet box above Panel A in the workshop. This box contains the grounding-type receptacle A1. The conduit contains three No. 12 TW conductors. A two-wire No. 12 cable runs from this outlet box above the lavatory and passage ceilings and ends in a handy box containing grounding-type receptacle A2 in the utility room. This arrangement is a typical three-wire circuit as shown in figure 16-3.

BONDING AT GROUNDING-TYPE RECEPTACLES

To wire grounding-type receptacles with armored cable, the metal armor and the integral bonding strip under the armor are adequate to ground the outlet box and the grounding terminal of the receptacle. If nonmetallic-sheathed cable can be used, the cable must contain a separate grounding conductor, *Sections 210-7, 250-45,* and *250-57.* Figure 16-4 shows the grounding conductor of nonmetallic-sheathed cable attached to the receptacle.

Almost all of the grounding terminals of the type shown in figure 16-4 are bonded to the metal yoke of the receptacle. There is a question, however, of whether this grounding terminal provides grounding continuity between a grounded outlet box and the grounding circuit of the receptacle.

An outlet box is said to be grounded when it is connected to grounded cable armor or metal raceway, *Section 250-57.* When a grounding wire is not connected to the receptacle grounding terminal and the receptacle is fastened to a grounded outlet box, then the grounding terminal is automatically grounded through the threads of the No. 6-32 mounting screws. For a 4-inch square outlet box, grounding to the raised cover takes place through the threads of the No. 6-32 screws. Grounding to the outlet box occurs through the threads of the No. 8-32 cover mounting screws. However, it is doubtful that the continuity of the grounding circuit is reliable under these conditions.

Section 250-74 requires the connection of a separate grounding conductor to the grounding screw of the grounding-type receptacle. This requirement holds regardless of the wiring method used – metal conduit, armored cable, or nonmetallic-

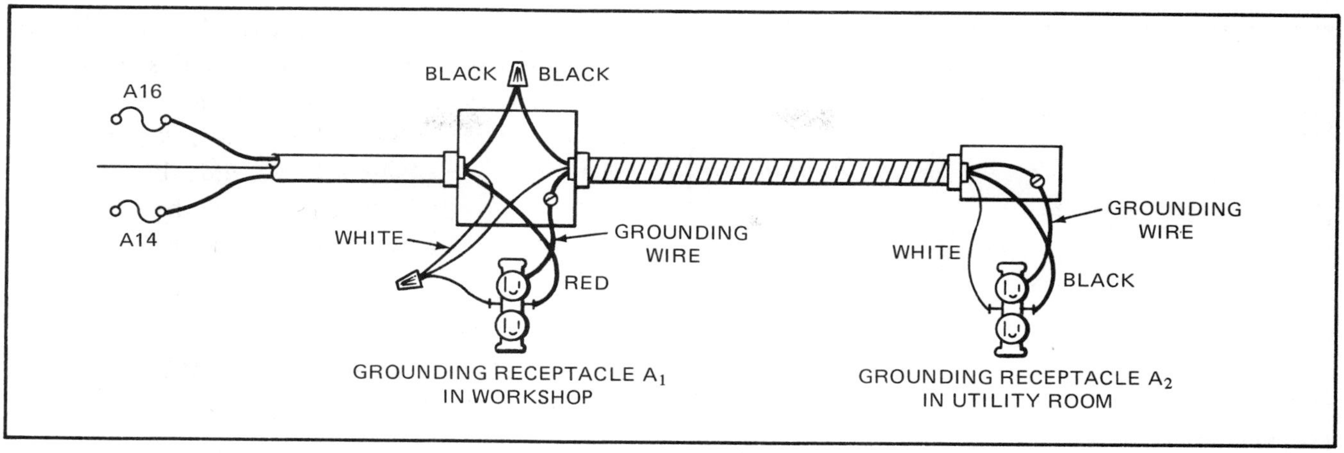

Fig. 16-3 Three-wire circuit connections for grounding-type receptacles.

sheathed cable with a ground conductor. In other words, the metal outlet box must be grounded in addition to using the grounding terminal of the receptacle, figure 16-4.

A bonding jumper, figure 16-4, is not necessary when the receptacle is fastened to a metal surface-mounted box such as the one in figure 15-2. In this case, there is direct metal-to-metal contact between the receptacle yoke and the box (*Section 250-74, Exception No. 1*). The No. 6-32 mounting screws of most receptacles and switches are held captive in the yoke by a small fiber or cardboard washer. Removing this washer gives metal-to-metal contact.

Some outlet and switch box manufacturers use washer head screws so that a grounding conductor can be placed under them. This method is useful when installing armored cable or nonmetallic-sheathed cable. That is, a means is provided to terminate the grounding conductor in the metal outlet or switch box. Grounding clips listed by Underwriters' Laboratories can be used to clamp the small equipment grounding conductor to the edge of the outlet box or switch box, figure 16-5. Grounding clips are convenient since special tools are not required to fasten them.

Exception No. 2 to *Section 250-74* states that the bonding jumper need not be used when the receptacle has a special self-grounding strap, figure 16-6. When such a receptacle is to be installed, a bonding jumper is not wired from the receptacle to the grounded outlet box.

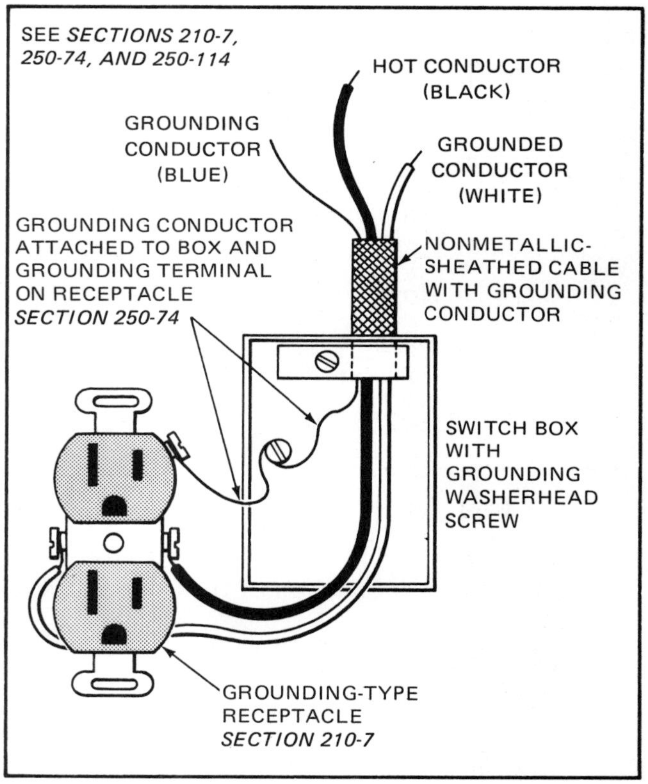

Fig. 16-4 Three-wire circuit connections for grounding-type receptacles.

GROUNDING PROVISIONS ON HEATER UNITS

Sections 250-45 and *250-59* require that the cord supplying the portable electric heater (or other portable electrical equipment) contain a separate grounding conductor.

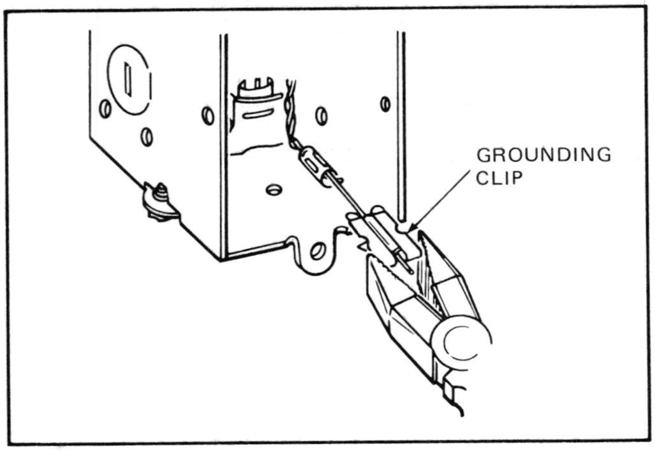

Fig. 16-5 Method of attaching grounding clip to switch box.

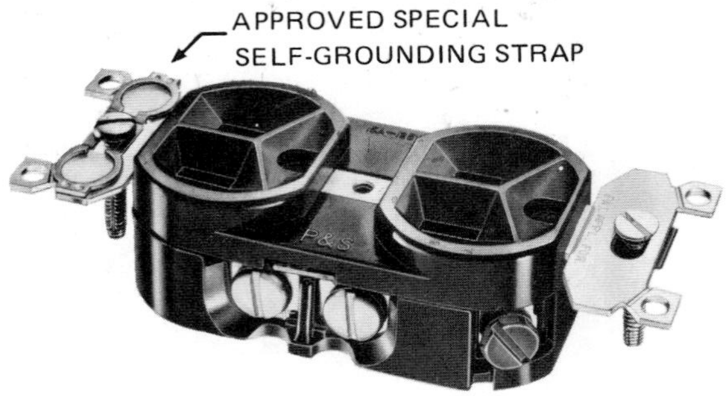

Fig. 16-6 Convenience receptacle with special self-grounding strap.

REVIEW

Note: Refer to the Code or the plans where necessary.

1. What type of convenience receptacles are used for the electric heaters? __________
2. a. Are portable electric heaters to be connected permanently to outlets A1 and A2?
   ____

   b. How are the heaters to be connected? __________
3. An electric heater is rated at 1400 watts, 120 volts. What is its current rating? Show all calculations. __________
4. a. What limitation is placed on the rating of portable appliances with regard to the branch-circuit rating? __________

   b. What Section of the Code states this limitation? __________
5. According to the Code, what is the maximum current rating of portable appliances that may be connected to:

   a. a 15-ampere branch circuit? __________

   b. a 20-ampere branch circuit? __________
6. a. Where is outlet ▲A1 mounted in the workshop? __________

   b. Where is outlet ▲A2 located? __________
7. List three ways in which outlet boxes can be grounded. __________
8. List three acceptable means of insuring the grounding continuity between a grounded outlet box and the grounding circuit of the receptacle. __________

9. List the methods by which a grounding conductor is connected to a switch box.

____

____

10. To permit the insertion of a three-wire grounding plug cap into a two-wire receptacle, some manufacturers supply the three-wire plug cap with a ______________________.

unit 17

Special-Purpose Outlets for a Water Pump and a Water Heater for Residential Use

OBJECTIVES

After studying this unit, the student will be able to

- list the requirements for a deep-well jet pump motor.
- calculate (given the rating of the motor) the conductor size, conduit size, and overcurrent protection required for the pump circuit.
- list the methods that can be used to heat water and the steps in each method.
- describe the basic operation of the water heating system.
- list the functions of the tank, the heating elements, and the thermostats of the water heating system.
- make the proper grounding connections for the water heater.

WATER PUMP CIRCUIT ⊕B

All dwellings need a good water supply. City dwellings generally are connected to the city water system. In rural areas, where there is no public water supply, each dwelling has its own water supply system in the form of a well. The residence plans show that a deep-well jet pump is used to pump water from the well to the various plumbing outlets. The outlet for this jet pump is shown by the symbol ⊕B. An electric motor drives the jet pump. A circuit must be installed for this system.

JET PUMP OPERATION

Figure 17-1 shows the major parts of a typical jet pump. The pump impeller wheel (1) forces water down a drive pipe (2) at a high velocity and pressure to a point just above the water level in the well casing. (Some well casings may be driven to a depth of more than 100 feet (30.5 m) before striking water.) Just above the water level, the drive pipe curves sharply upward and enters a larger vertical suction pipe (3) . The drive pipe terminates in a small nozzle or jet (4) in the suction pipe. The water emerges from the jet with great force and flows upward through the suction pipe.

Water rises in the suction pipe, drawn up through the tailpipe (5) by the action of the jet. The water rises to the pump inlet and passes through the impeller wheel of the pump. Part of the water is forced down through the drive pipe again. The remaining water passes through a check valve and enters the storage tank (6).

The tailpipe is submerged in the well water to a depth of about 10 feet (3.05 m). A foot valve (7) and strainer (8) are provided at the end of the tailpipe. The foot valve prevents water in the pumping equipment from draining back into the well when the pump is not operating.

When the pump is operating, the lower part of the storage tank fills with water and air is trapped in the upper part. As the water rises in the tank, the air is compressed. When the air pressure is 40 pounds per square inch, a pressure switch (9) disconnects the pump motor. The pressure switch is adjusted so that as the water is used and the air pressure falls to 20 pounds per square inch, the pump restarts and fills the tank again. One pound (lb) = 0.453 6 kilograms (kg). One kilogram (kg) = 2.2046 pounds (lb).

The Pump Motor

A jet pump for residential use may be driven by a 1-horsepower (hp) motor at 3400 r/min. A dual-voltage motor is used so that it can be connected to 120 or 240 volts. For a 1-hp motor, the higher voltage is preferred since the current at this voltage is half the value used by the lower voltage. It is recommended that a single-phase, capacitor-start motor be used for a residential jet pump. Such a motor is designed to produce a high starting torque. In pumping operations, a high starting torque helps to overcome the back pressure within the tank and the weight of the water being lifted from the well.

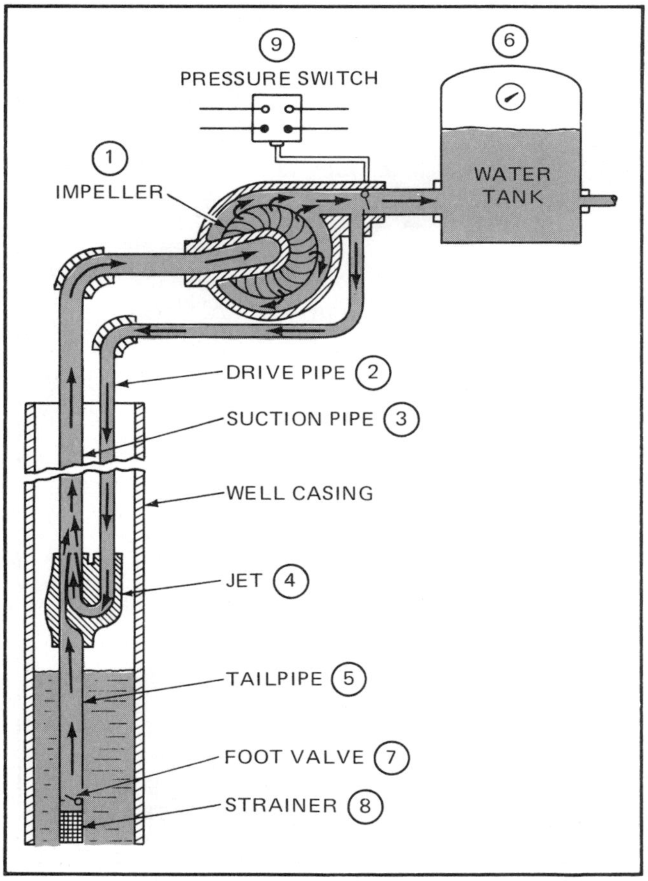

Fig. 17-1 Components of a jet pump.

The Pump Motor Circuit

Figure 17-2 is a diagram of the electrical circuit that operates the jet pump motor and pressure switch. This circuit is taken from electrical Panel B in the utility room. *Table 430-148* indicates that a 1-hp single-phase motor has a rating of 8 amperes at 240 volts. Using this information, the conductor size, conduit size, branch-circuit overcurrent protection, and running overcurrent protection can be determined for the motor.

Conductor Size (*Section 430-22*)

1.25 × 8 = 10 amperes

Table 310-16 indicates that the minimum conductor size permissible is No. 14 AWG. However, the specifications state that No. 12 AWG conductors are to be used for all circuits in the residence. Thus, No. 12 conductors are used for the pump circuit.

Conduit Size (*Table 3A, Chapter 9*)

Two No. 12 conductors require 1/2-inch conduit.

Branch-circuit Overcurrent Protection (*Section 430-52, Table 430-152*)

The jet pump motor is a single-phase motor without code letters. Therefore, *Table 430-152*,

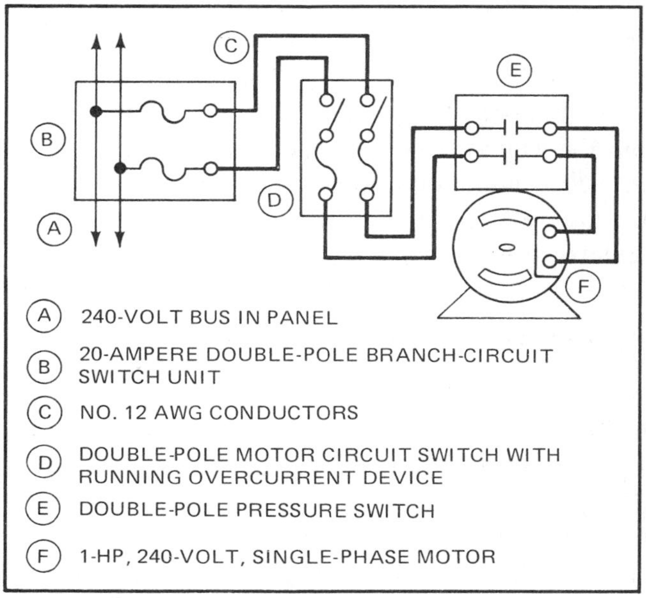

Fig. 17-2 The pump circuit.

line 1, can be used. If the branch-circuit protective device is an inverse time circuit breaker, then the maximum rating or setting of the breaker shall not exceed 250 percent of the full-load running current of the motor: 8 × 2.5 = 20 amperes. If the protective device is a nontime-delay fuse, the fuse rating may not exceed 300 percent of the full-load running current of the motor: 8 × 3 = 24 amperes. For dual-element time-delay fuses, the rating may not exceed 175 percent of the full-load current of the motor: 8 × 1.75 = 14 amperes.

Where the values for branch-circuit protective devices determined by *Table 430-152* do not correspond to the standard sizes or ratings of such devices, the next higher size, rating, or setting shall be permitted, *Section 430-52*.

Running Overcurrent Protection for the Motor [*Section 430-32(c)(1)*]

Dual-element, time-delay fuses must be rated at not more than 125 percent of the full-load current of the motor to provide running overcurrent protection according to *Section 430-32(c)(1)*. The maximum rating of thermal overload devices is also 125 percent of the full-load current rating of the motor: 1.25 × 8 = 10 amperes.

Wiring for the Pump Motor Circuit

The pump motor circuit is run in 1/2-inch conduit embedded in the utility room concrete floor. The conduit terminates in a metal-enclosed, double-pole thermal overload switch mounted on the wall in the pump room. Thermal overload devices (sometimes called *heaters*) are installed in the switch to provide running overcurrent protection. Two No. 12 conductors are connected to the double-pole, 20-ampere branch-circuit overcurrent protective device (in Panel B) to form Circuit B13-15. Type S Fustat time-delay fuses may be installed as well to provide running overcurrent protection. These fuses may be rated at not more than 125 percent of the full-load running current of the motor.

A short length of flexible metal conduit is connected between the load side of the thermal overload switch and the pressure switch on the pump. This pressure switch is usually connected to the motor at the factory. In figure 17-2, a double-pole pressure switch is shown at (E). When the contacts of the pressure switch are open, all conductors feeding the motor are disconnected.

SUBMERSIBLE PUMP

A submersible pump, figure 17-3, consists of a centrifugal pump driven by an electric motor. The pump and motor are contained in one housing, submersed below the permanent water level within the well casing. When running, the pump raises the water, upward through the piping, to the water tank. Proper pressure is maintained in the system by a pressure switch. The disconnect switch, pressure and limit switches, and controller are installed in a logical and convenient location near the water tank.

Some water tanks have a precharged air chamber that contains a vinyl bag that separates the air from the water. This assures that air is not absorbed by the water. The absorption of air causes a slow

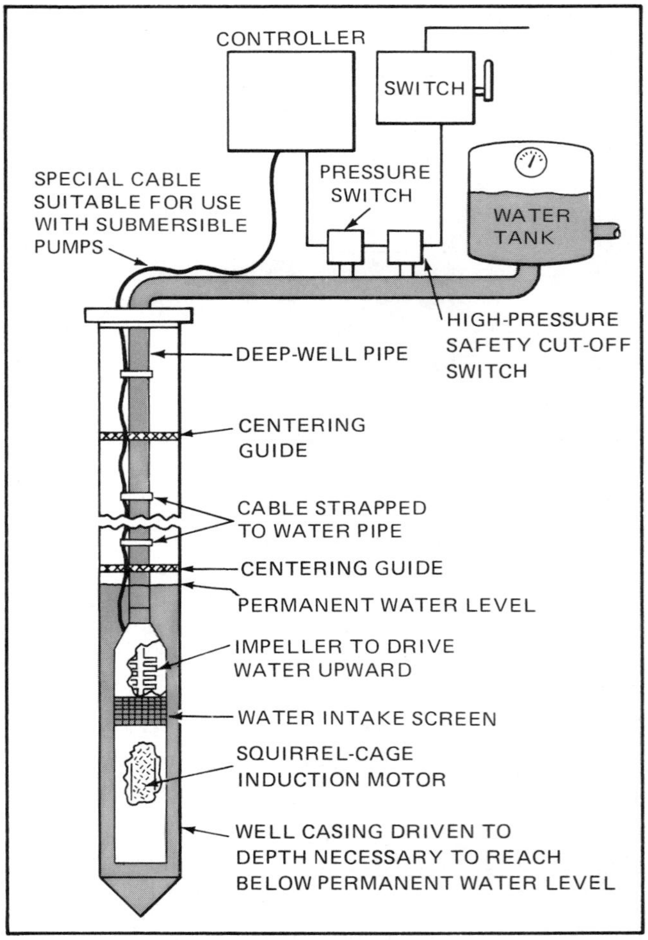

Fig. 17-3 Submersible pump.

reduction of the water pressure, and, ultimately, the tank will fill completely with water with no room for the air. The initial air charge is always maintained.

Power to the motor is supplied by a cable especially designed for use with submersible pumps. This cable is supplied with the pump and its various components. The cable is cut to the proper length to reach between the pump and its controller. When needed, the cable may be spliced according to the manufacturer's specification.

The controller contains the motor's starting relay, overload protection, starting and running capacitors, lightning arrester, and terminals for making the necessary electrical connections. Thus, there are no moving electrical parts within the pump itself, such as would be found in a typical single-phase, split-phase induction motor that would require a centrifugal starting switch.

The calculations for sizing the conductors, and the requirements for the disconnect switch, the motor's branch-circuit fuses, and the grounding connections are the same as those for the jet pump motor. The nameplate data and instructions furnished with the pump must be followed.

WATER HEATER CIRCUIT Ⓐ$_C$

All homes require a continuous supply of hot water. To meet this need, one or more automatic water heaters are installed close to the areas having the greatest need for hot water. Piping is installed to carry the heated water from the heater to the various plumbing fixtures and appliances that require hot water, such as clothes washers and dishwashers.

Modern water heaters have automatic temperature controls which maintain the water temperature within a range from 60°C to 71°C. The Code and Underwriters' Laboratories require that electric water heaters be equipped with a high-temperature cutoff. UL states that the maximum cutoff temperature is 90°C. The plumber may install a pressure relief valve in the water line if the appliance manufacturer or local electrical, building, and/or plumbing codes require such a valve.

Electric water heaters for residential use are available in sizes ranging from 20 to 82 gallons. (One gallon equals 3.785 liters.) Typical sizes are 20, 30, 40, 42, 45, 50, 52, 66, and 82 gallons. Many heaters contain one or two magnesium anodes (rods) which are permanently submerged in the water. These rods help to reduce corrosion.

The heating elements are generally rated at 236 volts. However, they will operate properly in the range from 220–240 volts. All approved electrical equipment and devices are designed to operate satisfactorily at plus or minus 10% of their respective operating voltages.

WATER HEATER METERING

The energy used by electric water heaters can be recorded in several ways. One method is to use a separate meter in addition to the regular watthour meter.

Off-peak Heating

The *off-peak* method of heating the water in the storage tank uses a time switch to control the heating. This switch has an electric clock which operates a switch. The clock operates on a separate heater circuit between the off-peak hours of 11:00 P.M. to 6:00 A.M. As a result, the water is heated at this time only. For the remainder of the 24-hour day, depending upon requirements, either the separate circuit or the main circuit may be used. Large capacity tanks should be used with this system.

A slight variation of the off-peak plan has most of the heating taking place at night. The rest of the heating occurs during the day. The rate charged to the customer for heating water is lower than the standard rate charged for the general use of electrical energy. Figure 17-4 shows the wiring for a typical off-peak water heater circuit.

The overcurrent protective devices for both the regular and off-peak circuits can be placed in the same enclosure. It is not necessary to have a separate switch for the off-peak circuitry. The off-peak protective device is called a *feedthrough unit*, figure 17-5. This device is not connected to the main bus of the panel even though it is placed in the same enclosure. The two wires from the off-peak meter are connected to one side of the feedthrough unit. The two wires supplying the element of the water heater are connected to the other side of the feedthrough unit.

Off-peak water heating can also be provided by connecting one heating element to the regular house meter and the second element to the off-peak meter. Thus, when the time clock of the off-peak meter is off, the water is heated through the house meter circuit at the regular rate for energy use. This method requires separate overcurrent protective devices for both elements.

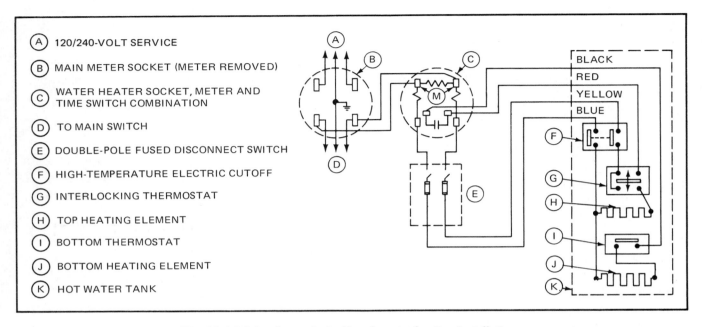

(A) 120/240-VOLT SERVICE
(B) MAIN METER SOCKET (METER REMOVED)
(C) WATER HEATER SOCKET, METER AND TIME SWITCH COMBINATION
(D) TO MAIN SWITCH
(E) DOUBLE-POLE FUSED DISCONNECT SWITCH
(F) HIGH-TEMPERATURE ELECTRIC CUTOFF
(G) INTERLOCKING THERMOSTAT
(H) TOP HEATING ELEMENT
(I) BOTTOM THERMOSTAT
(J) BOTTOM HEATING ELEMENT
(K) HOT WATER TANK

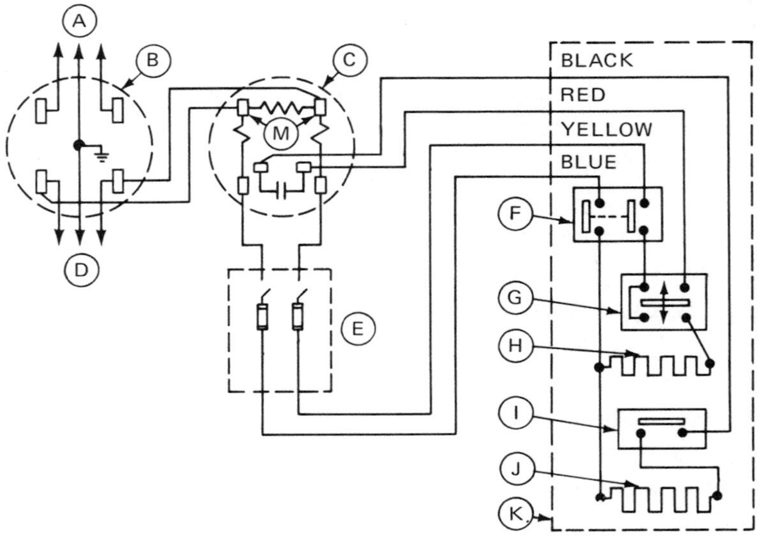

Fig. 17-4 Wiring for typical off-peak water heating installation.

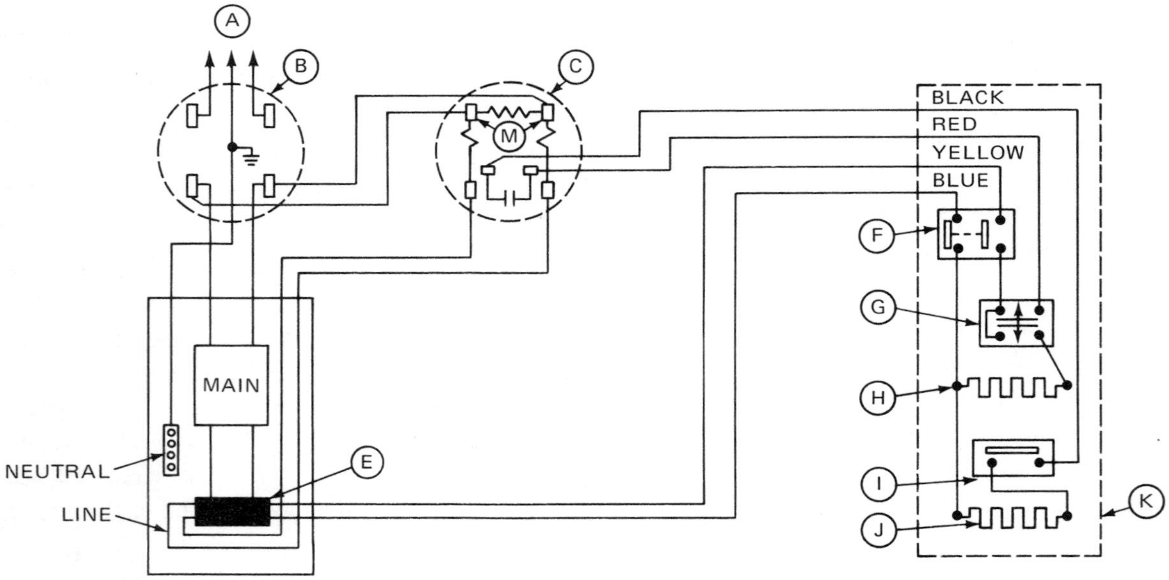

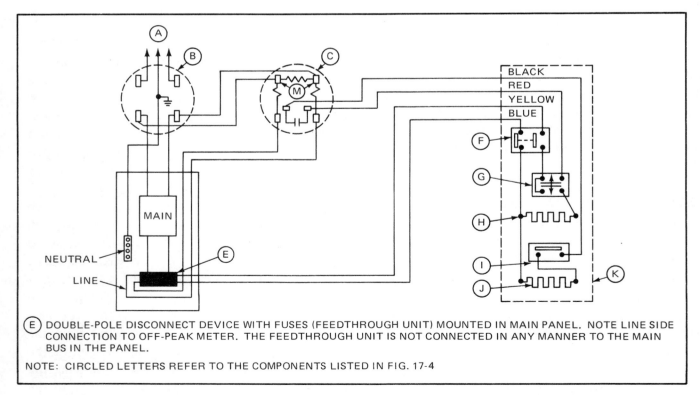

(E) DOUBLE-POLE DISCONNECT DEVICE WITH FUSES (FEEDTHROUGH UNIT) MOUNTED IN MAIN PANEL. NOTE LINE SIDE CONNECTION TO OFF-PEAK METER. THE FEEDTHROUGH UNIT IS NOT CONNECTED IN ANY MANNER TO THE MAIN BUS IN THE PANEL.

NOTE: CIRCLED LETTERS REFER TO THE COMPONENTS LISTED IN FIG. 17-4

Fig. 17-5 Feedthrough unit protective device for off-peak metering.

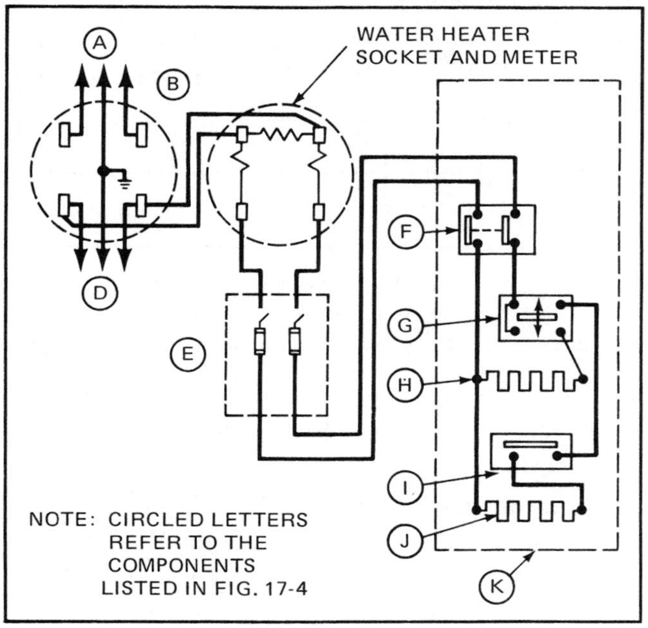
WATER HEATER SOCKET AND METER
A
B
C
D
E
F
G
H
I
J
K
NOTE: CIRCLED LETTERS REFER TO THE COMPONENTS LISTED IN FIG. 17-4

Fig. 17-6 24-hour water heating installation.

24-hour Heating

The water can be heated by providing electrical energy on a 24-hour basis. In this method, a time switch is not used. The water heating meter is tapped from the load side of the main meter. Figure 17-6 shows a 24-hour water heating circuit.

Still another method of connecting the water heater is used for the residence in the plans. In this method, the water heater is connected to one of the two-pole, 240-volt circuits in the main distribution panel, figure 17-7. This is a method used commonly for dwellings because the wiring is simple and extra switches, breakers, and off-peak meters are not required.

HEATING ELEMENT RATINGS

Most electrical heating elements are rated at 236 volts and can be used in 220-volt to 240-volt circuits.

The voltage ratings of residential electric water heaters vary greatly depending upon the size of the heater in gallons, the speed of recovery needed, local electric utility regulations, and local plumbing and building codes. (The *speed of recovery* is defined as the time required to bring the water temperature to satisfactory levels.) Water heaters may contain one or two heating elements. The rating of each heating element may be 750, 1250, 1500, 1650, 2000, 2500, 3000, 3800, or 4500 watts.

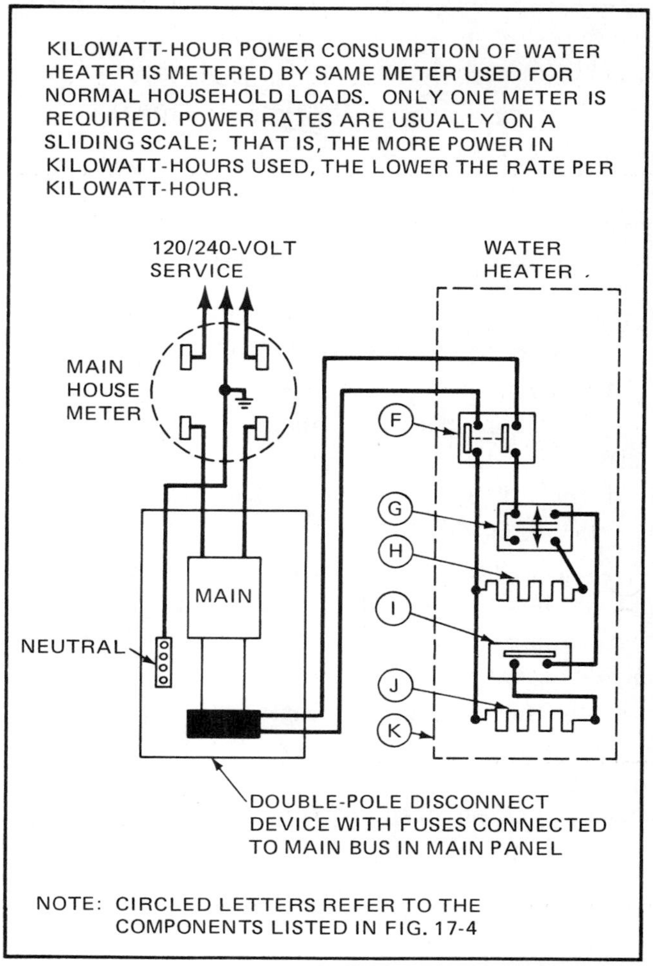
KILOWATT-HOUR POWER CONSUMPTION OF WATER HEATER IS METERED BY SAME METER USED FOR NORMAL HOUSEHOLD LOADS. ONLY ONE METER IS REQUIRED. POWER RATES ARE USUALLY ON A SLIDING SCALE; THAT IS, THE MORE POWER IN KILOWATT-HOURS USED, THE LOWER THE RATE PER KILOWATT-HOUR.
120/240-VOLT SERVICE
WATER HEATER
MAIN HOUSE METER
MAIN
NEUTRAL
F
G
H
I
J
K
DOUBLE-POLE DISCONNECT DEVICE WITH FUSES CONNECTED TO MAIN BUS IN MAIN PANEL
NOTE: CIRCLED LETTERS REFER TO THE COMPONENTS LISTED IN FIG. 17-4

Fig. 17-7 Water heater connected to two-pole circuit in main panel.

According to *Section 422-14*, electric water heaters shall be equipped with a temperature-limiting means in addition to its control thermostat to disconnect all ungrounded conductors.

Electric water heaters generally have two heating elements and two thermostats. The heating elements, figure 17-8, consist of nickel-chrome (nichrome) alloy wire. The proper size and length of wire are used to conform to the voltage and wattage ratings of the element.

The nichrome wire is coiled and embedded (compacted) in magnesium oxide in a copper-clad, tubular housing. The magnesium oxide is heat resistant and is a good insulator. Terminals are provided on one end of the heating element. A

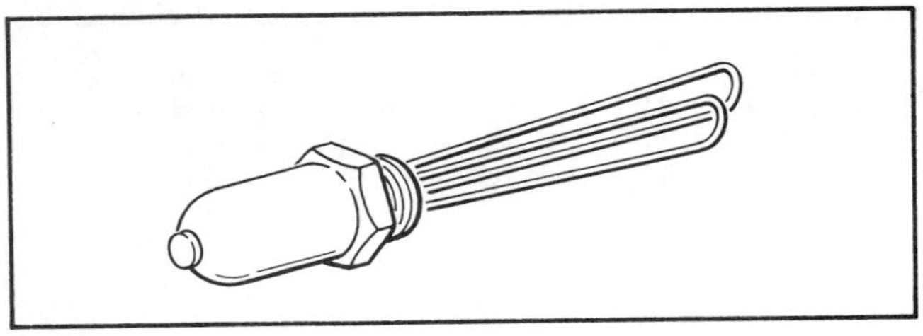

Fig. 17-8 Heating element.

threaded hub insures that the heating element can be screwed securely into the side of the water heating tank.

The top heating element is located about halfway up the tank (when the tank is in a vertical position). The bottom element is located close to the bottom of the tank.

The thermostats, figure 17-9, are usually placed slightly above the heating elements. These thermostats turn the heat on or off. The top thermostat is an interlocking or snap-over type. The bottom thermostat is a single-pole, on-off type. Both thermostats have silver contacts and quick make-and-break characteristics to prevent arcing. The installation diagrams show that both heating elements cannot be energized at the same time. This restriction is called *limited demand.*

Sequence of Thermostat Operation

When all of the water in the storage tank is cold, the top element heats only the water in the upper half of the tank. The top element is energized through the lower contacts of the top thermostat. The bottom thermostat also indicates that heat is needed, but heat cannot be supplied because of the open condition of the upper contacts of the top thermostat.

When the water in the upper half of the tank reaches a temperature of about 150°F, the top thermostat snaps over. That is, the upper contacts are closed and the lower contacts are opened. The lower element now begins to heat the cold water in the bottom of the tank. When the entire tank is heated to 150°F, the lower thermostat shuts off. Electrical energy is not being used by either element at this point. The heat-insulating jacket around the water tank is designed to keep the heat loss to a minimum. Thus, it takes many hours for the water to cool only a few degrees.

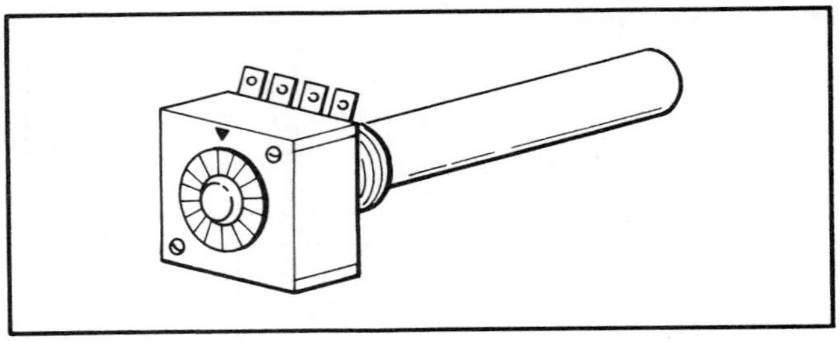

Fig. 17-9 Thermostat.

WATER HEATER LOAD DEMAND

The following Code sections apply to the load demands placed on a branch circuit by a water heater.

Section 422-14(b) applies to all fixed storage water heaters having a capacity of 120 gallons (454.2 L) or less. Such heaters are required to have a branch-circuit rating not less than 125 percent of the nameplate rating of the water heater.

Section 422-5(a), Exception 2, requires that a continuously loaded appliance must have a branch-circuit rating not less than 125 percent of the marked rating of the appliance.

Section 210-22(c) requires that continuous loads shall not exceed 80 percent of the rating of the branch circuit.

The water heater for the residence is to be connected for *limited demand.* This means that the 2000-watt element and the 3000-watt element (refer to the specifications) cannot be energized at the same time. The maximum load demand is:

$$\text{amperes} = \frac{\text{watts}}{\text{volts}} = \frac{3000}{236} = 12.7 \text{ amperes}$$

For a load of 12.7 amperes, two No. 12 AWG conductors (with a derated ampacity of 16A) are installed and connected to Circuit B18-20. This is a 20-ampere, double-pole overcurrent device located in Panel B in the utility room. (Note that some heating units with higher wattage ratings may require No. 10 AWG conductors.) Since the heating elements are rated at 236 volts, a neutral conductor is not required.

Table 3A, Chapter 9 indicates that 1/2-inch conduit (EMT) is the minimum size permitted for two No. 12 AWG conductors. To simplify the installation, a short length (18 to 24 inches) (457 to 610 mm) of 1/2-inch flexible metal conduit is run from the junction box on the water heater to the 1/2-inch conduit near the water heater.

Water Heater Grounding

The water heater is grounded through the 1/2-inch conduit and the flexible metal conduit, *Section 250-57*. The grounding of equipment by means of flexible connections is covered in detail in unit 4.

Effect of Voltage Variation

It was stated that the heating elements used in the residence are rated at 236 volts but function properly in the range from 220 to 240 volts. Ohm's Law and the wattage formula show how the wattage and current depend on the applied voltage.

$$R = \frac{E^2}{W} = \frac{236 \times 236}{3000} = 18.56 \text{ ohms}$$

$$I = \frac{E}{R} = \frac{236}{18.56} = 12.7 \text{ amperes}$$

If a different voltage is substituted, the wattage and current values change accordingly.

At 240 volts:

$$W = \frac{E^2}{R} = \frac{240 \times 240}{18.56} = 3100 \text{ watts}$$

and

$$I = \frac{W}{E} = \frac{3100}{240} = 12.9 \text{ amperes}$$

At 230 volts:

$$W = \frac{E^2}{R} = \frac{230 \times 230}{18.56} = 2850 \text{ watts}$$

and

$$I = \frac{W}{E} = \frac{2850}{230} = 12.4 \text{ amperes}$$

At 120 volts:

$$W = \frac{E^2}{R} = \frac{120 \times 120}{18.56} = 776 \text{ watts}$$

and

$$I = \frac{W}{E} = \frac{776}{120} = 6.46 \text{ amperes}$$

In general, when the voltage on a heating element is doubled, the current doubles and the wattage increases four times. When the voltage is reduced to one-half its original value, the current is halved and the wattage is reduced to one-fourth its original value.

REVIEW

Note: Refer to the Code or the plans where necessary.

WATER PUMP CIRCUIT ▲B

1. Does a jet pump have any electrical moving parts below the ground level?______
2. Which is larger, the drive pipe or the suction pipe?______
3. Where is the jet of the pump located?______
4. What does the impeller wheel move?______
5. Where does the water flow after leaving the impeller wheel?

   a. ______ b. ______
6. What prevents water from draining back into the pump from the tank?______
7. What prevents water from draining back into the well from the equipment?______
8. What is compressed in the water storage tank?______
9. How is the motor disconnected if pumping is no longer required?______
10. What is a common speed for jet pump motors?______
11. Why is a 240-volt motor preferable to a 120-volt motor for use in this residence?

    ______

    ______
12. How many amperes does a 1-hp, 240-volt, single-phase motor draw? (See *Table 430-148*).

    ______

13. What is the size of the conductors used for this circuit? ______________

14. What is the branch-circuit protective device? ______________

15. What furnishes the running protection for the pump motor? ______________

____

16. What is the maximum ampere setting required for running protection of the 1-hp, 240-volt pump motor? ______________

17. Submersible water pumps operate with the electrical motor and actual pump located ______________.

    a. above permanent water level.
    b. below permanent water level.
    c. half above and half below permanent water level.

18. Because the controller contains the motor starting relay and the running and starting capacitors, the motor itself contains ______________.

19. What type of pump moves the water upward inside of the deep-well pipe? ______________

20. Proper pressure of the submersible pump system is maintained by a ______________.

WATER HEATER CIRCUIT ▲C

1. At what temperature is the water in a residential hot water heater usually maintained?

   ____

2. Magnesium rods are installed inside the water tank to reduce ______________.

3. Is a separate meter required to record the amount of energy used to heat the water?

   ____

4. What is meant by the phrase *off-peak metering?* ______________

   ____

   ____

5. What is the most common method of metering water heater loads? ______________

   ____

6. Two thermostats generally are used in an electric water heater.
   a. What is the location of each thermostat? ______________
   b. What type of thermostat is used at each location? ______________

7. a. How many heating elements are provided in the heater? ______________
   b. Are these heating elements allowed to operate at the same time? ______________

8. When does the bottom heater operate? ______________

   ____

9. The Code states that water heaters having a capacity of 120 gallons (454.2 L) or less shall be considered ______________ duty and, as such, the circuit must have a rating of not less than ______________ percent of the rating of the water heater.

10. Why does the storage tank hold the heat so long? ______________

11. a. What size wire is used to connect the water heater? ______________
    b. What size overcurrent device is used? ______________

12. a. If both elements of the water heater are energized at the same time, how much current will they draw? (Assume the elements are rated at 236 volts.) ____________

    b. What size wire is required for the load of both elements? Show all calculations.

    ____

    ____

    ____

13. a. How much current do the elements in question 12 draw if connected to 220 volts?

    ____

    ____

    b. What is the wattage value at 220 volts? Show all calculations. ____________

    ____

unit 18

Special-Purpose Outlets for the Dryer and the Overhead Garage Door Openers

OBJECTIVES

After studying this unit, the student will be able to

- make the proper wiring and grounding connections for large appliances, based on the type of wiring method used.
- list the requirements for using service-entrance cable to connect large appliances.
- make the proper connections to install an overhead garage door opener based on the manufacturer's installation recommendations.
- select the proper overcurrent protective devices based on the amperage rating of the device.

DRYER CIRCUIT (▲)$_D$

A separate circuit is provided in the utility room for the electric clothes dryer. This appliance demands a large amount of power. The special circuit provided for the dryer is indicated on the plans by the symbol (▲)$_D$.

The Clothes Dryer

Clothes dryer manufacturers make many dryer models with different wiring arrangements and connection provisions. All dryers have electric heating units and a motor-operated drum which tumbles the clothes as the heat evaporates the moisture. Dryers also have thermostats to regulate the temperature of the air inside the dryer. Timers are used to regulate the lengths of the various drying cycles. Drying time is determined by the type of cloth and can be set to a maximum of about 85 minutes. Many dryers have fans or blowers, internal lights, ozone lamps, or germicidal lamps. In addition, dampness controls are often provided to stop the drying process at a preselected stage of dampness so that it is easier to iron the clothes.

Most electric clothes dryers operate from a 120/240-volt, single-phase, three-wire circuit. Some dryers are designed for two-wire, 120-volt operation. Some manufacturers claim that their 240-volt dryers can be connected to 120 volts. Unit 17 shows that when the voltage is reduced to one-half its rated value, the wattage is reduced to one-fourth its rated value. Because of this fact, the drying time of a 240-volt dryer connected to 120 volts is increased accordingly.

Dryer Connection Methods

Electric clothes dryers can be connected in several ways. The method used depends largely upon local codes. The local electrical inspector can provide information about the proper wiring procedure. Figure 18-1 shows the internal components and wiring of a typical dryer circuit.

One method of connecting the dryer is to run a three-wire armored cable to a junction box on the dryer provided by the manufacturer, figure 18-2. The cable conductors are connected to their corresponding dryer terminals in this junction box.

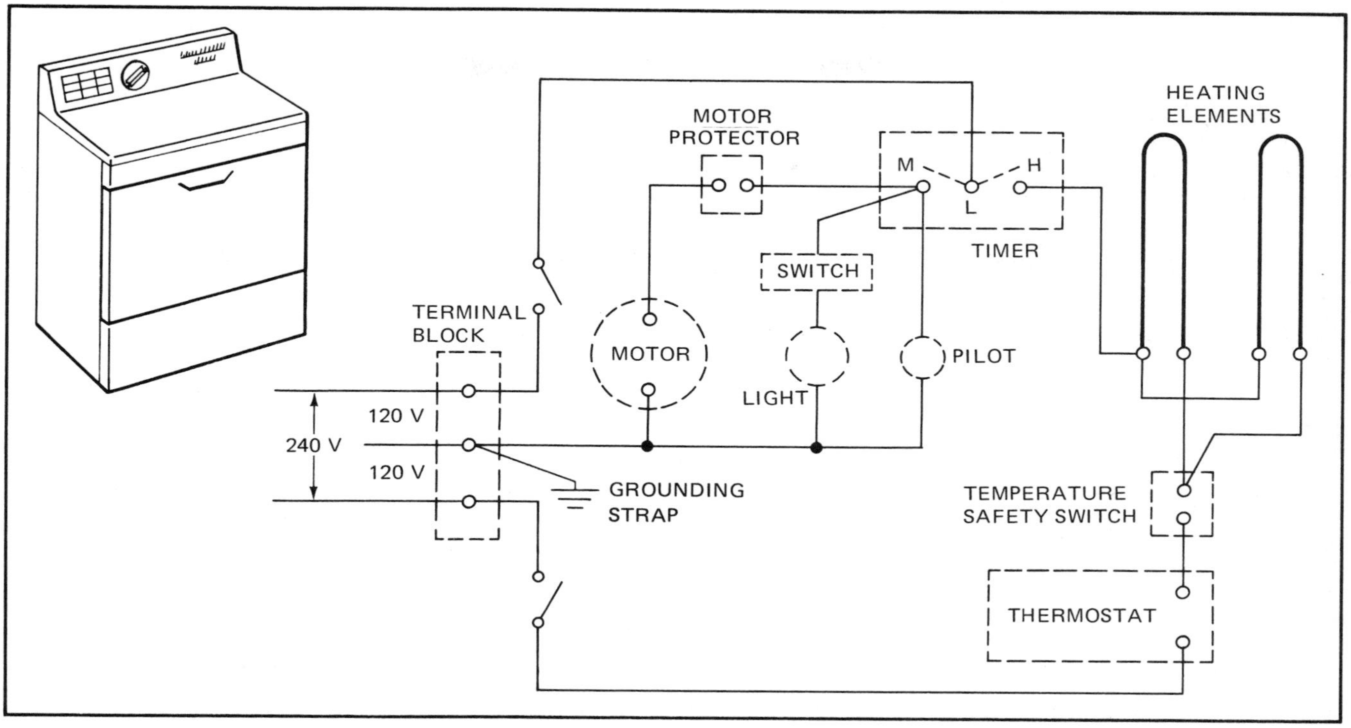

Fig. 18-1 Laundry dryer – wiring and components.

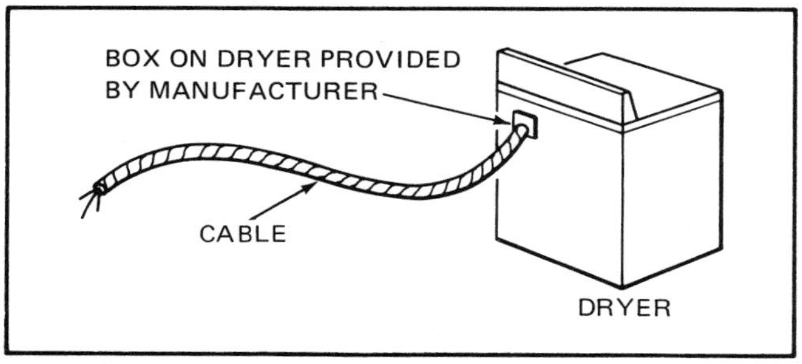

Fig. 18-2 Dryer connected by armored cable.

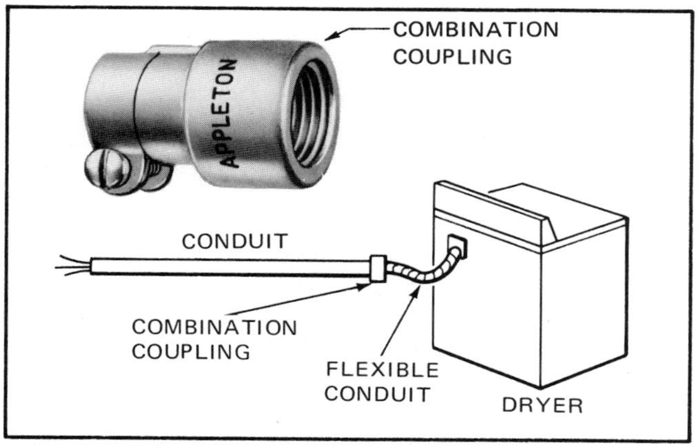

Fig. 18-3 Dryer connected by EMT and flexible conduit.

Another method of connecting a clothes dryer is to run conduit (EMT) to a point just behind the dryer. Flexible conduit from 24 to 36 inches (610 mm to 914 mm) in length is installed between the dryer junction box and the EMT, figure 18-3. The transition from the conduit (EMT) to the flexible conduit is made using a fitting such as the combination coupling shown in figure 18-3. This flexible connection allows the dryer to be moved for cleaning or servicing without disconnecting the wiring.

Another method of connecting a clothes dryer is to install a receptacle on the wall at the rear of the dryer. A cord set is connected to the dryer, figure 18-4, and is plugged into the receptacle. This method satisfies the requirements of *Section 422-22(a)* for a disconnecting means.

The receptacle can be wired with cable or conduit if concealed wiring is desired. Conduit (EMT) must be used if the wiring is to be exposed. In the residence, the wiring for the dryer is conduit which runs from Panel B to a dryer receptacle on the wall behind the dryer. A cord set is attached to the dryer and is plugged into this receptacle. The receptacle and cord must have a current-carrying capacity not less than that of the attached appliance.

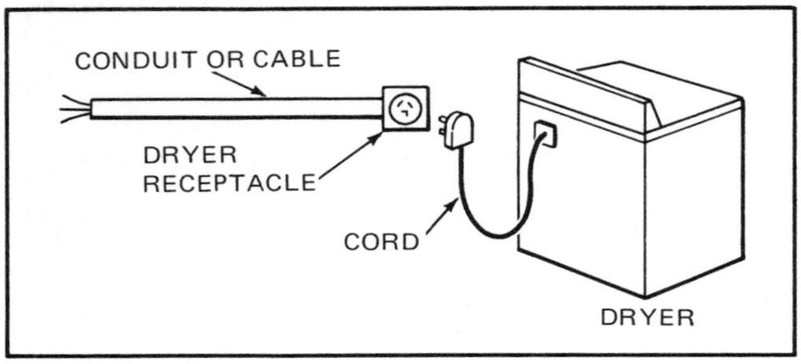

Fig. 18-4 Dryer connection using a cord set.

Receptacle and Cord Set

A standard dryer receptacle and cord set is shown in figure 18-5. Such a set is usually rated at 30 amperes, 250 volts. Cords and receptacles rated for higher currents are available. The L-shaped grounding and neutral slot of the outlet does not accept a 50-ampere cord set.

The surface mounting dryer outlet, figure 18-5A, can be wired directly with cable or conduit. The flush mounting outlet, figure 18-5B, will fit single-gang or two-gang sectional switch boxes. Figure 18-5C shows the cord set used with this receptacle. The outlet will also fit a 4-inch square or 4 11/16-inch square box if the proper single-gang or two-gang covers are used.

CODE REQUIREMENTS FOR CONNECTING A DRYER (*ARTICLE 422*)

The information necessary to install a dryer, or any appliance, is contained in *Article 422* of the National Electrical Code.

In the residence, the dryer is installed in the utility room next to the electric water heater. The schedule of special-purpose outlets in the specifications shows that the dryer is rated at 4700 watts and 120/240 volts. The schematic wiring diagram, figure 18-1, shows that the heating elements are connected to the 240-volt terminals of the terminal block. The dryer motor and light are connected between the hot wire and the neutral terminal.

The nameplate on the dryer lists the minimum circuit size permitted and the maximum rating of the circuit overcurrent device, *Section 422-32*. In the residence, a 30-ampere circuit is used for the dryer. Thirty-ampere fuses are selected as the overcurrent protective devices.

The motor of the dryer has integral thermal protection. This thermal protector prevents the motor from reaching dangerous temperatures as the result of an overload or failure to start. *Section 422-27* lists overcurrent protection requirements. This section also refers to parts of *Article 430.*

To determine the feeder or branch-circuit rating, *Section 220-18* requires the use of a load of 5000 watts or the nameplate rating of the dryer (whichever is larger).

$$I = \frac{W}{E} = \frac{5000}{240} = 20.83 \text{ amperes}$$

Conductor Size

According to *Table 310-16*, the dryer circuit requires No. 10 TW conductors. *Table 3A, Chapter 9*, shows that three No. 10 TW conductors require 1/2-inch conduit. The neutral conductor may be smaller than the phase wires because it carries the current of the motor and light only.

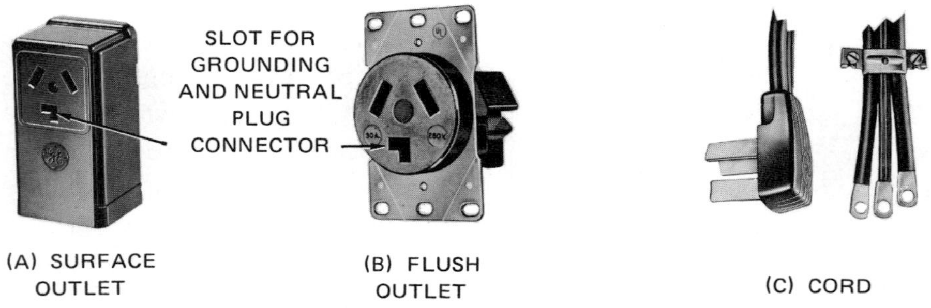

Fig. 18-4 Typical dryer receptacle and cord.

GROUNDING FRAMES OF RANGES AND DRYERS

Section 250-60 states that a dryer may be grounded to the grounded neutral conductor if the neutral is no smaller than No. 10 AWG, figure 18-6. If a three-wire cord set is used to install a dryer, the cord and conductors supplying the receptacle must have a neutral conductor no smaller than No. 10 AWG. If flexible conduit or armored cable is connected to the junction box of the dryer, then a smaller neutral may be installed. That is, two No. 10 AWG conductors and one No. 12 AWG neutral may be used because the metal conduit or armor provides the necessary equipment ground.

Two-conductor, No. 10 AWG nonmetallic-sheathed cable containing a bare grounding conductor may not be used for a dryer connection because the grounding conductor in this cable is smaller than No. 10 AWG.

The branch-circuit conductors for the dryer are connected to Panel B, Circuit B17–19. This is a double-pole circuit which breaks only the two ungrounded conductors. The neutral conductor remains solid and unbroken.

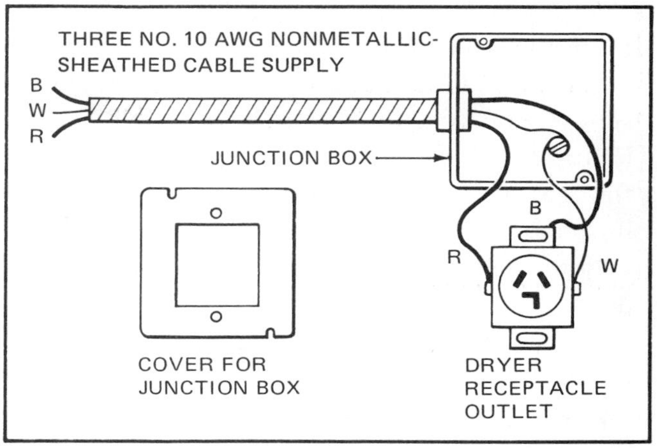

Fig. 18-6 It is permitted to ground the junction box to the neutral only when the box is part of the circuit for an electrical dryer or a range *(Section 250-60).*

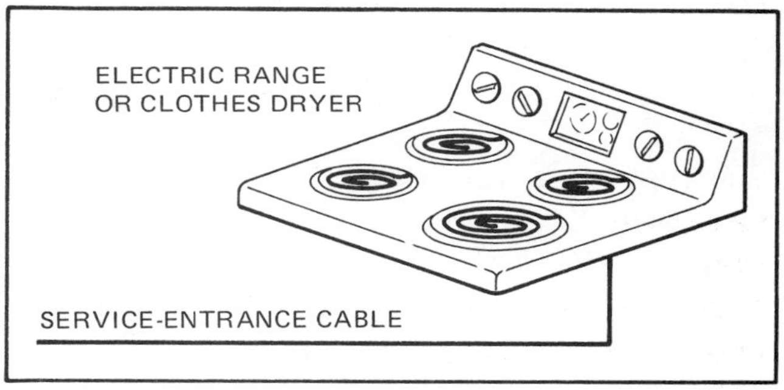

Fig. 18-7 An electric range or a clothes dryer may be connected to service-entrance cable.

Alternate Method: Service-entrance Cable Installation

In previous units, when cable was used as the means of wiring the general lighting and small appliance branch circuits, it was assumed that the cable was either armored or nonmetallic-sheathed cable. The residence specifications require that conduit be installed in certain areas. *Section 338-3* of the National Electrical Code recognizes another wiring method that may be used under certain conditions.

To install service-entrance cable with an *insulated* neutral, the cable may start from the service-entrance equipment or a subpanelboard. If the other requirements of *Section 250-60* are met, then the metal frame of a range or clothes dryer may be grounded to this neutral.

When service-entrance cable with an *uninsulated* neutral is used to connect a range or clothes dryer, the cable must originate from either the service-entrance equipment or a subpanel. The metal frame of a range or clothes dryer may be grounded to this neutral if the other requirements of *Section 250-60* are met, figure 18-7.

The student should always check with the local electrical inspector to determine if there are any variations between local code rules and the National Electrical Code.

OVERHEAD GARAGE DOOR OPENER CIRCUIT Ⓐ$_E$

A separate circuit must be installed for the two overhead garage door openers in the residence. The garage doors, controls, and all of the required hardware will be installed by the carpenter or a specially trained overhead door installer. The symbol Ⓐ$_E$ in figure 18-8 shows the location of the outlet for the door openers.

Principles of Operation

The overhead door opener contains a motor, gear reduction unit, motor reversing switch, and electric brake. This unit is preassembled and wired by the manufacturer.

Almost all residential overhead door openers use split-phase, capacitor-start motors in sizes from 1/6 hp to 1/3 hp. The motor size selected depends upon the size (weight) of the door which is to be raised and lowered. By using springs in addition to the gear reduction unit to counterbalance the weight of the door, a small motor can lift a fairly heavy door.

The two ways to change the direction of rotation of a split-phase motor are: (1) reverse the starting winding with respect to the running winding, or (2) reverse the running winding with respect to the starting winding. The motor can be run in either direction by properly connecting the starting and running winding leads of the motor to the reversing switch. When the motor shaft is connected to a gear reduction unit, which is connected to a chain drive or worm gear drive, the door can be raised or lowered according to the direction in which the motor is rotating.

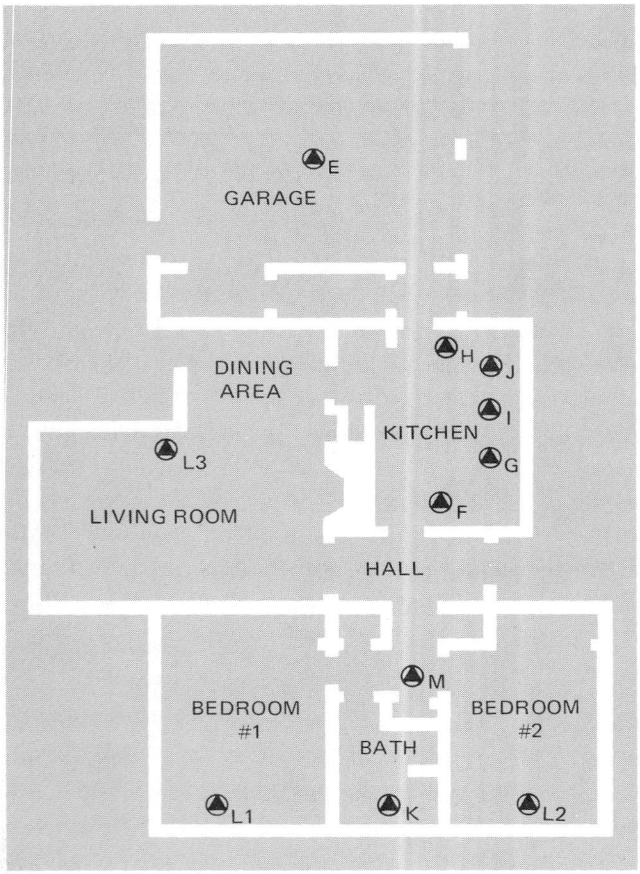

Fig. 18-8 Special purpose outlets – first floor.

The electrically operated reversing switch can be controlled in a number of ways. These methods include push button stations marked "open-close-stop," "up-down-stop," or "open-close"; indoor and outdoor weatherproof push buttons, and key-operated stations. Radio-operated controllers are popular because they permit overhead doors to be controlled from an automobile.

Wiring of Door Openers

The wiring of overhead door openers for residential use is quite simple, since these units are completely prewired by the manufacturer. The electrician must provide a 120-volt circuit close to where the overhead door operators are to be installed. This 120-volt circuit can be wired directly into the overhead door unit, or provided with a receptacle into which the cord on the overhead door operator is plugged.

The buttons that are pushed to operate the overhead door opener can be placed in any convenient, desirable location. These push buttons are wired with low-voltage wiring, usually 24 volts, figure 18-9. This is the same type of wire used for wiring bells and chimes. From each button, the electrician runs a 2-wire, low-voltage cable back to the control unit. Since the residential openers operate on the *momentary contact* principle for the push buttons, a 2-wire cable is all that is necessary between the operator and the push

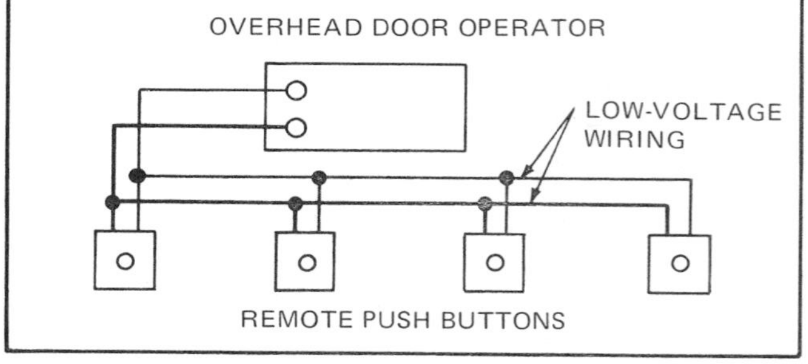

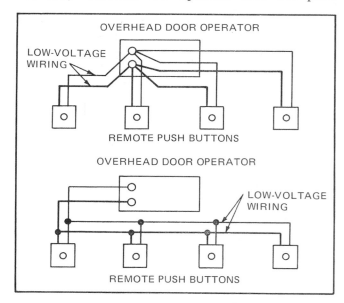

Fig. 18-9 Push button wiring for overhead door operator.

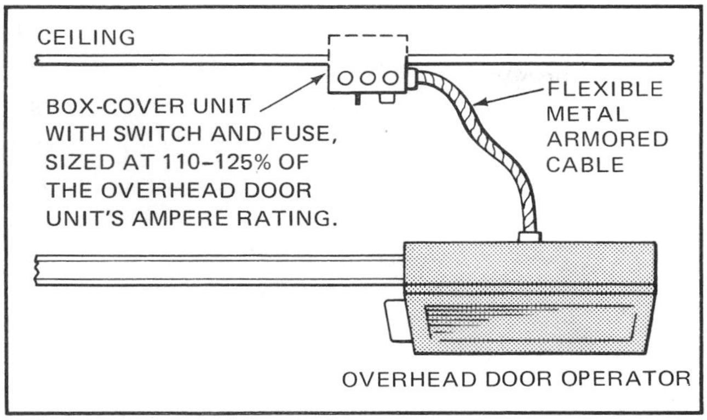

Fig. 18-10 Connections for overhead door operator.

buttons. Commercial overhead door openers can require three, or more, conductors.

It is desirable to install push buttons at each door leading into the garage.

The actual connection of the low-voltage wiring between the controller and push buttons is a parallel circuit connection.

Overhead door openers can be connected directly by using flexible metal conduit or flexible metal armored cable. The illustration in figure 18-10 shows how this can be accomplished. Note that the flexible cable is run between the door opener and an electrical box that has been mounted on the ceiling. A box-cover unit, containing a switch and fuse of a size rated at approximately 110 percent to 125 percent of the ampere rating of the overhead door unit, is fastened to the outlet box in the ceiling.

Install a box-cover unit close to the overhead door unit operator. Install a dual-element fuse sized approximately 1.25 times the full-load ampere rating of the overhead door unit. This receptacle does *not* have to be protected by a GFCI (*Section 210-8(a)(2)*.

Many overhead door openers come complete with a 3-wire cord for ease in making the electrical connection and ease in servicing the unit. The unit is merely plugged into the receptacle, figure 18-11. The type shown can be fused to provide additional overcurrent protection to the overhead door operator, which has integral, built-in overload protection. If a short circuit should occur in the opener, only the fuse in the box-cover unit will blow. Therefore, the branch-circuit fuse or breaker that supplies the overhead door unit is unaffected.

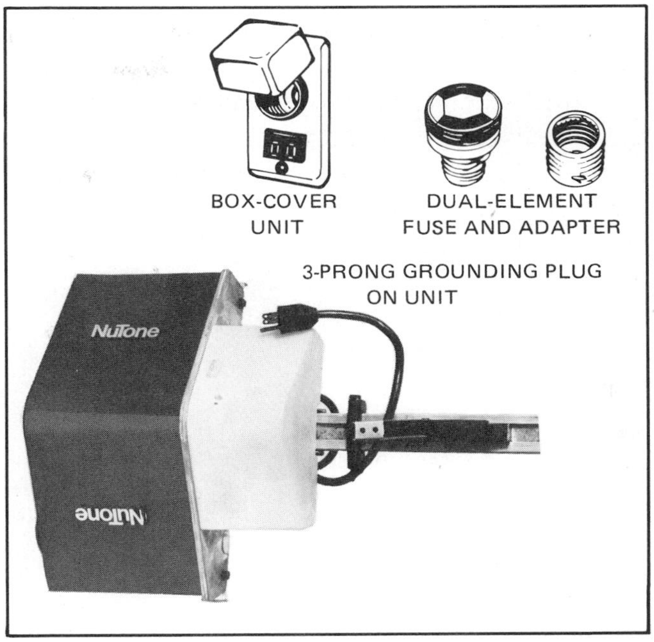

Fig. 18-11 Overhead door opener with 3-wire cord connector.

Double fusing not only provides added protection, it also makes troubleshooting much easier.

For security reasons, it is desirable for a homeowner to disconnect the overhead door opener when away for extended periods of time.

In the residence, a two-wire circuit at 120 volts is installed using No. 12 AWG cable. The circuit runs from Panel A, Circuit A17, to a 4-inch square, 2 1/8-inch deep outlet box mounted near the overhead door operator.

Electricians often install box-cover units as a disconnecting means as required by the Code. The unit also serves as a disconnecting means to provide security when the owner is away for more than a day. These units also provide running overcurrent protection for small motors. A box-cover unit, figure 18-12, is economical to use.

A box-cover unit is used in the residence. Both switches of the unit are fed by the black hot conductor at the outlet box near the garage door operators. A two-wire cable runs from the outlet box to a junction box on each of the two overhead door operators. The black hot conductors of each of these cables are connected on each of the two overhead door operators. The black hot conductors of each of these cables are connected to the load side of the box-cover unit. All three white conductors are connected together, figure 18-13.

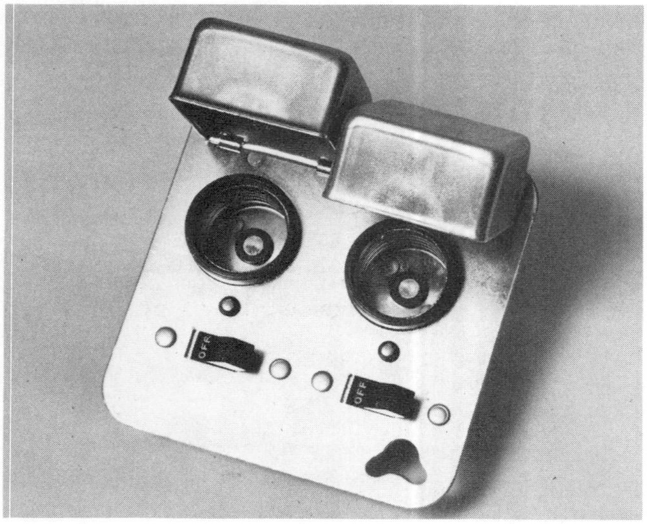

Fig. 18-12 Box-cover unit.

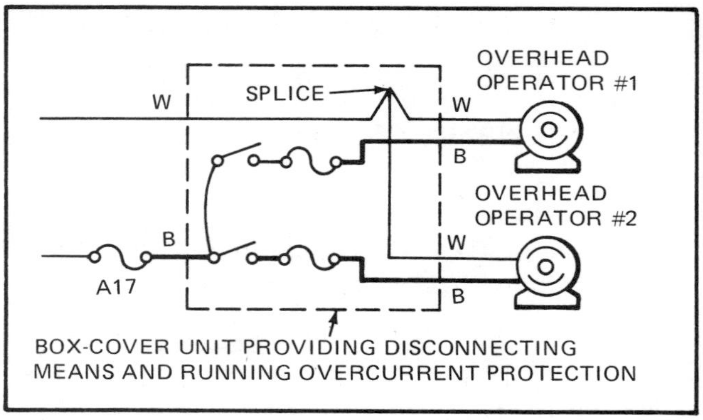

Fig. 18-13 Wiring circuit for overhead garage door opener.

The walls and ceilings of the garage are finished with 3/16-inch flexboard. Thus, all wiring is concealed except where the cables from the box-cover unit drop through the ceiling to the junction boxes on the overhead door operators.

Overcurrent Protection

Running overcurrent protection is usually an integral part of the motor, *Section 430-32(c)(2)*, or is provided by a separate overcurrent device, *Section 430-32(c)(1)*.

The protective devices installed in the box-cover units are dual-element, Type S fuses. The fuses are selected for their time-delay characteristics. In other words, they do not open on heavy inrush starting current surges. Overcurrent devices are described in units 19 and 27.

As noted in the schedule of special-purpose outlets in the specifications, each overhead door operator draws 5.8 amperes. *Table 430-148* shows that a motor that draws 5.8 amperes has a rating of 1/4 hp. According to *Section 430-32(c)*, running overcurrent protection shall be based on not more than 125 percent of the full-load current rating of the motor, or

$$5.8 \times 1.25 = 7.25 \text{ amperes}$$

Since 7.25-ampere, dual-element fuses are not available, 6 1/4- or 7-ampere dual-element fuses may be used. It is *not* a violation of the Code to provide circuit protection smaller than 125 percent of the full-load current rating of the motor. Sizing time-delay fuses at 110 percent to 125 percent of the motor's rating is a good practice. However, it is a Code violation to provide protection *exceeding* 125 percent of the full-load running current of the motor.

Section 430-34 does permit running overcurrent protection up to 140 percent of the full-load running current of the motor when the 125 percent value does not correspond to standard overcurrent device sizes or ratings. However, many sizes of overcurrent devices are readily available, including adjustable overcurrent devices. As a result, an electrician seldom installs overcurrent devices rated at greater than 125 percent of the full-load current rating of the motor. The maximum overload protection results when the rating of the overcurrent device is as close as possible to the actual full-load current rating of the motor without opening on harmless momentary overloads.

REVIEW

Note: Refer to the Code or the plans where necessary.

DRYER CIRCUIT ⓓ D

1. a. What regulates the temperature in the dryer? ______________________
   b. What regulates drying time? ______________________

2. List the various methods of connecting an electric clothes dryer.

3. What is the unique shape of the neutral blade of a dryer cord set?

4. a. What is the minimum power demand for the dryer?
   b. How much current is used?

5. What is the maximum permitted current rating of a portable appliance on a 30-ampere branch circuit?

6. What provides motor running overcurrent protection for the dryer?

7. a. Must an electric clothes dryer be grounded?
   b. May a dryer be grounded to a neutral conductor?
   c. Under what condition?

8. A dryer is rated at 7.5 kW and 240 volts.
   a. What is the wattage rating?
   b. What is the current rating?
   c. What size conductors are required?
   d. What size conduit is required?

9. When a metal junction box is installed as part of the cable wiring to a clothes dryer or electric range, may this box be grounded to the circuit neutral?

OVERHEAD GARAGE DOOR OPENER CIRCUIT E

1. What type of motors are generally used for garage door openers?

2. How is the direction of a split-phase motor reversed?

3. a. What circuit supplies the garage door openers?
   b. What size conductors are used in this circuit?

4. What provision is made for the overcurrent protection and disconnecting means for the garage door opener?

5. Motor running overcurrent protection is usually based on what percentage of the full-load running current of the motor?

6. When is running overcurrent protection based on 140 percent of the full-load current rating of the motor applicable to a motor circuit?

unit 19

Special-Purpose Outlets for the Refrigerator-Freezer, Counter-Mounted Cooking Unit, and Wall-Mounted Oven Circuits

OBJECTIVES

After studying this unit, the student will be able to

- interpret electrical plans to determine special installation requirements for refrigerator-freezers, counter-mounted cooking units, wall-mounted ovens, and free-standing ranges.
- select proper conductor sizes for wiring installations based on the ratings of appliances.
- ground all appliances properly regardless of the wiring method used.
- describe a seven-heat control and a three-heat control for a dual-element heating unit for a cooktop.
- install a counter-mounted cooking unit and a wall-mounted oven using the group circuit arrangement.
- install a subfeeder circuit to a load center.

REFRIGERATOR-FREEZER CIRCUIT (▲)$_F$

Special-purpose outlet (▲)$_F$ is connected to a 20-ampere circuit (Circuit B3). This outlet serves only the built-in refrigerator-freezer and is located on the southeast wall of the kitchen, (see figure 10-3). The receptacle is a grounding type as required by *Section 210-7*.

Panel B is mounted in the basement utility room directly below the refrigerator-freezer in the kitchen. Thus, Circuit B3 is relatively short. A separate circuit is recommended for the refrigerator-freezer so that trouble on any other circuit in the dwelling will not affect this appliance. A clock connected to this circuit will indicate if the power is off or was off previously.

Wiring

Most refrigerators or refrigerator-freezers are furnished with a three-wire cord (two conductors plus a ground) and a three-wire attachment plug cap.

Electrical plans generally show the refrigerator-freezer outlet as a special-purpose outlet. In this way, the electrician is alerted that there may be special installation instructions for the convenience receptacle. For example, the receptacle may not be installed at the standard height. In this dwelling, the standard height is 12 inches to center from the finish floor. Specific instructions must be given to the electrician during the roughing-in period so that the outlet can be located correctly.

COUNTER-MOUNTED COOKING UNIT CIRCUIT (▲)$_G$

Counter-mounted cooking units are available in many styles. Such units may have two, three, or four heating elements. The heating elements of some cooking units are completely covered by a sheet of high-temperature ceramic. These units have the same controls as standard units and are wired according to the general Code guidelines for cooking units. When the elements of these ceramic covered are turned on, the ceramic above each element changes color slightly. Most of these cooking units have a faint design in the ceramic to locate the heating elements. The ceramic is rugged, attractive, and easy to clean.

In the residence in the plans, a separate circuit is provided for a standard, exposed element, counter-mounted cooking unit. This unit is rated at 7450 watts and 120/240 volts. The ampere rating is:

$$I = \frac{W}{E} = \frac{7450}{240} = 31.04 \text{ amperes}$$

The cooking unit outlet is shown by the symbol G.

NEC *Table 220-19* shows the demand loads for household cooking appliances. A free-standing range may be supplied by conductors having an ampacity for the demand load shown in the table for the specific range.

However, the demand load for individual wall-mounted oven circuits or individual counter-mounted cooking unit circuits must be calculated at the full nameplate rating. A reduction in demand is not permitted (see *Note 4* to *Table 220-19*).

Installing the Cooking Unit

Table 310-16 shows that No. 8 TW conductors may be used as the circuit conductors for the cooking unit installation. These circuit conductors are connected to Circuit B14-16. This is a two-pole, 40-ampere circuit in Panel B.

The electrician must obtain the roughing in dimensions for the counter-mounted cooking unit so that the cable can be brought out of the wall at the proper height. In this way, the cable is hidden from view. To simplify both servicing and installation, a separable connector or a plug and receptacle combination may be used in the supply line, *Section 422-17*.

Most cooktop manufacturers provide a junction box on their units. The electrician can terminate the circuit conductors of the appliance in this box and make the proper connections to the residence wiring. Instead of the junction box, the manufacturer may connect a short length of flexible conduit to the appliance. This conduit contains the conductors to which the electrician will connect the supply conductors. These connections must be made in a junction box furnished by the electrician or the manufacturer of the appliance.

Temperature Effects on Conductors

It is a good practice to check the appliance instructions provided by the manufacturer before beginning any installation. The manufacturer may require the supply conductors to have an insulation rated for temperatures higher than the standard insulation rating of 60°C.

Underwriters' Laboratories require appliance manufacturers to attach some form of label or otherwise mark their units to indicate that the supply conductors will be subjected to temperatures higher than 60°C. If such information is not provided, then it is safe to assume that the conductors used may have insulation rated for 60°C. *Table 310-13* gives the conductor insulation temperature limitations and the accepted applications for the conductors.

The surface heating elements and the controls of cooking units are usually prewired by the manufacturer with asbestos (Type A) insulated conductors. These conductors are made of copper or nickel, depending upon the temperature they must withstand. The maximum operating temperature of Type A conductors is 200°C or 392°F (as compared to the maximum operating temperature of standard Type T or TW conductors which is 60°C or 140°F).

General Wiring Methods

Any of the standard wiring methods may be used to install counter-mounted cooking units, wall-mounted ovens, free-standing ranges, and clothes dryers. That is, armored cable, nonmetallic-sheathed cable, and conduit may be used. These appliances can also be installed with service-entrance cable, *Article 338* (refer to unit 18).

Local electrical codes and utility companies can supply information as to restrictions on the use of any of the wiring methods listed.

Cooking Unit Grounding Methods

The subject of the various methods of grounding frames of counter-mounted cooktops, free-standing ranges, and clothes dryers is discussed in unit 18.

Surface Heating Elements for Cooking Units

The heating elements used in surface-type cooking units are manufactured in several steps. First, spiral-wound nichrome resistance wire is impacted in magnesium oxide (a white, chalklike powder). The wire is then encased in a nickel-steel alloy sheath, flattened under very high pressure, and formed into coils. The flattened surface of the coil makes good contact with the bottoms of cooking utensils. Thus, efficient heat transfer takes place between the elements and the utensils.

Cooktops with a smooth ceramic surface offer the user ease in cleaning, and a neater, more modern appearance. The electric heating elements are in direct contact with the underside of the ceramic top.

Seven-heat Control for Dual-element Heating Unit

The most effective method of obtaining several levels of heat from one heating coil unit is to use a dual-element unit. Figure 19-1 shows that seven levels of heat can be obtained from such a unit if a seven-point heater switch is used. Note that the two elements of the unit are connected together at the inner end of the steel coil. A return wire (not a resistance wire) is carried back to the terminal block. This arrangement means that the elements can be operated singly, together, or in series.

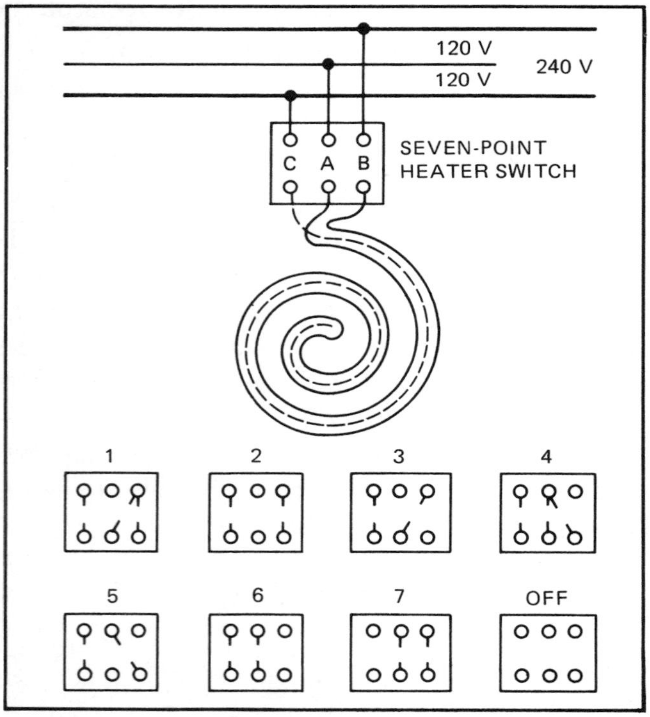

Fig. 19-1 Surface heating element with seven levels of heat.

Figure 19-2 shows the heating element connected with element A as lead 2, element B as lead 1, and the return wire as the common (C). Resistance readings can be taken between these leads as follows:

Test between C and 2	90 ohms (element A)
Test between C and 1 (single operation)	74 ohms (element B)
Test between 2 and 1 (elements in series)	164 ohms (elements A and B in series)
Test between C and 1–2 connected together (elements connected in parallel)	40.6 ohms (elements A and B in parallel)

Element B allows more current to flow than does element A because element B has 74 ohms of resistance as compared to 90 ohms of resistance for element A. As the current increases, the wattage also increases. The eight switch positions shown in figure 19-1 result in the wattages shown in the following list. For each switch position, the formula $W = E^2/R$ is used to find the wattage value.

1.	A and B parallel, 240 volts	40.6 ohms	1418 watts
2.	B only, 240 volts	74 ohms	778 watts
3.	A only, 240 volts	90 ohms	640 watts
4.	A and B parallel, 120 volts	40.6 ohms	_____ watts*
5.	B only, 120 volts	74 ohms	_____ watts*
6.	A only, 120 volts	90 ohms	_____ watts*
7.	A and B series, 120 volts	164 ohms	_____ watts*
8.	Off position	Infinity	0 watts

*These values are omitted because the student is required to calculate the values in the Review section. A sample calculation is completed as follows:

$$W = \frac{E^2}{R} = \frac{240 \times 240}{40.6} = \frac{57\ 600}{40.6} = 1418 \text{ watts}$$

Three-heat Control for Dual-element Heating Unit

A simple, three-heat series-parallel switch is shown in figure 19-3. This switch is similar to the type used on many electric ranges.

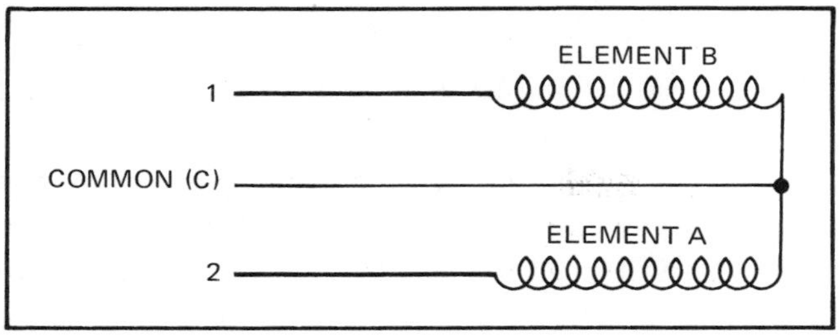

Fig. 19-2 Typical surface unit wiring.

Another type of heating element is the 120-volt flash or high-speed element. When connected to a special control switch, this type of element is connected briefly across 240 volts. In a few seconds, the element is returned automatically to the 120-volt supply for continued operation. A 1250-watt flash unit connected to 240 volts produces 5000 watts of heat in a matter of seconds. This brings the unit to maximum heat quickly. When the flash element is returned to the 120-volt supply, the unit continues to produce 1250 watts, the rated value.

To calculate the maximum wattage at 240 volts, assume that a 1250-watt heating element is rated at 120 volts. This unit has a resistance of:

$$R = \frac{E^2}{W} = \frac{120 \times 120}{1250} = \frac{14\,400}{1250} = 11.52 \text{ ohms}$$

When this 11.52-ohm element is connected to a 240-volt supply, the wattage is:

$$W = \frac{E^2}{R} = \frac{240 \times 240}{11.52} = \frac{57\,600}{11.52} = 5000 \text{ watts}$$

It is obvious that the wattage quadruples when the supply voltage is doubled. Or, the wattage is reduced to one-fourth its original value when the supply voltage is reduced to one-half its original value.

Heat Generation by Surface Heating Elements

The surface heating elements generate a large amount of radiant heat. For example, a 1000-watt heating unit generates approximately 3412 Btu of heat per hour. Heat is measured in British thermal units (Btu). A *Btu* is defined as the amount of heat required to raise the temperature of one pound of water by one degree Fahrenheit.

WALL-MOUNTED OVEN CIRCUIT Ⓐ$_H$

A separate circuit is provided for the wall-mounted oven. The circuit is connected to Circuit B10-12, a two-pole, 20-ampere circuit in Panel B. The receptacle for the oven is shown by the symbol Ⓐ$_H$.

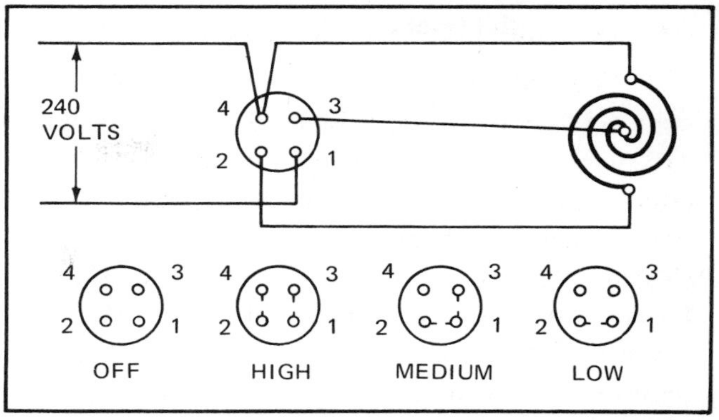

Fig. 19-3 Surface unit with three-heat series-parallel switch.

The wall-mounted oven installed in the residence is rated at 6.6 kW or 6600 watts at 120/240 volts. The current rating is:

$$I = \frac{W}{E} = \frac{6600}{240} = 27.5 \text{ amperes}$$

Note 4 to *Table 220-19* states that the branch-circuit load for the wall-mounted oven must be based on the actual nameplate rating of the appliance. With this in mind, *Table 310-16* indicates that No. 10 TW conductors may be installed. Under certain conditions, the use of a conductor having an insulation with a higher temperature rating will permit smaller conductors to be installed.

The information given on supply conductor connections, conductor temperature limitations, grounding requirements, and general installation procedures for counter-mounted cooking units also applies to wall-mounted ovens.

Self-cleaning Oven

Self-cleaning ovens are popular appliances because they automatically remove cooking spills from the inside of the oven. Lined with high-temperature material, a self-cleaning oven can be set for a cleaning temperature that is much higher than baking and broiling temperatures. When the oven is set for at least 800°F, all residue in the oven (such as grease and drippings) is burned until it is a fine ash which can be removed easily. Self-cleaning ovens do not require special wiring and are installed in the same manner as standard ovens.

Operation of the Oven

The oven heating elements are similar to the surface elements. The oven elements are mounted in removable frames. A thermostat controls the temperature of these heating units. Any oven temperature can be obtained by setting the thermostat knob to the proper point on the dial. The thermostat controls both the baking element and the broiling element. These elements can be used together to preheat the oven.

A combination clock and timer is used on most ovens. The timer can be preset so that the oven turns on automatically at a preset time, heats for a preset duration, and then turns off.

Microwave Oven

Modern microwave ovens, a result of electronic technology, have introduced many features never before possible. Such features include:

- control of time and temperature.
- programmable multiple-stage settings.
- temperature probes for insertion into the food to assure desired temperature.
- ability to "hold" food temperatures for up to one hour following cooking cycle.
- touch-sensitive controls.
- programmable defrost settings, having lower temperatures than those required for actual cooking or baking, so that premature cooking does not occur.
- computer calculations to automatically increase the time of operation when recipe quantities are doubled, tripled, or more.
- delayed starting of oven for up to 12 hours.
- preprogrammed temperature and time for recipes.

The electrician's primary concern is how to calculate for these features, and how to install the circuit for microwave appliances so as to conform to the Code. The same Code rules apply for both electrical and electronic microwave cooking appliances.

Ovens are insulated with fiberglass or polyurethane foam insulation placed within the oven walls to prevent excessive heat leakage.

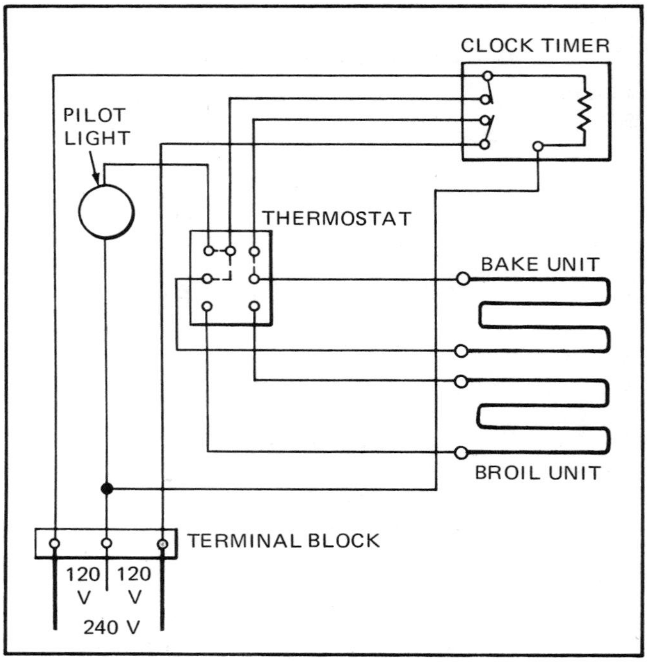

Fig. 19-4 Wiring diagram for an oven unit.

Figure 19-4 shows a typical wiring diagram for a standard oven unit.

CIRCUIT REQUIREMENTS FOR GROUP INSTALLATIONS

The Code states that a counter-mounted cooking unit and a wall-mounted oven may be connected to one circuit. When only one circuit is used, its capacity must be large enough to serve both appliances.

Group Circuit Arrangement

The Code permits ranges and other household cooking equipment to be group connected to one feeder. However, certain restrictions must be met, figure 19-5. These Code rules are contained in *Section 210-19(b)*.

Figure 19-5 shows how *Exception No. 2* to *Section 210-19(b)* is applied in the residence in the plans.

Note 4 of *Table 220-19* states that the size of the tap conductors must be based on the actual nameplate rating of the loads they serve. The tap conductors may be smaller than the 50-ampere branch-circuit conductors. The Code does not establish a definite length for the taps, but requires that they be not longer than necessary for servicing.

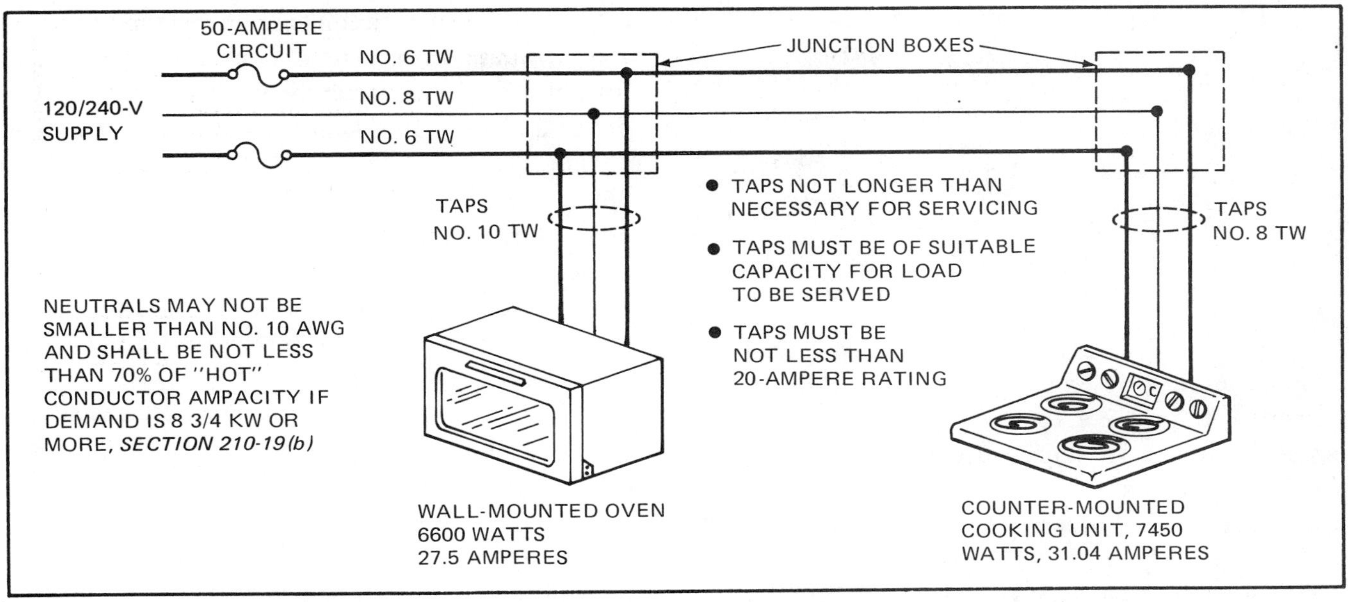

Fig. 19-5 Counter-mounted cooktop and wall-mounted oven connected to one circuit.

This length is usually determined by the local electrical inspector.

The group circuit method requires that smaller conductors be tapped from the 50-ampere circuit. This means that extra junction boxes, cable connectors, conduit fittings, and wire connectors are used to complete the installation. Thus, the initial cost of installing a 50-ampere circuit may be higher than the cost of installing separate circuits having smaller capacities. Another factor must be considered by the electrician when deciding which method of installing cooking units is more economical. Since the individual overcurrent device for each appliance on an individual circuit has a lower rating, there is greater protection if a short circuit or ground fault occurs on the circuit.

Grounding Requirements

The subject of the various methods of grounding the frames of cooking appliances is discussed in unit 18.

Circuit Calculations for the Group Circuit Arrangement

The load demand of a single, free-standing electric range (combined cooking unit and oven) is similar to the load demand of a separate wall-mounted oven and counter-mounted cooking unit. *Note 4* to *Table 220-19* points out special requirements for a branch-circuit load for a counter-mounted cooking unit and not more than two wall-mounted ovens which are all supplied from a single branch circuit and located in the same room. The branch-circuit load for this arrangement must be computed by adding the nameplate ratings on the individual appliances, and treating the total as equivalent to that of one range.

To compute the load demand for the counter-mounted cooking unit and the wall-mounted oven, the nameplate ratings are used.

Counter-mounted cooktop	7450 watts
Wall-mounted oven	6600 watts
Total	14 050 watts

The total connected load of 14 050 watts at 240 volts is

$$I = \frac{W}{E} = \frac{14\,050}{240} = 58.5 \text{ amperes}$$

According to *Note 1* of *Table 220-19*, for ranges rated over 12 kW but not over 27 kW, the demand load given in column A of *Table 220-19* must be increased by 5% for each kilowatt or major fraction of a kilowatt in excess of 12 kW.

Assume that the combined load of the counter-mounted cooking unit and the wall-mounted oven is equivalent to the load of a free-standing range rated at 14 050 watts. This value exceeds 12 000 watts (12 kW) by 2050 watts, or roughly 2 kW.

The demand is determined as follows:

2 kW (kW over 12 kW) × 5% per kW = 10%
8 kW (from Column A of *Table 220-19*) × 0.10 = 0.8 kW
Calculated demand = 8 kW + 0.8 kW = 8.8 kW or = 8800 watts

and,

$$I = \frac{W}{E} = \frac{8800}{240} = 36.7 \text{ amperes}$$

Section 210-23(c) states that fixed cooking appliances may be connected to 40- or 50-ampere circuits. If taps are to be made from 40- or 50-ampere circuits, the taps must be able to carry the load to be served. In no case may the taps be less than 20 amperes, *Section 210-19(b)*.

For example, the 50-ampere circuit shown in figure 19-5 has No. 10 TW and No. 8 TW tap conductors. Each of these taps has an ampacity of more than 20 amperes. The branch circuit requires No. 6 Type TW conductors. *Section 210-19(b), Exception No. 2* states that the load on the neutral conductor supplying the wall-mounted oven and the counter-mounted cooking unit may be based on 70 percent of the ampacity of the ungrounded conductors.

$$55 \times 0.70 = 38.5 \text{ amperes}$$

According to *Table 310-16*, the neutral may be a No. 8 TW conductor.

Using a Load Center

A *load center* can be used to make the installation of the cooking units easier. If the main power panel and the built-in range components are far apart, a subfeeder circuit can be installed to a load center, figure 19-6. The maximum power demand for the appliances is obtained from Column A of *Table 220-19*. It was shown that the calculated demand for the 7450-watt cooking unit and the 6600-watt oven is 8.8 kW (36.7 amperes). The two ungrounded conductors are No. 8 TW wire (40 amperes).

Section 210-19(b) states that the neutral conductor must have a rating of not less than 40 × 0.70 = 28 amperes.

Table 310-16 shows that No. 10 TW wire is the minimum size of copper conductor that can be used for the neutral of this subfeeder circuit.

The No. 8 TW conductors are run from a 40-ampere circuit to the load center. The circuits supplying each appliance are protected at the load center according to their individual current ratings. The fuses used have lower ratings based on the nameplate rating of the appliance (see *Note 4, Table 220-19*).

In figure 19-6, a No. 8 TW neutral feeder conductor is shown even though the conductor may be reduced to No. 10 TW, *Section 210-19(b)*. Cable having three conductors of the same size is available.

FREE-STANDING RANGE

The demand calculations for a single, free-standing range are simple. For example, a single electric range is to be installed. This range has a rating of 14 050 watts (exactly the same as the combined built-in units).

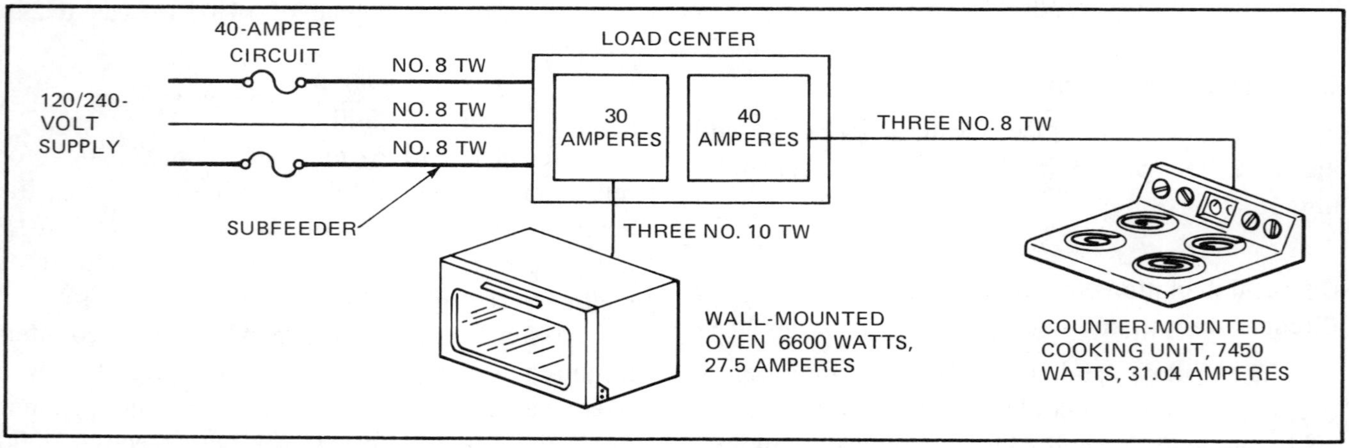

Fig. 19-6 Appliances connected to a load center.

The load for a single range is the same as that for the two built-in units:

1. 14 050 watts - 12 000 watts = 2050 watts or 2 kW.
2. According to *Note 1, Table 220-19*, a 5 percent increase for each kW in excess of 12 kW must be added to the load demand in Column A.
   2 kW (kW over 12 kW) × 5% per kW = 10%
3. 8 kW (from Column A) × 0.10 = 0.8 kW
4. Calculated load = 8 kW + 0.8 kW = 8.8 kW
5. In amperes, the calculated load is:

$$I = \frac{W}{E} = \frac{8800}{240} = 36.7 \text{ amperes}$$

No. 8 TW conductors should be used for the ungrounded conductor. The size of the neutral is based on 70 percent of the current-carrying capacity of the No. 8 TW conductors, or

$$40 \times 0.70 = 28 \text{ amperes}$$

A No. 10 TW conductor is to be used for the neutral.

If a range is rated at not over 12 kW, then the maximum power demand is based on Column A of *Table 220-19*. For example, if the range is rated at 11.4 kW, the maximum demand is based on 8 kW:

$$I = \frac{W}{E} = \frac{8000}{240} = 33.3 \text{ amperes}$$

As in the previous example, two No. 8 TW ungrounded conductors and one No. 10 TW neutral conductor can be installed for this range. For ranges of 8 3/4 kW or higher rating, the minimum branch-circuit rating shall be 40 amperes, *Section 210-19(b)*.

REVIEW

Note: Refer to the Code or the plans where necessary.

REFRIGERATOR-FREEZER CIRCUIT ▲F

1. The refrigerator-freezer is connected to what circuit? ________________
2. What is the rating of this circuit? ________________
3. Why is it advisable to supply the refrigerator-freezer with a separate circuit?
   ____
   ____
   ____

COUNTER-MOUNTED COOKING UNIT CIRCUIT ▲G

1. a. What circuit supplies the counter-mounted cooking unit in this residence? ________
   b. What is the rating of this circuit? ________________
2. What three methods may be used to connect counter-mounted cooking units? ________
   ____
   ____
3. Is it permissible to use standard 60°C insulated conductors to connect all counter-mounted cooking units? Why? ________________
   ____
4. What is the maximum operating temperature (in degrees Celsius) for a
   a. Type RH conductor? ________________
   b. Type RHH conductor? ________________
   c. Type T conductor? ________________
5. a. May service-entrance cable be used to connect counter-mounted cooking units? ______
   b. What Code section answers question a? ________________

6. a. What section of the Code applies to grounding a counter-mounted cooking unit?

   ____

   b. What methods may be used to ground the counter-mounted cooking unit? __________

   ____

7. An electric range or clothes dryer may be grounded to the ______________ conductor provided the ______________ conductor is not smaller than No. ________ AWG.

8. In the illustration of a typical seven-heat, eight-position switch (figure 19-1), the wattage values for positions 4, 5, 6, and 7 are omitted. Calculate these values and insert them in the spaces provided below and on page 166. Show all calculations.

   Position 4 ______________________________

   ____

   Position 5 ______________________________

   ____

   Position 6 ______________________________

   ____

   Position 7 ______________________________

   ____

9. A 120-volt flash or high-speed unit rated at 1000 watts produces __________ watts when connected briefly to the 240-volt source.

10. When the voltage to an element is doubled, the wattage: (underline one)

    a. increases b. decreases

11. By how much is the wattage in question 10 increased or decreased? (underline one)

    a. doubled b. tripled c. quadrupled d. halved

12. One kilowatt equals ______________ Btu per hour.

WALL-MOUNTED OVEN CIRCUIT Ⓐ H

1. To what circuit is the wall-mounted oven connected? ______________________________

2. An oven is rated at 7.5 kW. This is equal to:

   a. How many watts? ______________________________

   b. How many amperes at 240 volts? ______________________________

3. a. What section of the Code governs the grounding of a wall-mounted oven? __________

   b. By what methods may wall-mounted ovens be grounded? ______________

   ____

   ____

4. What is the type and rating of the overcurrent device protecting the wall-mounted oven? ______________________________

   ____

5. How many feet (meters) of cable are required to connect the oven in the residence? ______

6. When connecting a wall-mounted oven and a counter-mounted cooking unit to one feeder, how long are the taps to the individual appliances? ______________

7. The branch-circuit load for a single wall-mounted self-cleaning oven or counter-mounted cooking unit shall be the ______________ rating of the appliance.

8. A 6-kW counter-mounted cooking unit and a 4-kW wall-mounted oven are to be installed in a residence. Calculate the maximum demand according to Column A, *Table 220-19*. Show all calculations. ______________

____

____

9. The size of the neutral conductor supplying an electric range may be based on ______ percent of the ampacity of the ungrounded conductor.

10. a. A free-standing range is rated at 11.8 kW, 240 volts. According to Column A, *Table 220-19*, what is the maximum demand? ______________
    b. What size wire (Type TW) is required? ______________

    ____

    c. What size neutral is required? ______________

    ____

11. A double-oven electric range is rated at 18 kW, 240 volts. Calculate the maximum demand according to *Table 220-19*. Show all calculations. ______________

____

____

12. a. What size conductors are required for the range in question 11? Show all calculations.

    ____

    ____

    ____

    b. What size neutral is required? ______________

13. For ranges of 8 3/4 kW or higher rating, the minimum branch-circuit rating is ______ amperes.

unit 20

Special-Purpose Outlets for a Food Waste Disposer and a Dishwasher

OBJECTIVES

After studying this unit, the student will be able to

- install circuits for kitchen appliances such as a food waste disposer and a dishwasher.
- provide adequate running overcurrent protection for an appliance when such protection is not furnished by the manufacturer of the appliance.
- describe the difference in wiring for a food waste disposer with and without an integral on-off switch.
- determine the maximum power demand of a dishwasher.
- make the proper grounding connections to the appliances.
- describe disconnecting means for kitchen appliances.

FOOD WASTE DISPOSER Ⓐ_I

The food waste disposer outlet is shown on the plans by the symbol Ⓐ_I. The food waste disposer is rated at 7.2 amperes. This appliance is connected to Circuit B7, a separate 20-ampere, 120-volt branch circuit in Panel B. To meet the specifications, No. 12 TW conductors are used. The conductors terminate in a junction box provided on the disposer. Food waste disposers normally are driven by a 1/3-hp, split-phase, 120-volt motor.

The food waste disposer is a motor-operated appliance. Therefore, running overcurrent protection must be provided. Such protection must not exceed 125 percent of the full-load current rating of the motor. Most food waste disposer manufacturers install a built-in thermal protector to meet Code requirements. Either a manual reset or an automatic reset thermal protector may be used.

When running overcurrent protection is not built into the disposer, the electrician must install separate protection. For example, a box-cover unit may be installed under the sink near the food waste disposer. A dual-element fuse of the proper size is then mounted in the box-cover unit. This unit also serves as the disconnecting means.

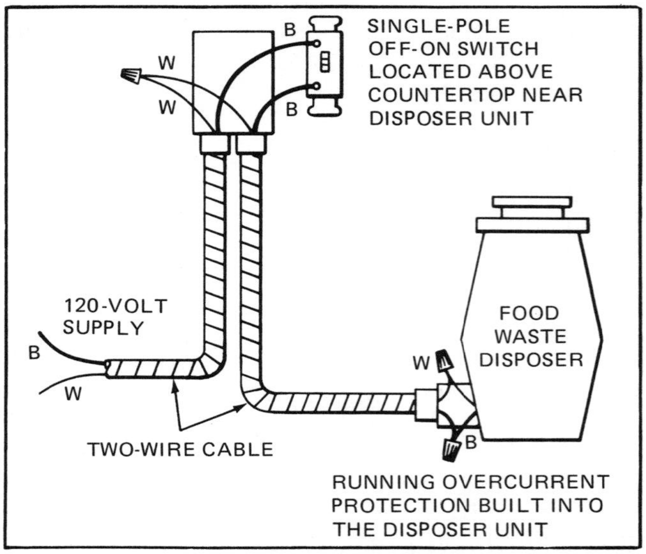

Fig. 20-1 Wiring for a food waste disposer operated by a separate switch.

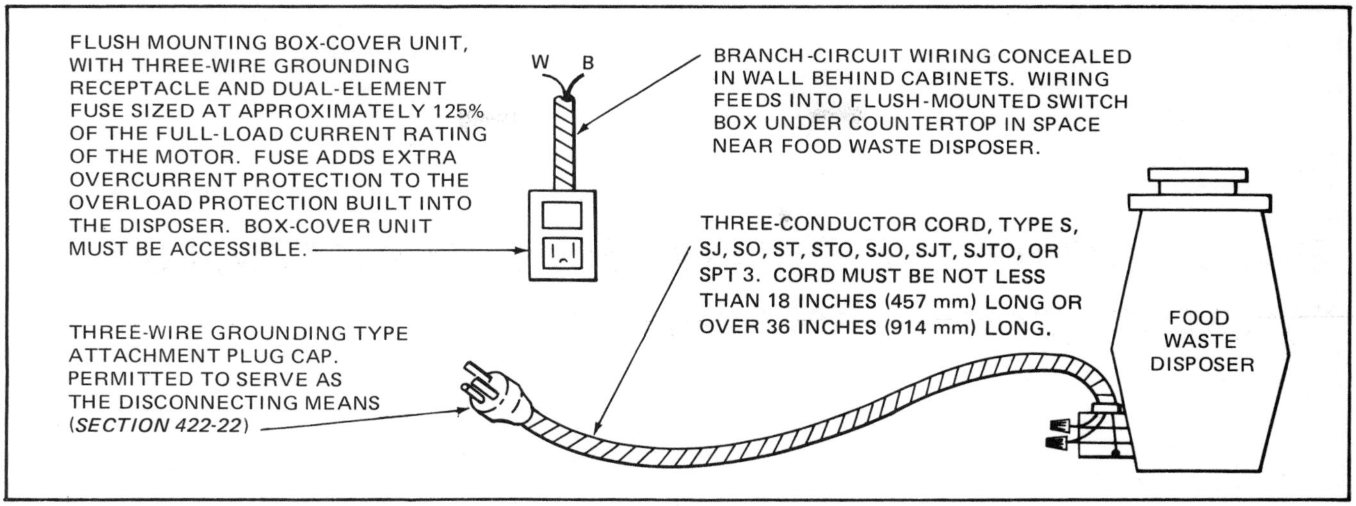

FLUSH MOUNTING BOX-COVER UNIT, WITH THREE-WIRE GROUNDING RECEPTACLE AND DUAL-ELEMENT FUSE SIZED AT APPROXIMATELY 125% OF THE FULL-LOAD CURRENT RATING OF THE MOTOR. FUSE ADDS EXTRA OVERCURRENT PROTECTION TO THE OVERLOAD PROTECTION BUILT INTO THE DISPOSER. BOX-COVER UNIT MUST BE ACCESSIBLE.
W
B
BRANCH-CIRCUIT WIRING CONCEALED IN WALL BEHIND CABINETS. WIRING FEEDS INTO FLUSH-MOUNTED SWITCH BOX UNDER COUNTERTOP IN SPACE NEAR FOOD WASTE DISPOSER.
THREE-CONDUCTOR CORD, TYPE S, SJ, SO, ST, STO, SJO, SJT, SJTO, OR SPT 3. CORD MUST BE NOT LESS THAN 18 INCHES (457 mm) LONG OR OVER 36 INCHES (914 mm) LONG.
THREE-WIRE GROUNDING TYPE ATTACHMENT PLUG CAP. PERMITTED TO SERVE AS THE DISCONNECTING MEANS (SECTION 422-22)
FOOD WASTE DISPOSER

Fig. 20-2 Typical cord connection for a food waste disposer, ***Section 422-8(b).***

DISCONNECTING MEANS

All electrical appliances must be provided with some means of disconnecting the appliance, *Article 420, Part D*. The disconnecting means may be

- a separate on-off switch, figure 20-1,
- a cord connection, figure 20-2, or
- a branch-circuit switch or circuit breaker, such as that supplying the installation in figure 20-3. See NEC *Sections 422-21, 422-24, 422-25,* and 422-26.

Some local codes require that food waste disposers, dishwashers, and trash compactors be cord-connected to make it easier to disconnect the unit, to replace it, to service it, and to reduce noise and vibration, *Sections 422-8(c)* and *400-7*.

Food Waste Disposer Wiring

Many types of food waste disposer units are available. Most of these units have a junction box or wiring space in which the supply conductors can be terminated.

Circuit for Separate On-off Switch

When a separate on-off switch is used, a simple circuit can be made by running a two-wire supply cable to a switch box located next to the disposer unit, figure 20-1. A second cable is run from the switch box to the junction box of the food waste disposer. A switch in the switch box provides off-on control of the disposer.

In the kitchen of the residence, there is a switch next to the sink for the lighting fixture, and a convenience receptacle to be used for small portable appliances. An additional single-pole switch for the disposer can be installed here as well. In this case, a three-gang faceplate must be installed instead of the two-gang faceplate specified on the electrical plans. Another, less convenient, arrangement is to install the switch and the convenience receptacle to the left of the kitchen sink. A two-gang combination switch and receptacle faceplate is then installed at this location. A third possibility is to install the switch for the disposer inside the cabinet space directly under the kitchen sink near the food waste disposer. The switch for a disposer may also be installed in the face of the counter cabinet.

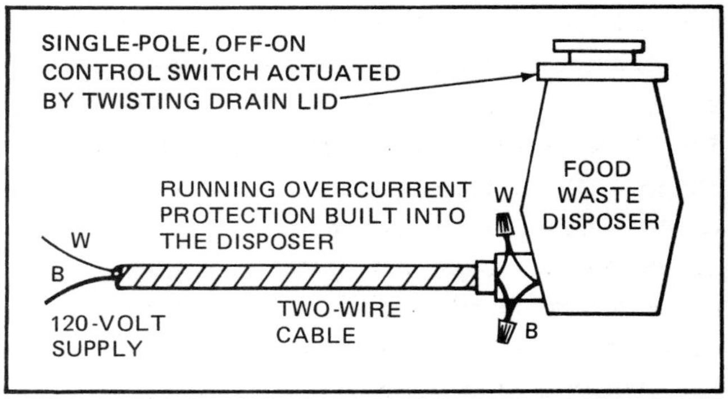

Fig. 20-3 Wiring for a food waste disposer with an integral on-off switch.

Most electricians prefer to mount the food waste disposer switch on the wall near the sink (with other wall switches or receptacles). This location is convenient for the user and places the switch out of the reach of children.

Circuit for Integral On-off Switch

A food waste disposer may be equipped with an integral prewired control switch. This integral control starts and stops the disposer when the user twists the drain cover into place. An extra on-off switch is not required with this type of control. To install the disposer, the electrician runs the supply conductors directly to the junction box on the disposer and makes the proper connections, figure 20-3.

Disposer with Flow Switch in Cold Water Line

Some manufacturers of food waste disposers recommend that a *flow switch* be installed in the cold water line under the sink, figure 20-4. This switch is connected in series with the disposer motor. The flow switch prevents the disposer from operating until a predetermined quantity of cold water is flowing through the disposer. The cold water helps prevent clogged drains by solidifying any grease in the disposer. The water also washes away the shredded waste. If the disposer is to operate properly, a certain amount of water must be in the disposer. Thus, the addition of a flow switch means that the disposer cannot be operated without water.

Figure 20-4 shows one method of installing a food waste disposer with a flow switch in the cold water line.

Grounding

Section 422-16 states that where required by *Article 250*, all exposed, noncurrent-carrying parts are to be grounded in the manner specified in *Article 250*.

The presence of any of the conditions outlined in *Sections 250-42* through *250-45* require that all electrical appliances in the dwelling be grounded. Food waste disposers can be grounded using any of the methods covered in the previous units, such as the metal armor of armored cable or the separate grounding conductor of nonmetallic-sheathed cable, *Section 250-57*.

DISHWASHER Ⓐ$_J$

The dishwasher is supplied by a separate 20-ampere circuit connected to Circuit B5. No. 12 TW conductors are used to install the appliance. The dishwasher has a 1/3-hp motor rated at 7.2 amperes and 120 volts. The outlet for the dishwasher is shown by the symbol Ⓐ$_J$. During the drying cycle of the dishwasher, a thermostatically controlled, 1000-watt electric heating element turns on.

The actual connected load is:

Motor	7.2 amperes
Heater	8.3 amperes
Total	15.5 amperes

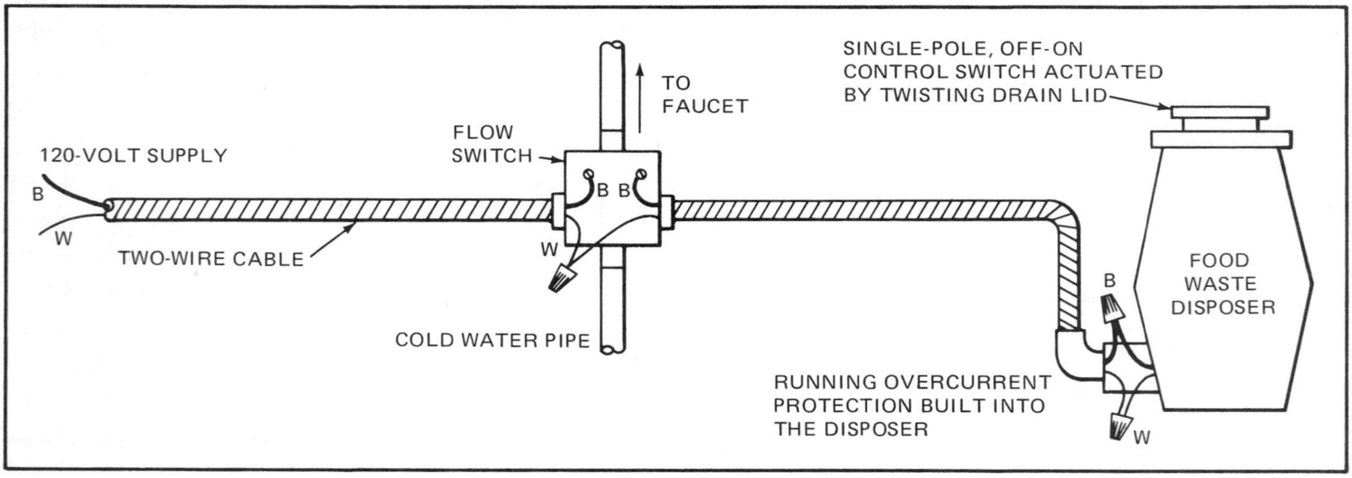

Fig. 20-4 Disposer with a flow switch in the cold water line.

The motor does not run during the drying cycle. Thus, the actual maximum demand on Circuit B5 is that of the larger of the two loads – the 1000-watt heating element.

Energy-saving dishwashers are equipped with a built-in booster water heater which allows the homeowner to adjust the regular water heater temperature to a temperature of 120 degrees, or lower, when desired. The booster heater then boosts the water temperature from 120 degrees to approximately 155 degrees, the temperature considered necessary to sanitize dishes.

Wiring

The dishwasher manufacturer normally supplies a terminal or junction box on the appliance. The electrician connects the supply conductors in this junction box. The electrician must verify the dimensions of the dishwasher so that the supply conductors are brought through the wall at the proper point.

The dishwasher is a motor-operated appliance and requires running overcurrent protection. Such protection prevents the motor from burning out if it becomes overloaded. Normally, the required protection is supplied by the manufacturer as an integral part of the dishwasher motor. If the dishwasher lacks protection, the electrician must provide it as part of the installation of the unit. For example, a box-cover unit with appropriate fuses can be installed under the sink. This unit serves as the disconnecting means according to *Sections 422-21, 422-24* and *422-26.*

Grounding

The dishwasher is required to be grounded. The grounding of appliances is described fully in previous units.

PORTABLE DISHWASHERS

In addition to built-in dishwashers, portable dishwashers are available. These portable units have one hose which is attached to the water faucet and a water drainage hose that hangs in the sink. The obvious location for a portable dishwasher is near the sink. Thus, the dishwasher probably will be plugged into the convenience outlet nearest the sink.

NEC *Section 210-7* requires that receptacles installed on 15- and 20-ampere branch circuits shall be of the grounding type.

Portable dishwashers are supplied with a three-wire cord containing two circuit conductors and one grounding conductor and a three-wire grounding-type attachment plug cap. If the three-wire plug cap is plugged into the three-wire grounding-type receptacle, the dishwasher is adequately grounded. Whenever there is a chance that the user may touch the appliance and a grounded surface at the same time – such as a water pipe or a baseboard heating unit – **the equipment must be grounded to reduce the shock hazard.** The exception to *Section 250-45(c)* permits the double insulation of appliances and portable tools instead of grounding. *Double insulation* means that the appliance or tool has two separate insulations between the hot conductor and the person using the device. Although refrigerators, cooking units, water heaters, and other large appliances are not double insulated, most portable hand-held appliances and tools have double insulation. All double-insulated tools and appliances must be marked by the manufacturer to indicate this feature.

DISHWASHER RATINGS

Many dishwashers have heating elements with lower wattage ratings than the example used in this unit. Some dishwashers do not have any heating elements. For these units, the total connected load is less than that of the dishwasher in the dwelling.

The combined loads of the dishwasher and the food waste disposer may be connected to a single 20-ampere circuit. The electrician must split the circuit in a junction box mounted in the cabinet space below the food waste disposer. If a box-cover unit is used and has the proper fuse sizes, then separate overcurrent protection and separate disconnecting means are available. The connections for this circuit are similar to those for the overhead garage door opener circuit (unit 18).

REVIEW

Note: Refer to the Code or the plans where necessary.

FOOD WASTE DISPOSER CIRCUIT $\blacktriangle_{I}$

1. How many amperes does the food waste disposer draw? ________________
2. To what circuit is the food waste disposer connected? ________________
3. Means must be provided to disconnect the food waste disposer. A plumber need not be involved in the electrical connections if the disconnecting means for the disposer is ________________.
4. How is running overcurrent protection provided in most food waste disposer units? ________________
5. When running overcurrent protection is not provided by the manufacturer, dual-element time-delay fuses may be installed in a separate box-cover unit. These fuses are sized at not over ________ percent of the full-load current rating of the motor.
6. Why are flow switches installed on food waste disposers? ________________
7. What Code sections relate to the grounding of appliances? ________________
8. Do the plans show a wall switch for controlling the food waste disposer? ________
9. If a separate circuit supplies the food waste disposer, how many feet (meters) of cable will be required to connect the disposer? Remember that cable may not be run exposed in the utility room. ________________

DISHWASHER CIRCUIT $\blacktriangle_{J}$

1. a. To what circuit is the dishwasher in the residence connected? ________________
   b. What size wire is used to connect the dishwasher? ________________
2. The motor on the dishwasher is: (underline one)
   a. 1/4 hp b. 1/3 hp c. 1/2 hp
3. The heating element is rated at: (underline one)
   a. 750 watts b. 1000 watts c. 1250 watts
4. How many amperes at 120 volts do the following heating elements draw?
   a. 750 watts ________________
   b. 1000 watts ________________
   c. 1250 watts ________________
5. How is the dishwasher in this residence grounded? ________________
6. What type of cord is used on most portable dishwashers? ________________
7. How is a portable dishwasher grounded? ________________
8. What is meant by double insulation? ________________
9. Who is to furnish the dishwasher? ________________

unit 21

Special-Purpose Outlets for the Bathroom Ceiling Heater, Air Conditioner, and Attic Exhaust Fan

OBJECTIVES

After studying this unit, the student will be able to

- explain the operation and switching sequence of the Heat-A-Vent light.
- discuss why heating and cooling systems can be connected to the same circuit.
- install an air conditioner to meet Code requirements for the branch circuit, grounding, and disconnecting means.
- describe the operation of a humidistat.
- install attic exhaust fans with humidistats – both with and without a relay.
- list the various methods of disconnecting appliances.

BATHROOM CEILING HEATER CIRCUIT (▲)$_K$

The first floor bathroom contains an electric baseboard heating unit. In addition, a combination heater, light, and exhaust fan is installed in the bathroom ceiling. The *Heat-A-Vent light* is such a device and is shown in figure 21-1. The symbol (▲)$_K$ represents the outlet for this unit.

The Heat-A-Vent light contains a heating element similar to the surface burners of electric ranges. The device also has a single-shaft motor with one blower wheel, a lamp with a diffusing lens, and a means of discharging air to the outside of the dwelling.

The Heat-A-Vent light specified for the residence is rated at 1475 watts, 120 volts. It is protected by Circuit A23. To meet the manufacturer's recommendations, Circuit A23 is a 15-ampere overcurrent device. Heat-A-Vent units are available with higher ratings in watts that require 20-ampere circuits.

The current rating of the 1475-watt unit is:

$$I = \frac{W}{E} = \frac{1475}{120} = 12.29 \text{ amperes}$$

Wiring

In figure 21-2, the circuit consists of a two-wire, No. 12 TW cable. This cable runs from Panel A to the control switch in the bathroom. A 4-inch square, 2 1/8-inch deep outlet box is installed at this point. The box has a deep two-gang plaster cover and can hold the switch assembly and the required splices. A four-wire cable is connected between the switch and the Heat-A-Vent light.

The Heat-A-Vent unit has a rotary, six-position switch. The following operations are possible once the proper connections are made.

Rotary Switch Position	**Operation**
First	Light only
Second	Exhaust only
Third	Light and exhaust only
Fourth	Heat only
Fifth	Light and heat only
Sixth	Off

Note in the list of operations that it is not possible to heat and exhaust at the same time. A pilot light at the rotary switch location indicates when the heating element is on. A thermostat can be

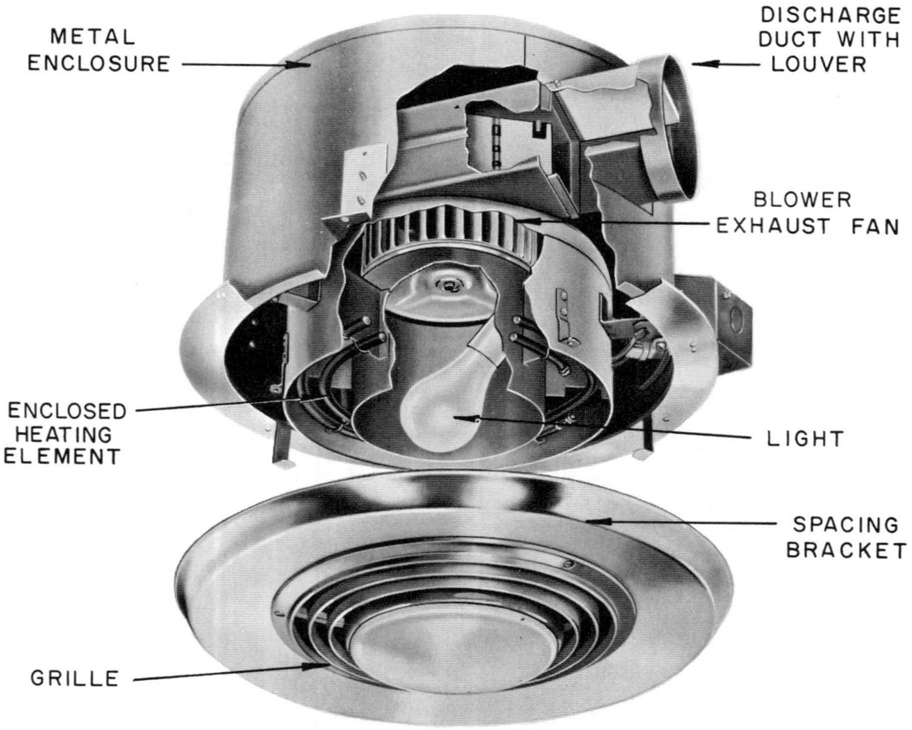

Fig. 21-1 Heat-A-Vent light.

connected at the rotary switch to provide automatic control of the heater. In this dwelling, however, the baseboard electric heater is controlled by a thermostat. Thus, a thermostat is not needed on the Heat-A-Vent unit.

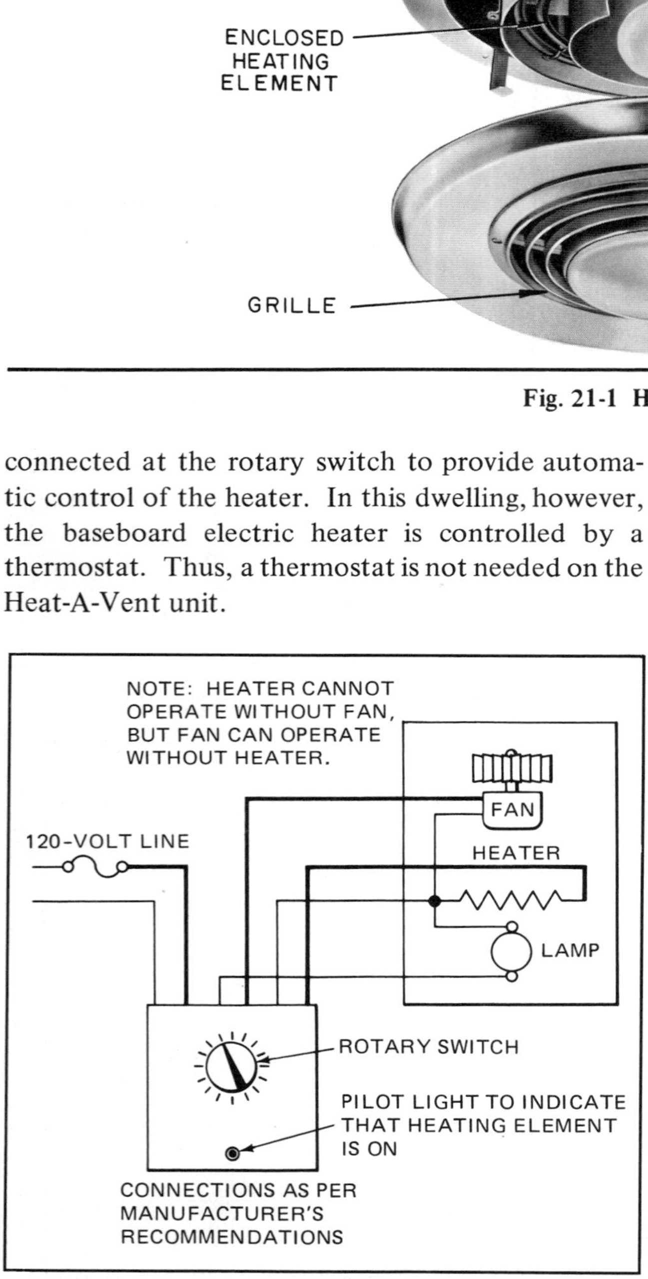

Fig. 21-2 Wiring for Heat-A-Vent light.

Operation of the Unit

The heating element begins to give off heat as soon as the unit is turned on. The heat activates a bimetallic coil attached to a damper section in the housing of the unit. The heat-sensitive coil expands until the damper closes the discharge opening of the exhaust fan. Air is taken in through the outer grille of the unit. The air is blown downward over the heating element. The blower wheel circulates heated air back into the room area.

If the heater is turned off so that the exhaust fan alone is on, air is pulled into the unit and is exhausted to the outside by the blower wheel.

AIR CONDITIONER CIRCUITS ▲L_1 ▲L_2 ▲L_3

Special-purpose outlets ▲L_1 ▲L_2 and ▲L_3 are installed in the residence for three through-the-wall air-conditioning units. Each air conditioner is rated at 8 amperes and 240 volts. The special-purpose outlets also can be used to supply heat pumps.

A *heat pump* cools and dehumidifies on hot days and heats on cool days. To provide heat, the pump picks up heat from the outside air. This heat is added to the heat due to the compression of the refrigerant. The resultant heat is used to warm the air flowing through the unit. This warm air is then delivered to the inside of the building. The direction of flow of the refrigerant is reversed for the heating operation.

Wiring

Each air-conditioner outlet is connected to a separate 15-ampere, 240-volt circuit. No. 12 TW conductors are used. The special-purpose outlets are located as follows:

Ⓐ L_1	Bedroom No. 1	Circuit A9–11
Ⓐ L_2	Bedroom No. 2	Circuit A10–12
Ⓐ L_3	Living Room	Circuit A13–15

The circuit schedule for Panel A shows that the 240-volt, two-pole circuits also supply electric baseboard units. *Section 220-21* permits this arrangement since it is unlikely that the heating and cooling units will be used at the same time. If *Section 220-21* is applied, the total cost of the wiring is less because fewer overcurrent devices are used. The space saved by combining the circuits often means that a smaller electrical panel or load center can be installed.

Circuit Requirements

The Code requirements for room air-conditioning units are found in *Sections 440-60* through *440-64*.

In this residence, three window or through-the-wall air conditioners are to be installed. The basic Code rules for installing these units and their receptacle outlets are as follows:

- the air conditioners must be grounded.
- the air conditioners may be connected using a cord and an attachment plug.
- the air-conditioner rating may not exceed 40 amperes at 250 volts, single phase.
- the rating of the branch-circuit overcurrent device must not exceed the branch-circuit conductor rating, or the receptacle rating – whichever is less.
- the air-conditioner load shall not exceed 80% of the branch-circuit ampacity if no other loads are served.
- the air-conditioner load shall not exceed 50% of the branch-circuit ampacity if other loads are served.
- the attachment plug cap may serve as the disconnecting means.

Circuit Loading

Each air-conditioner outlet is protected by a 15-ampere overcurrent device. *Section 440-62(b)* permits the maximum load on each circuit to be equal to 80 percent of 15 amperes or 12 amperes. Since the air conditioners used are rated at 8 amperes each, they are well within the limits of a 15-ampere branch circuit. See *Table 210-21(b)(2)*.

Receptacles for Air Conditioners

Figure 21-3 shows two types of receptacles that can be used for 240-volt installations. A combination receptacle is shown in figure 21-3B. The upper portion of the receptacle is for 120-volt use and the lower portion is for 240-volt use. Note that the tandem and grounding slot arrangement for the 240-volt receptacles meets the requirements of *Section 210-7(f)*. This section states that receptacle outlets for different voltage levels must be noninterchangeable with each other.

CENTRAL AIR CONDITIONING AND HEATING

The plans show that this dwelling is cooled by three through-the-wall air conditioners. It is heated

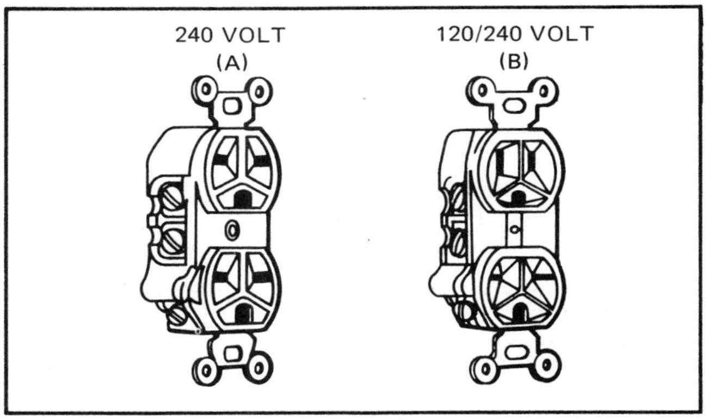

Fig. 21-3 Types of 240-volt receptacles.

by electric baseboard heaters and by several cord-connected heaters used in some of the basement areas.

For discussion purposes, assume that the dwelling has central heating and air conditioning consisting of an electric furnace, a central air-conditioning unit, and/or a heat pump. For this system, the total load is much the same as that of the individual baseboard units and window-mounted air conditioners. The wiring for the central heating and cooling system is shown in figure 21-4. Note that one feeder runs to the electric furnace and another feeder runs to the air conditioner or heat pump outside the dwelling. Low-voltage wiring is used between the inside and outside units to provide control of the systems.

Heat pumps are available as self-contained packages that provide both cooling and heating. In certain climates, the heat pump alone may not provide enough heat. To overcome this condition, the heat pump may contain supplementary heating elements.

NEC *Article 440* covers air-conditioning equipment. It is required by Underwriters' Laboratories that all air conditioners, heat pumps, and electric furnaces be marked with the branch-circuit selection current. In this way, the electrician will be able to determine the conductor sizes and the switch and overcurrent device ratings required by the appliance.

NEC *Section 110-3(b)* requires that listed or labeled equipment be used or installed in accordance with any instructions included in the listing or labeling. For example, the nameplate of the air-conditioning unit for the dwelling reads "Maximum Size Fuse-40 Amperes." Thus, *only* 40-ampere fuses can be used. Forty-ampere circuit breakers are *not* permitted. If the nameplate calls for "Maximum Overcurrent Protection," then either fuses or circuit breakers may be used.

DISCONNECT TO BE WITHIN SIGHT AND READILY ACCESSIBLE

NEC *Section 440-14* requires that the disconnect for the air conditioner must be within sight of the unit and readily accessible. Figures 21-5 and 21-6 illustrate the requirements of *Section 440-14*. For cord- and plug-connected appliances, the cord and plug are considered to be the disconnecting means.

ATTIC EXHAUST FAN CIRCUIT Ⓐ$_M$

An exhaust fan is mounted in the hall ceiling between the bedrooms, figure 21-7. This fan removes humid air from the dwelling. Such an exhaust fan can be used to lower the temperature of the dwelling by as much as 15 to 20 degrees. The exhaust fan in the residence is connected to Circuit A24. This is a 120-volt, 15-ampere circuit in Panel A.

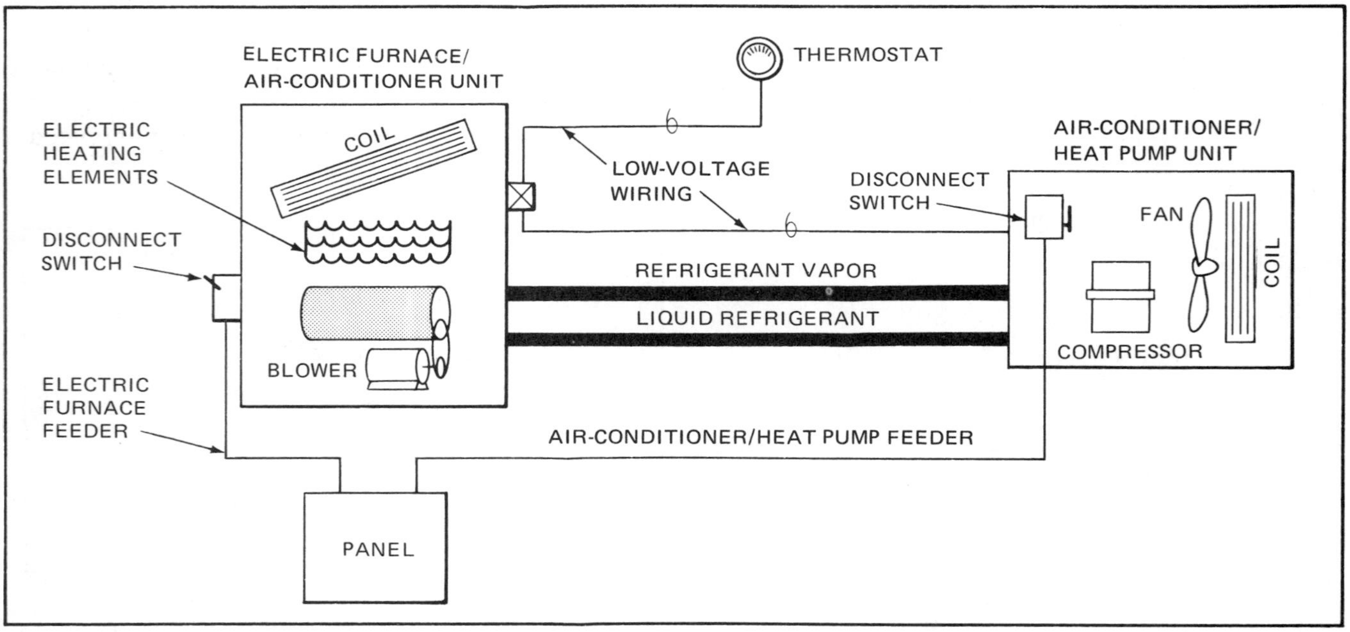

Fig. 21-4 Connection diagram showing typical electric furnace and air-conditioner/heat pump installation.

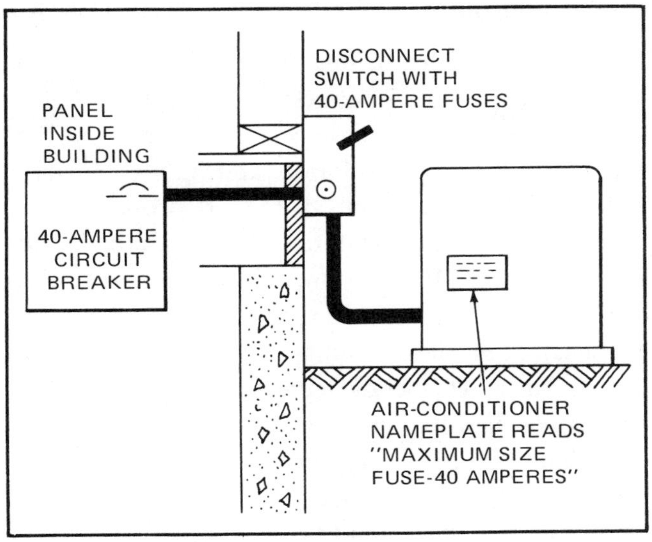

Fig. 21-5 This installation *conforms* to the Code, *Section 440-14*. The disconnect switch is within sight of the unit and contains the 40-ampere fuses called for on the air-conditioner nameplate as the branch-circuit protection.

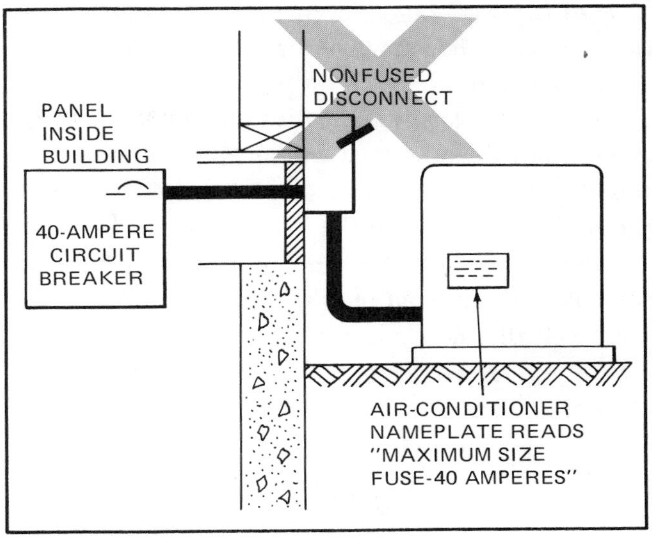

Fig. 21-6 This installation *violates* the Code, *Section 110-3(b)*. Although the disconnect is within sight of the air conditioner, it does not contain fuses. Note that the branch-circuit protection is provided by the 40-ampere circuit breaker inside the building. Note also that the nameplate requires that the branch-circuit protection be 40-ampere fuses maximum. If fused branch-circuit protection were provided at the panel inside the building, the installation would meet Code requirements.

Fan Operation

A 1/4-hp motor is normally used to drive an exhaust fan. Smaller or larger motors are available if the 1/4-hp motor is unsuitable for some reason. A rubber cushion or other shock absorbing material is placed under the motor. To gain quiet operation, a V-belt is used to connect the motor shaft to the fan shaft. Direct drive units are available also.

A prefabricated motor and fan unit may be installed in a dwelling in the attic space directly above a ceiling grille, in the gables, or through the roof. The grille has adjustable louvers to prevent the escape of heat from the living quarters. When the fan is operating, air is drawn upward from the interior of the house and basement and is exhausted through the attic space. Louvers in the gables of the attic permit the air to escape to the outside.

On hot days, the air in the attic may reach a temperature of 150°F or more. Thus, it is desirable to provide a means of taking this attic air to the outside. The heat from the attic air can radiate through the ceiling into the living areas. As a result, the temperature of these areas is raised and the air-conditioning load increases. Personal discomfort increases as well. These problems are minimized by a properly vented exhaust fan.

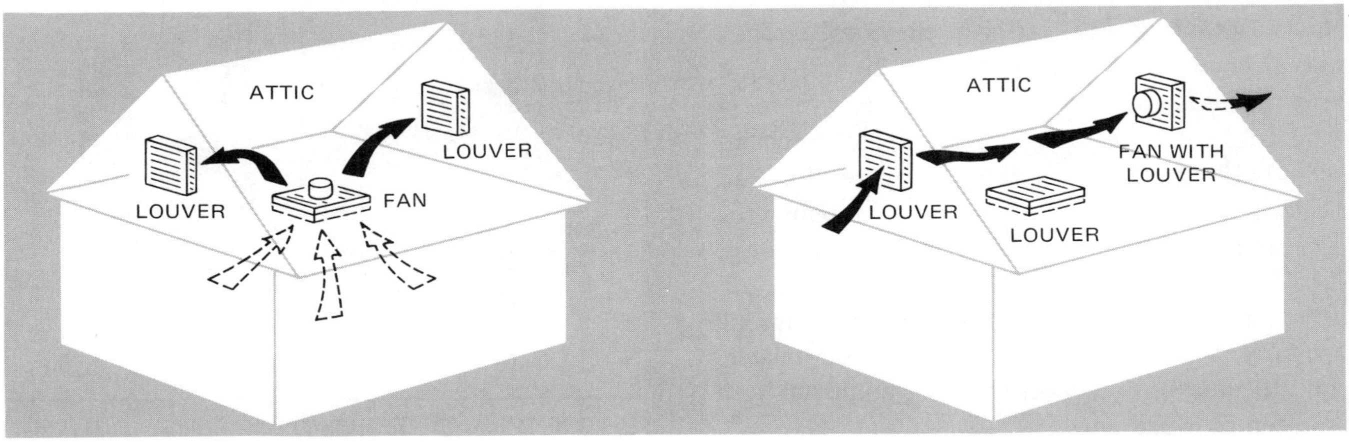

Fig. 21-7 Exhaust fan installation in an attic.

A separate on-off wall switch is installed to control the exhaust fan if auxiliary controls, such as a humidistat, are not required. The switch for the exhaust fan is usually placed about six feet (1.83 m) above the floor so that it will not be confused with other wall switches. The exact height of the switch should be verified with the owner or architect.

Some exhaust fan and motor units have an automatic thermostat mounted on the frame of the fan. The thermostat turns the motor on when the attic temperature reaches 105°F. The fan is turned off when the attic temperature drops to 95°F.

Overcurrent Protection for the Fan Motor

Several methods can be used to provide running overcurrent protection for the 1/4-hp attic fan motor. For example, an overload device can be built into the motor or a combination overload tripping device and thermal element can be added to the switch assembly. Several manufacturers of motor controls also make switches with overcurrent devices. Such a switch may be installed in any standard flush switch box opening (2″ x 3″). A pilot light may be used to indicate that the fan is running. Thus, a two-gang switch box or 1 1/2″ x 4″ square outlet box with a two-gang raised plaster cover may be used for both the switch and the pilot light.

NEC *Section 430-32(c)* states that the overcurrent device must not exceed 125 percent of the full-load running current of the motor. *Table 430-148* shows that the exhaust fan motor has a full-load current rating of 5.8 amperes. Thus, the overcurrent device rating is 5.8 × 1.25 = 7.25 amperes. For a full-load current rating of 5.8 amperes, *Table 430-152* shows that the rating of the branch-circuit overcurrent device shall not exceed 20 amperes if fuses are used (5.8 × 3 = 17.4 amperes; the next standard fuse size is 20 amperes). The overcurrent device rating shall not exceed 15 amperes if circuit breakers are used (5.8 × 2.5 = 14.5 amperes; the next standard circuit breaker size is 15 amperes). Time-delay fuses sized at 125 percent of the full-load ampere rating of the motor provide running overcurrent protection and branch-circuit protection. In addition, the motor may be equipped with a built-in overload device.

Humidistat Control

An electrically heated dwelling usually has a problem with excess humidity due to the "tightness" of the house, because of the care taken in the installation of the insulation and vapor barriers. High humidity is uncomfortable. It promotes the growth of mold and the deterioration of fabrics and floor coverings. In addition, the framing members, wall panels, and plaster of a dwelling may deteriorate because of the humidity. Insulation must be kept dry or its efficiency decreases. A low humidity level can be maintained by automatically controlling the exhaust fan. One type of automatic control is the *humidistat*. This device starts the exhaust fan when the relative humidity reaches a high level. The fan exhausts air until the relative humidity drops to a comfortable level. The comfort level is about 50 percent relative humidity.

The electrician must check the maximum current and voltage ratings of a humidistat before the device is installed. Some humidistats are low-voltage devices and require a relay. Other humidistats are rated at line voltage and can be used to switch the motor directly (a relay is not required). However, a relay must be installed on the line voltage humidistat if the connected load exceeds the maximum allowable current rating of the humidistat.

The switching mechanism of a humidistat is controlled by a nylon element, figure 21-8. This element is very sensitive to changes in humidity. A bimetallic element cannot be used because it reacts to temperature changes only.

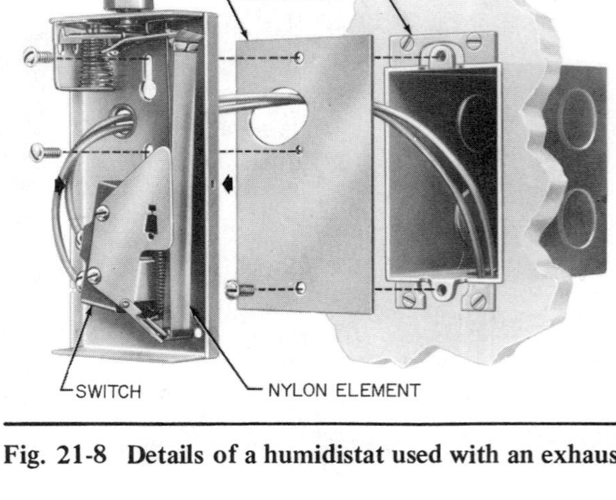

Fig. 21-8 Details of a humidistat used with an exhaust fan.

Wiring

The humidistat and relay in the dwelling are installed using two-wire No. 12 AWG cable. The cable runs from Circuit A24 to a 4-inch square, 1 1/2-inch deep outlet box near the fan, figure 21-9B. A box-cover unit is mounted on this outlet box. The box-cover unit serves as a disconnecting means within sight of the motor as required by *Sections 422-26* and *430-102*. Running overcurrent protection is provided by the dual-element time-delay fuses installed in the box-cover unit.

Mounted next to the box-cover unit is a relay that can carry the full-load current of the motor. Low-voltage thermostat wire runs from this relay to the humidistat. (The humidistat is located in the hall between the bedrooms.) The thermostat cable is a two-conductor cable since the humidistat has a single-pole, single-throw switching action. The line voltage side of the relay provides single-pole switching to the motor. Normally, the white grounded conductor is not broken.

APPLIANCE DISCONNECTING MEANS

To clean, adjust, maintain, or repair an appliance, it must be disconnected to prevent personal injury. *Sections 422-20* through *422-26* outline the basic methods for disconnecting fixed, portable, and stationary electrical appliances. Note that each appliance in the dwelling conforms to one or more of the disconnecting methods listed. The more

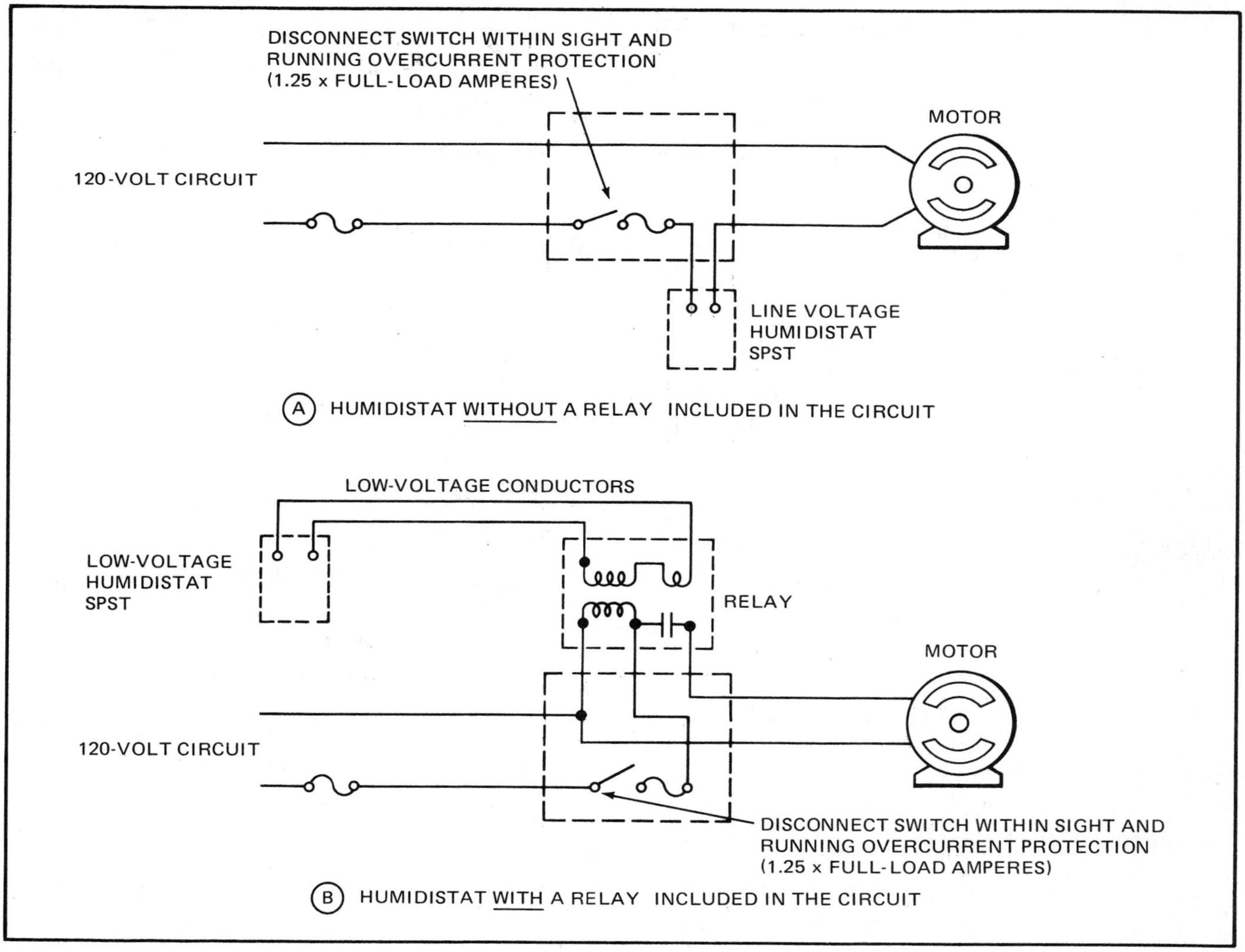

Fig. 21-9 Wiring for the humidistat control.

important Code rules for appliance disconnects are as follows:

- each appliance must have a disconnecting means.
- the disconnecting means may be a separate disconnect switch if the appliance is permanently connected.
- the disconnecting means may be the branch-circuit switch or circuit breaker if the appliance is permanently connected and not over 1/8 hp or 300 volt-amperes.
- the disconnecting means may be an attachment plug cap if the appliance is cord connected.
- the disconnecting means may be the unit switch on the appliance only if other means for disconnection are also provided. That is, in a single-family residence, the service disconnect serves as the other means.
- the disconnect must have a positive on-off position.
- the disconnect must be within sight of a motor-driven appliance where the motor is more than 1/8 hp.

Read *Sections 422-20* through *422-26* for the details of appliance disconnect requirements.

REVIEW

Note: Refer to the Code or the plans where necessary.

BATHROOM CEILING HEATER CIRCUIT ▲$_K$

1. What is the wattage rating of the Heat-A-Vent light? ______________________
2. To what circuit is the Heat-A-Vent light connected? ______________________
3. Why is a 4-inch square, 2 1/8-inch deep box with a two-gang raised plaster ring used for the switch assembly for the Heat-A-Vent light? ______________________

   ____

   ____
4. a. How many wires are required to connect the control switch and the Heat-A-Vent light? ______________________

   b. What size wires are used? ______________________
5. Can the heating element be energized when the fan is not operating? ____________
6. Can the fan be turned on without the heating element? ______________________
7. What device can be used to provide automatic control of the heating element and the fan of the Heat-A-Vent light? ______________________
8. a. What device controls the baseboard electric heating unit? ____________

   b. Where is this device located? ______________________
9. Where does the air enter the Heat-A-Vent light? ______________________
10. Where does the air leave this unit? ______________________
11. Who is to furnish the Heat-A-Vent light? ______________________

AIR-CONDITIONER CIRCUITS ▲L_1 ▲L_2 ▲L_3

1. How many air-conditioner outlets are to be installed in the residence? ____________
2. The air-conditioner outlets are: (underline one)

   a. 120 volts   b. 240 volts
3. Where are the air-conditioner outlets located? ______________________

4. a. When calculating air-conditioner load requirements and electric heating load requirements, is it necessary to add the two loads together to determine the combined load on the system? ______
   b. Explain the answer to part a. ______

5. The total load of an air conditioner shall not exceed what percentage of a separate branch circuit? (underline one)
   a. 75%   b. 80%   c. 125%

6. The total load of an air conditioner shall not exceed what percentage of a branch circuit that also supplies lighting? (underline one)
   a. 50%   b. 75%   c. 80%

7. a. Must an air conditioner installed in a window opening be grounded if a person on the ground outside the building can touch the air conditioner? ______
   b. What Code section governs the answer to part a? ______

8. A 120-volt air conditioner draws 13 amperes. What size is the circuit to which the air conditioner will be connected? ______

9. What is the Code requirement for receptacles connected to circuits of different voltages and installed in one building? ______

10. How many feet (meters) of cable are required to connect all of the air-conditioning outlets? ______

11. When a central air-conditioning unit is installed, and the label states "Maximum size fuse-50 amperes," is it permissible to connect the unit to a 50-ampere circuit breaker? ______

12. When a central air conditioner is connected to a branch-circuit breaker, what type of disconnect must be used within sight of the unit? ______

ATTIC EXHAUST FAN CIRCUIT ⓂM

1. What is the purpose of this fan? ______

2. At what voltage does the fan operate? ______
3. What is the horsepower rating of the fan motor? ______
4. Is the fan direct- or belt-driven? ______
5. How is the fan controlled? ______
6. What is the setting of the running overcurrent device? ______
7. What is the rating of the running overcurrent protection if the motor is rated at 10 amperes? ______
8. What is the basic difference between a thermostat and a humidistat? ______

9. What size conductors are to be used for this circuit? ______
10. How many feet (meters) of cable are required to complete the wiring for the attic exhaust fan circuit? ______

unit 22

Television, Telephone, and Signal Systems

OBJECTIVES

After studying this unit, the student will be able to

- install television outlets so that there is no line voltage interference with the television signal.
- install outlet boxes and outlets, and provide cable or conduit according to telephone company requirements.
- define what is meant by a signal circuit.
- describe the operation of a two-tone chime and a four-note chime.
- install a chime circuit with one main chime and one or more extension chimes.

TELEVISION

Television is a highly technical and complex field. A trained television servicer or installer should be consulted before a television system is installed.

General Wiring

According to the plans for the dwelling, television outlets are installed in the following rooms:

Living Room - 3
Recreation Room - 2
Bedroom No. 1 - 1
Bedroom No. 2 - 1

These outlets may be connected in several ways. The method selected depends upon the proposed locations of the outlets. In general, for installation in dwellings, the electrician uses standard sectional switch boxes or 4-inch square, 1 1/2-inch deep outlet boxes with single-gang raised plaster covers. A box is installed at each point on the system where an outlet is located. Nonmetallic boxes are to be used if the system is to be wired with nonshielded cable, figure 22-1.

Television Installation Methods

A typical television installation is shown in figure 22-2. A master amplifier distribution system is used. The lead-in wire from the antenna is connected to an amplifier placed in an accessible area, such as the basement. Twin-lead, 300-ohm cable runs from the amplifier to each of the outlet boxes. The cable is connected to tap-off units. These units have a 300-ohm input and a 300-ohm output. The tap-off units have terminals or plug-in arrangements so that the 300-ohm, twin-lead cable can be connected between the television outlet and the television receiver.

Faceplates are provided for tap-off units to match conventional switch and convenience outlet faceplates. Combination two-gang faceplates also are available for tap-off units, figure 22-3. Both a 120-volt convenience receptacle and a television outlet can be mounted in one outlet box using this combination faceplate.

However, a combination installation requires that a metal barrier be installed between the 120-volt section and the television section of the outlet box, *Section 810-18(b)*. This barrier prevents line

Fig. 22-1 Nonmetallic boxes and a nonmetallic raised cover.

voltage interference with the television signal. In addition, line voltage cables and lead-in wire must be separated to prevent interference when they are run in the same space within the walls and ceilings. The line voltage cables should be fastened to one side of the space and the lead-in cable should be fastened to the other side.

A residential television system can also be installed using a *multiset coupler*, figure 22-4. The lead-in wire from the antenna is connected to a specific pair of terminals on the coupler. The cables to each television receiver are connected to other pairs of terminals. Two, three, or four television receivers can be connected to a coupler, depending upon the number of terminal pairs provided. When a coupler is used, an amplifier is not required. A coupler system is less expensive than an amplifier distribution system.

A third method of installing a television system uses shielded 75-ohm coaxial cable. The cable is connected to impedance matching transformers with 75-ohm or 300-ohm outputs. Additional cable is then connected between the matching transformer and the television receiver. Many older television receivers have 300-ohm inputs. Newer models generally have 75-ohm inputs. Some manufacturers recommend that shielded coaxial cable be used to prevent interference and keep the color signal strong. When unshielded lead-in wire is used, distorted color television pictures may result. Both shielded and nonshielded lead-in wire deliver good television reception when properly installed.

Cable Installation

Lead-in cables must be installed with care to prevent damage to the cable. The various connec-

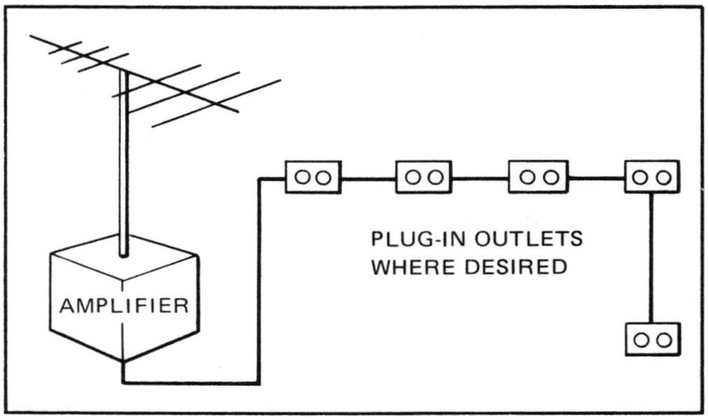

Fig. 22-2 Television master amplifier distribution system.

Fig. 22-3 Typical faceplates.

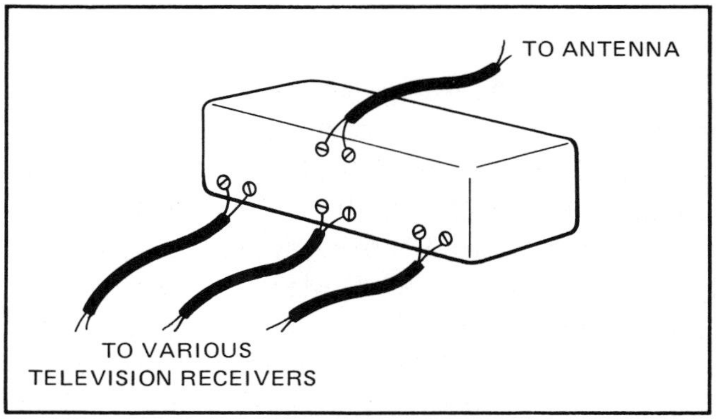

Fig. 22-4 Multiset coupler.

tions and splices must be made carefully so as not to change the spacing of the conductors within the cable. Such changes result in a distorted television signal. Shielded lead-in cables may be run in or near metal pipes and ducts, and close to metal objects without affecting television reception. Unshielded lead-in wires installed close to these items may affect reception. Television cables should be installed during the roughing-in phase of construction so that none of the cables is exposed in the residence.

Television Antennas

Either indoor or outdoor antennas may be used with black and white and color televisions. The front of an outdoor antenna is aimed at the television transmitting station. When there is more than one transmitting station and they are located in different directions, a rotor is installed. A rotor turns the antenna on its mast so that it can face in the direction of each transmitter. The rotor is controlled from inside the building. A four-wire cable is usually installed between the rotor motor and the control unit. The wiring for a rotor may be installed during the roughing-in phase of construction. Plug-in jacks (similar to television plug-in jacks) can be used to flush mount the rotor.

TELEPHONES

If several telephones are used in a residence, most telephone companies require that at least one of these telephones be permanently installed. The remaining phones may be portable. This means that they can be plugged into any of the phone jacks furnished by the telephone company. If a permanent stationary telephone is not installed, the telephone company must install a permanent ringer as an audible signal. For example, if all of the phones are portable, there may be a time when none of the phones is connected to phone jacks. As a result, there will be no audible signal.

The electrical plans show that five telephone locations are provided as follows:

Bedroom No. 1 (NE wall)
Bedroom No. 2 (SW wall)
Living room (in shelf next to fireplace)
Kitchen (SW wall)
Recreation room (NE wall)

The permanent phone is installed on the southwest wall of the kitchen. The phone is mounted 5′0″ (1.52 m) to center from the finish floor. The remaining four jacks are flush mounted and trimmed with faceplates, figure 22-5. These faceplates match the switch and convenience outlet faceplates used elsewhere in the residence.

Telephone System Installation

The installation of a residential telephone system is not regulated by the National Electrical Code. Installation requirements are provided by the local telephone company. Generally, the telephone company will provide all materials and perform all work involved in the installation. That is, the company furnishes and installs the switch and outlet boxes, multiconductor cables, phone jacks, faceplates, service drop, lightning arrester (protector), connecting blocks, and any other necessary material.

As an alternative, a telephone company engineer may inform the electrician of the number of outlets required and the preferred locations for these outlets. The electrician then mounts the switch or outlet boxes and the proper raised plaster covers. In some areas, a raised plaster cover alone can be used for the installation. The electrician drops a fish tape (fishing wire) from each box to the basement. Cables are attached to the fish tape which is then withdrawn from the outlet box. The cable is pulled up to the outlet box where final connections are made. This last step may take some time since the insulation tends to obstruct the fish tape.

Telephone outlets may be installed by running 1/2-inch conduit (EMT) or 3/4-inch conduit, depending upon telephone company recommendations, from each outlet location to an accessible point in the basement. This method is preferred since the installer can then pull the cables through

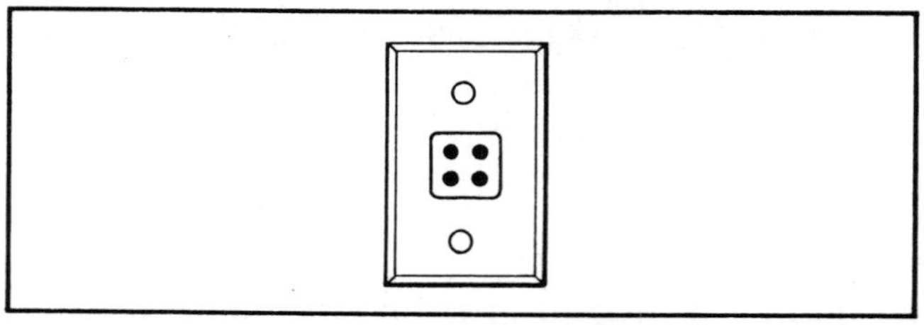

Fig. 22-5 Telephone jack plate.

the conduit, run them exposed in the basement, and terminate them at the proper location.

Conduit systems are used for commercial installations, but are rarely used for residential telephone systems. In some cases, however, the telephone company may require that a residential installation be roughed in using conduit alone. For example, conduit is used when the installation is in a building (of masonry construction) which lacks a basement. Regardless of how the system is installed, the telephone company must be consulted before work is started.

SIGNAL SYSTEM

A signal circuit is described in the National Electrical Code as any electrical circuit that energizes signaling equipment, or one which is used to transmit certain types of signaling messages. Signaling equipment includes such devices as doorbells, buzzers, code-calling systems, and signal lights.

Door Chimes

Present-day dwellings often use chimes rather than bells or buzzers to announce that someone is at a door. A musical tone is sounded rather than a harsh ringing or buzzing sound. Chimes are available in single-note, two-note, eight-note (four-tube), and repeater tone styles. In a repeater tone chime, both notes sound as long as the push button is depressed. In an eight-note chime, contacts on a motor-driven cam are arranged in sequence to sound the notes of a simple melody when the chime button is pushed. This type of chime is usually installed in dwellings having three entrances. The chime can be connected so that the eight-note melody sounds for the front door, two notes sound for the side door, and a single note sounds for the rear door. Chimes are also available with clocks and lights.

Electronic chimes may relay their chime tones through the various speakers of an intercom system. When any chime is installed, the manufacturer's instructions must be followed.

Residence Chimes

The plans show that two chimes are installed in the residence. Two-note chimes are used. Each chime has two solenoids and two iron plungers. When one solenoid is energized, the iron plunger is drawn into the opening of the solenoid. A plastic or wooden peg in the end of the plunger strikes one chime tone bar. When the solenoid is deenergized, spring action returns the plunger where it comes to rest against a soft felt pad so that it does not strike the other chime tone bar. Thus, a single chime tone sounds. As the second solenoid is energized, one chime tone bar is struck. When the second solenoid is deenergized, the plunger returns and strikes the second tone bar. A two-tone signal is produced. The plunger then comes to rest between the two tone bars.

Figure 22-6 shows four push button styles used for chimes. Many other styles are available. Figure 22-7 shows several wall-mounted cover plates for chime units. The symbols used to indicate push buttons and audible signals on the plans are shown in figure 22-8.

Transformers

The transformers used to operate door chimes have a greater capacity than the transformers used with bell and buzzer circuits, figure 22-9. The

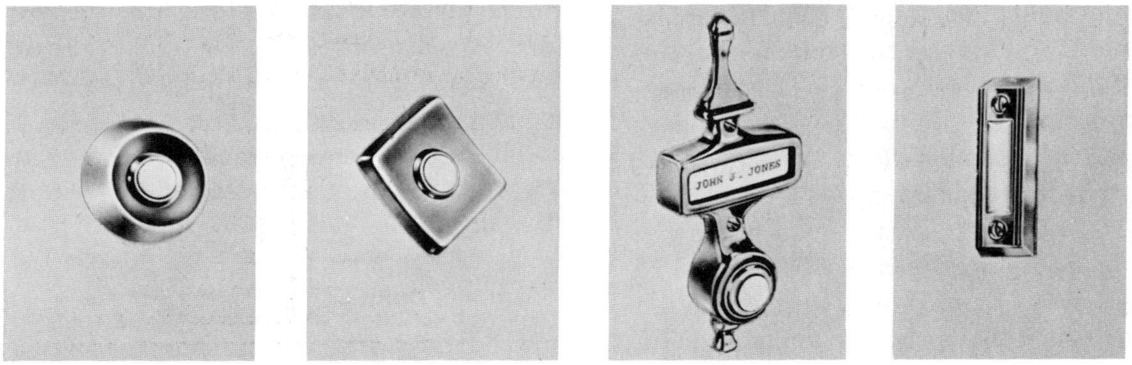

Fig. 22-6 Push buttons for door chimes.

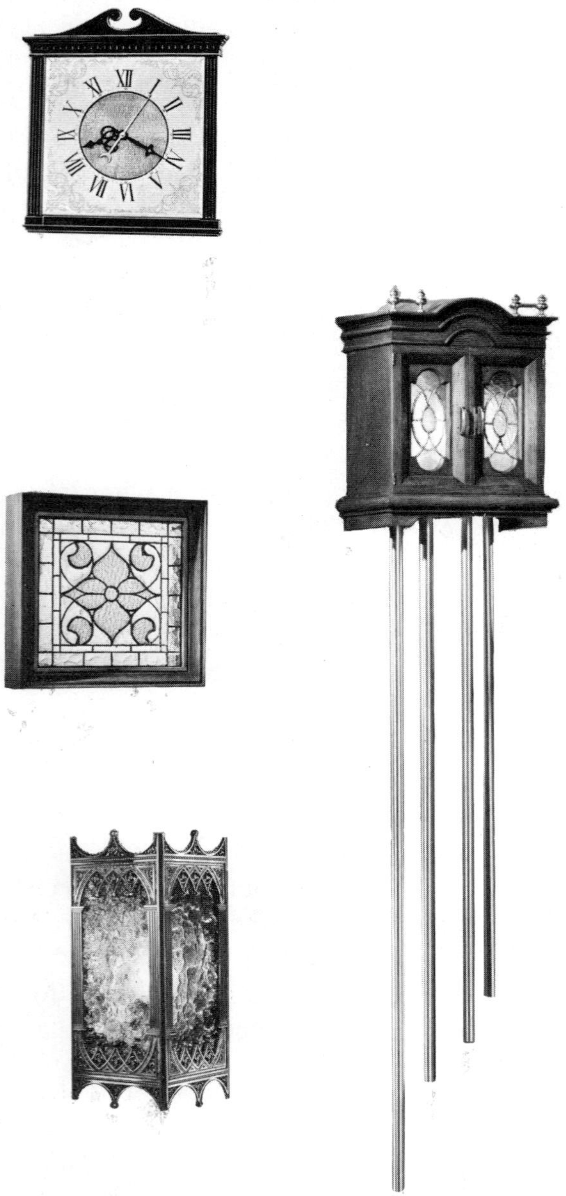

Fig. 22-7 Typical residential door chimes.

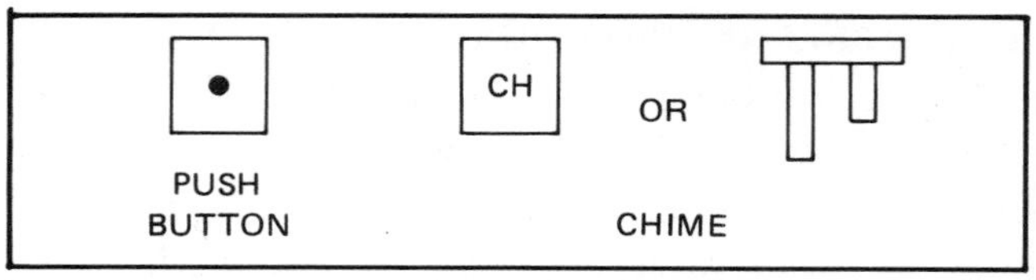

Fig. 22-8 Symbols for signal circuits.

Fig. 22-9 Chime transformer.

voltage output of chime transformers ranges between 10 and 24 volts. These transformers are rated from 5 to 30 volt-amperes (watts). Bell transformers have a voltage output range of 6 to 10 volts and a rating of 5 to 20 volt-amperes (watts).

Chime transformers used in dwellings are available with a 16-volt rating. Transformers which give a combination of voltages, such as 4, 8, 12, and 24 volts, also are used. Transformers with output ratings of up to 100 volt-amperes are available. NEC *Section 725-31* limits the ratings of transformers supplying Class 2 systems to 100 volt-amperes. (Class 2 circuits are defined in *Article 725.*)

Underwriters' Laboratories list two types of chime transformers that are normally used in Class 2 locations. The *energy-limiting transformer* is designed to limit the short-circuit current to a maximum of 8 amperes. The *nonenergy-limiting transformer* is rated at 100 volt-amperes or less. This transformer has an overcurrent protective device that limits the voltage and current to the values specified in *Section 725-31*. The open circuit voltage limitation of both types of transformers must not exceed 30 volts. Most transformers suitable for use in dwellings have a built-in thermal overload device. Whenever a short circuit occurs in the bell-wire circuit, the overload device opens and closes repeatedly until the short is cleared.

Additional Chimes

To extend a chime system to cover a larger area, a second or third chime may be added. In the residence, two chimes are used: one chime is mounted in the front hall and a second (extension) chime is mounted in the recreation room. The extension chime is wired in parallel to the first chime. The wires are run from one chime terminal board to the terminal board of the other chime. The terminals are connected as follows: transformer to transformer, front to front, and rear to rear, figure 22-10.

When chimes are added to a chime circuit, a transformer with a higher rating may be required to energize the greater number of solenoids being used at one time. Also, more wire is used in the circuit and the voltage drop and power loss are greater. This increase is the result of resistance in the wire between the chimes and the transformer. *Section 725-32* does not allow transformers supplying Class 2 systems to be connected in parallel unless listed for interconnection.

If a buzzer (or bell) and a chime are connected to a single transformer and are used at the same time, the transformer will put out a fluctuating voltage. This condition does not allow either the buzzer or the chime to operate properly. The use of a transformer with a larger rating may solve this problem.

The wattage consumption of chimes varies with the manufacturer. Typical ratings are as follows:

TYPE OF CHIME	POWER CONSUMPTION
Standard two-note	10 watts
Repeating chime	10 watts
Internally lighted, two lamps	10 watts
Internally lighted, four lamps	15 watts
Combination chime and clock	15 watts
Motor-driven chime	15 watts
Electronic chime	15 watts

Transformers with ratings of 5, 10, 15, 20, and 30 watts (VA) are available. For a multiple chime installation, wattage ratings for the individual chimes are added. The total value is the minimum transformer rating needed to do the job properly. If there are still technical questions or problems, check the literature supplied by the chime and transformer manufacturers.

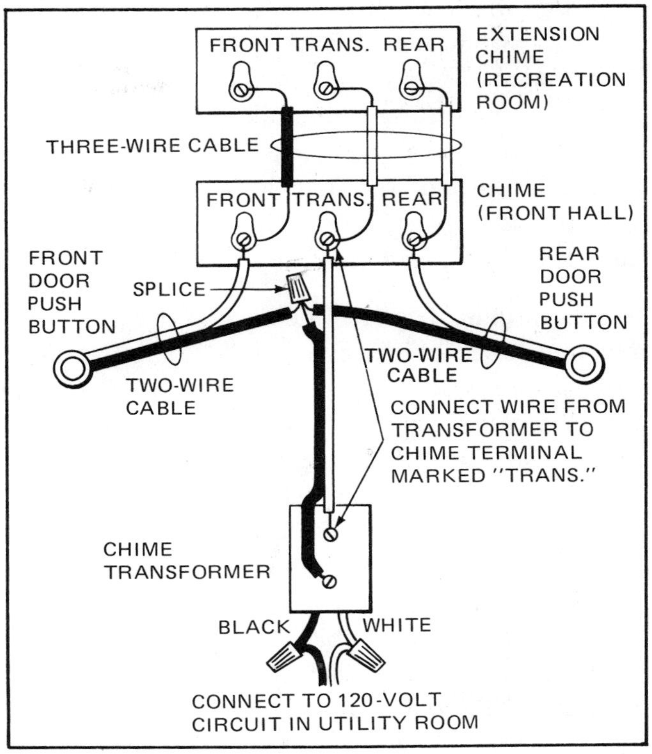

Fig. 22-10 Circuit for chime installation.

Wiring for Chime Installation

Bell Wire and Cable. The wire used for low-voltage bell and chime circuits is called bell wire, annunciator wire, or thermostat wire. One type of wire consists of a copper conductor covered with two layers of cotton wrapped in opposite directions. These layers can be tied off to prevent the insulation from unraveling at terminals or splices. Both layers of cotton are saturated with paraffin. Another type of wire is insulated with a thermoplastic compound (Type T). Because of the low voltages involved, the use of paraffin and thermoplastic insulations is satisfactory. The current required for bell and chime circuits is small and No. 18 AWG conductors are used.

Multiconductor cables consist of two, three, or more single wires covered with a single protective insulation. This type of cable is often used for electrical installations because there is less danger of damage to individual wires and it gives a neat appearance to the wiring. The conductors within the cables are color coded to make circuit identification easy.

Bell wire and cable may be fastened directly to surfaces with insulated staples or cleats, or they may be installed in raceways. The job requirements will determine how the conductors are to be attached. In the residence in the plans, the bell wire is run along the sides of the floor joists in the basement and on the sides of studs in the walls. The installation of low-voltage wiring is covered in unit 29.

Wiring Circuit. The circuit shown in figure 22-10 is recommended for this chime installation because it provides a hot low-voltage circuit at the front hall location. However, it is not the only way in which these chimes may be connected.

Figure 22-10 shows that a two-wire cable runs from the transformer in the utility room to the front hall chime. A two-wire cable then runs from the chime to both the front and rear door push buttons. A three-wire cable also runs between the front hall chime and the recreation room chime. Because of the hot, low-voltage circuit at the front hall location, a chime with a built-in clock can be used. A four-conductor cable may be run to the extension chime so that the owner may install a clock-chime at this location also.

The plans show that the chime transformer is mounted on one of the ceiling boxes in the utility room. The line voltage connections are easy to make here. Some electricians prefer to mount the chime transformer on the top or side of the distribution panel or load center. The electrician decides where to mount the transformer after considering the factors of convenience, economy, and good wiring practice.

National Electrical Code Rules for Signal Systems

Bell wire with low-voltage insulation must not be installed in the same enclosure or raceway with light or power conductors. Bell wires must be not closer than two inches (50.8 mm) to open light or power conductors unless the bell wires are permanently separated from the other conductors by some approved insulation used in addition to the insulation on the wire. Such insulation is provided by the outer jacket on nonmetallic-sheathed cable, UF cable, and armored cable. Furthermore, bell wire with low-voltage insulation may not enter an outlet box or switch box containing light or power conductors unless a metal barrier is used to separate the two types of wiring.

National Electrical Code requirements for low-voltage bell circuits are contained in *Article 725, Sections 725-31* through *725-42*.

REVIEW

Note: Refer to the Code or the plans where necessary.

TELEVISION CIRCUIT

1. How many television outlets are installed in this residence? ____________________
2. What type of boxes are recommended when nonshielded lead-in wire is used? ________

   ____

3. What determines the design of the faceplates used? ____________________
4. What must be provided when installing a television outlet and convenience receptacles in one wall box? Why? ____

   ____

   ____

5. From a cost standpoint, which system is more economical to install: a master amplifier distribution system or a multiset coupler? Explain the basic differences between these two systems. ____

   ____

   ____

6. How many wires are in the cable used between a rotor and its controller? ______

7. What precautions should be noted when installing cable for TV systems? ______

______

______

TELEPHONE SYSTEM

1. How many locations are provided for telephones in the residence? ______

2. At what height are the telephone outlets mounted? ______

3. Sketch the symbol for a telephone outlet. ______

4. Is the telephone installation regulated by the National Electrical Code? ______

5. a. Who is to furnish the outlet boxes required at each telephone outlet? ______
   b. Who is to furnish the faceplates? ______

6. Who is to furnish the telephone? ______

7. Who does the actual installation of the telephone equipment? ______

8. How are the telephone cables concealed in this residence? ______

9. a. What must the electrician provide in the conduits run to the basement from each telephone outlet location? ______
   b. What size conduit is installed? ______

10. a. How many 4 inch square, 1 1/2-inch deep outlet boxes are required for the telephone installation? ______
    b. How many raised plaster covers are required? ______
    c. How many faceplates are required? ______
    d. Must the faceplates match the switch and convenience outlet plates? ______

11. a. Approximately how many feet (meters) of conduit are needed to rough in the telephone outlets? ______
    b. How many connectors are required? ______
    c. How many conduit straps are required? ______

SIGNAL SYSTEM

1. What is a signal circuit? ______

2. What style of chimes is used in this residence? ______

3. a. How many solenoids are contained in a two-tone chime? ______
   b. What closes the circuit to the solenoid of a chime? ______

4. Explain briefly how two notes are sounded by depressing one push button (when two solenoids are provided). ______

______

______

______

______

5. a. Sketch the symbol for a push button. ______
   b. Sketch the symbol for a chime. ______

6. a. At what voltage do residence chimes generally operate? ____________
   b. How is this voltage obtained? ____________
7. What is the maximum volt-ampere rating of transformers supplying Class 2 systems? ____________
8. What two types of chime transformers for Class 2 systems are listed by Underwriters' Laboratories? ____________
9. How is the extension chime connected with the front hall chime, in series or in parallel? ____________
10. How many bell wires terminate at:
    a. the transformer? ____________
    b. the front hall chime? ____________
    c. the extension chime? ____________
    d. each push button? ____________
11. a. What change in equipment may be necessary when more than one chime is connected to sound at the same time on one circuit? ____________
    b. Why? ____________
12. How do electricians take advantage of the fact that the two layers of cotton covering on a bell wire are wrapped in opposite directions? ____________
13. What are the two basic types of insulation permitted on low-voltage wires? ____________
14. What size wire is installed for signal systems of the type in this residence? ____________
15. a. How many wires are run between the front hall chime and the extension chime in the recreation room? ____________
    b. How many wires are required to provide a hot low-voltage circuit at the extension chime? ____________
16. Why is it recommended that the low-voltage secondary of the transformer be run to the front hall chime location and separate two-wire cables be installed to each push button? ____________
17. Is it permissible to install low-voltage systems in the same raceway or enclosure with light and power wiring? ____________
18. a. Where is the transformer in the residence mounted? ____________
    b. To what circuit is the transformer connected? ____________
19. a. How many feet (meters) of two-conductor bell wire cable are required? ____________
    b. How many feet (meters) of three-conductor bell wire cable are required? ____________
20. How many insulated staples are needed for the bell wire if it is stapled every two feet (610 mm)? ____________

unit 23

Electric Heating

OBJECTIVES

After studying this unit, the student will be able to

- list the advantages of electric heating.
- describe the components and operation of electric heating systems.
- describe thermostat control systems for electric heating units.
- install electric heaters with appropriate temperature control according to National Electrical Code rules.

Electric heating units are installed in this residence. The installation meets the requirements of NEC *Article 424, Fixed Electric Space Heating Equipment*.

This text cannot cover in detail the methods used to calculate heat loss and the wattage required to provide a comfortable level of heat in the building. For this residence, the total required wattage for the residence is estimated at 15 500 watts. Depending upon the location of the residence (in the Northeast or Midwest, for example), the estimated heating load will vary.

Electric heating has gained wide acceptance when compared with other types of heating systems. It has a number of advantages over the other types of heating systems. Electric heating is flexible because each room can have its own thermostat. Thus, one room can be kept cool, while an adjoining room is warm. This type of temperature control for a gas- or oil-fired central heating system is complex and very expensive.

Electric heating is safer than heating with fuels. The system does not require storage space, tanks, chimneys, or filters. Electric heating is quiet. Electric heat does not add or remove anything from the air. As a result, electric heat is cleaner. This type of heating is considered to be healthier than fuel heating systems which remove oxygen from the air. The only moving part of an electric heating system is the thermostat. This means that there is a minimum of maintenance.

If an electric heating system is to be used, more insulation should be added. The increased insulation helps to keep the residence cool during the hot summer months. The cost of the extra insulation is offset through the years by the decreased burden on the air-conditioning equipment. Energy conservation measures require the installation of proper and adequate insulation.

TYPES OF ELECTRIC HEATING SYSTEMS

Electric heating units are available in baseboard, wall-mounted, and floor-mounted styles. These units may or may not have built-in thermal overload protection. The type of unit to be installed depends upon structural conditions and the purpose for which the room is to be used.

Another method of providing electric heating is to embed resistance-type cables in the plaster of ceilings, *Sections 424-34* through *424-48*.

Electric heat can also be supplied by an electric furnace and a duct system similar to the type used on conventional hot-air central heating systems. The heat is supplied by electric heating elements rather than the burning of fuel. Air conditioning, humidity control, air circulation, and zone control can be provided on an electric furnace system (see *Sections 424-57* through *424-66*). Heat pumps are another method of supplying heat. See unit 21 for a discussion of heat pumps, air conditioning, and electric furnace connections.

CONTROL OF ELECTRIC HEATING SYSTEMS

Line voltage thermostats can be used to control the heat load for most electric heating systems, figure 23-1. Common ratings for line voltage thermostats are 2500 watts, 3000 watts, and 5000 watts. (Other ratings are available, however.) The electrician must check the nameplate ratings of the heating unit and the thermostat to insure that the total connected load does not exceed the thermostat rating.

The amperage limit of a thermostat is found using the wattage formula. For example, if a thermostat has a rating of 2500 watts at 120 volts, or 5000 watts at 240 volts, the current value is:

$$I = \frac{W}{E} = \frac{2500}{120} = 20.8 \text{ amperes}$$

or

$$I = \frac{5000}{240} = 20.8 \text{ amperes}$$

The total connected load for this thermostat must not exceed 20.8 amperes.

When the connected load is larger than the rating of a line voltage thermostat, or when the control relays are to be placed in one location, low-voltage thermostats may be used. In this case, a relay must be connected ahead of the thermostat. The low-voltage contacts of the relay are connected to the thermostat. The line-voltage contacts of the relay are used to switch the actual heater load. Low-voltage thermostat cable is run between the low-voltage terminals of the relay and the thermostat.

When the connected load is larger than the maximum current rating of a thermostat and relay combination, the thermostat can be used to control a heavy-duty relay. An example of such a relay is the magnetic switch used for motor controls. The load side of the magnetic switch feeds a distribution panel containing as many 15- or 20-ampere circuits as needed for the connected load. A 40-ampere load may be divided into three 15-ampere circuits and still be controlled by one thermostat. The proper overcurrent protection is obtained by dividing the 40-ampere circuit into several circuits having lower ratings.

CIRCUIT REQUIREMENTS FOR BASEBOARD UNITS

Figure 23-2 shows the type of baseboard electric heating units used in this residence. These units are rated at 240 volts. Although these units

Fig. 23-1 Thermostats for electric heating systems.

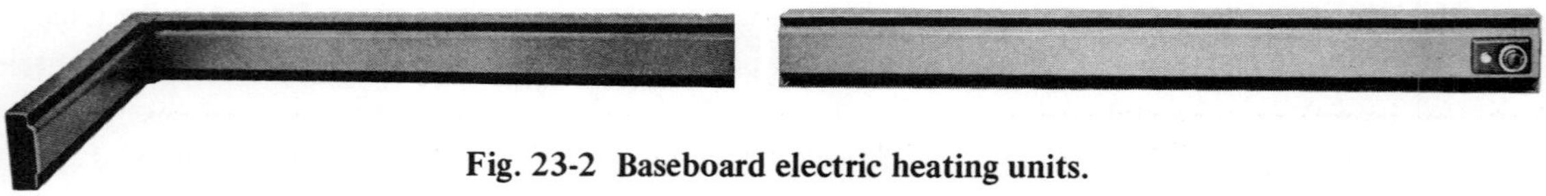

Fig. 23-2 Baseboard electric heating units.

are the same size for each room, it is permissible to install units having different ratings in different rooms, depending upon requirements.

The branch circuit for each heating unit or group of units is shown in figure 23-3. A two-wire cable (armored cable or nonmetallic-sheathed cable with ground) runs from a 240-volt, two-pole circuit in Panel A to the outlet box or switch box installed at the thermostat location. A second two-wire cable runs from the thermostat to the junction box on the heater unit. The proper connections are made in this junction box. Most heating unit manufacturers provide knockouts at the rear and on the bottom of the junction box. The supply conductors can be run through these knockouts. Most baseboard units also have a channel or wiring space running the full length of the unit, usually at the bottom. When two or more heating units are joined together, the conductors are run in this wiring channel. Most manufacturers indicate the type of wire required for these units because of conductor temperature limitations.

The heating units may be placed at different points in the room. The units in the living room and the dining room are examples of this arrangement. Generally, a separate cable is run from each heating unit to the outlet or switch box at the thermostat where the proper connections are made, figure 23-4.

If the connected heater load exceeds the rating of a line voltage thermostat, a combination low-voltage thermostat and relay must be installed, figure 23-5. The total length of the cable run between the main panel and the heater is usually less than when line voltage thermostats are installed.

Most wall and baseboard heating units are available with built-in thermostats (figure 23-2). The supply cable for such a unit runs from the main panel to the junction box on the unit.

Some heaters have convenience receptacles. Underwriters' Laboratories states that when receptacle sections are included with the other components of baseboard heating systems, they must be supplied separately using conventional wiring methods.

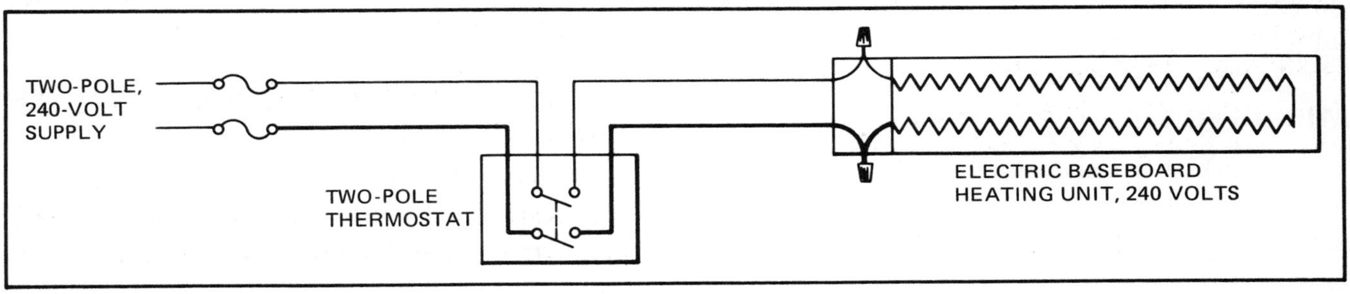

Fig. 23-3 Wiring for a single baseboard electric heating unit.

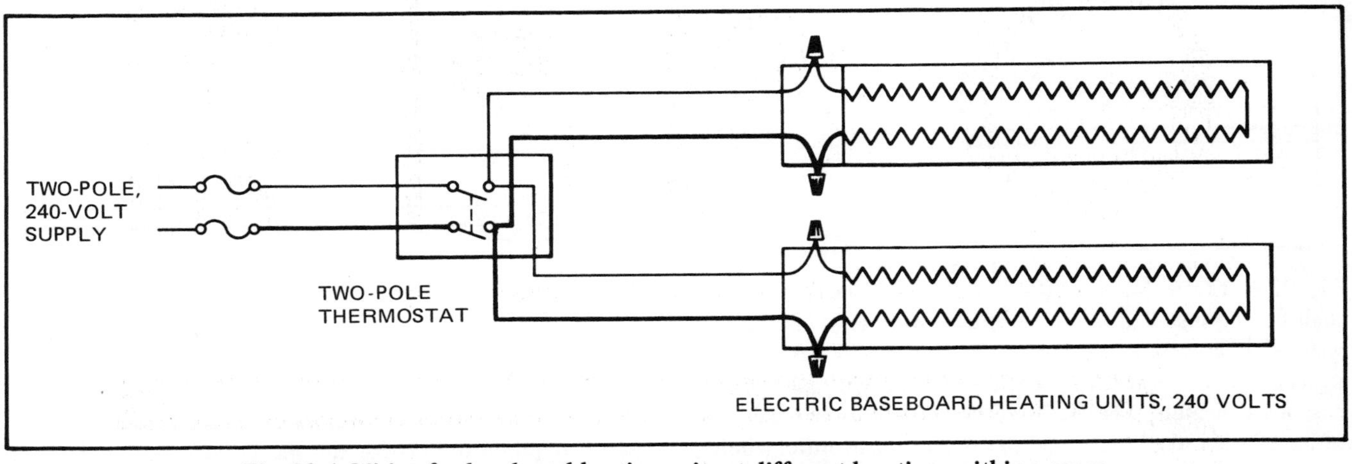

Fig. 23-4 Wiring for baseboard heating units at different locations within a room.

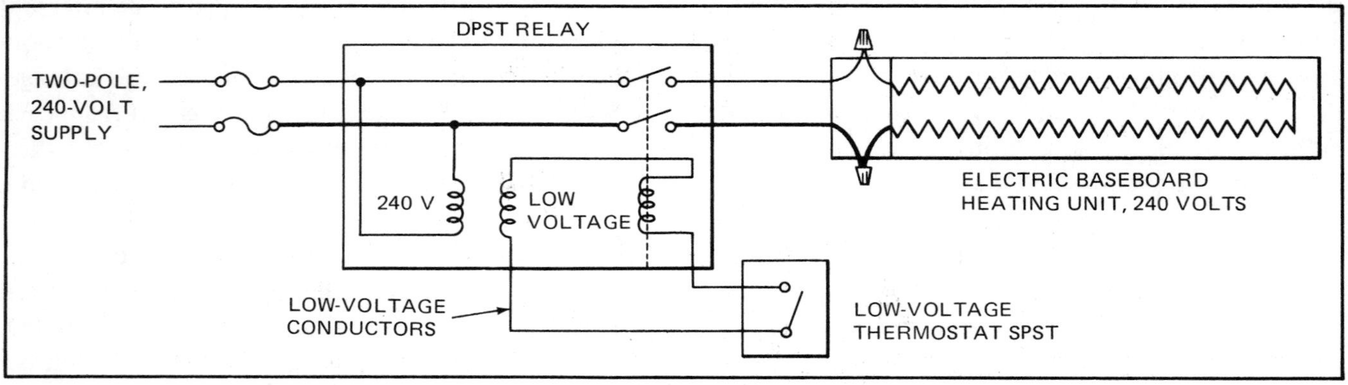

Fig. 23-5 Wiring for baseboard electric heating unit having a low-voltage thermostat and a relay.

LOCATION OF ELECTRIC BASEBOARD HEATERS

The electrician must study the plans and specifications to determine the best locations for the receptacle outlets and the baseboard heating units, figure 23-6. Some of the receptacle outlets to be installed are part of electric baseboard heating units. The use of these outlets provides the proper spacing of receptacle outlets in a given room as required by *Section 210-52(a)*, figure 23-7. Refer to unit 3 for more information on the installation of receptacle outlets.

WIRING FOR HEATER CIRCUITS

No. 12 TW conductors are used in the residence for each circuit that supplies electric heating units. This conductor size is more than adequate for the loads to be served. When larger conductors are used, the voltage drop on each circuit is much smaller than the voltage drop that would result from the use of No. 14 TW conductors.

The workshop and utility room have portable electric heaters with built-in thermostats. These heaters are rated at 1750 watts and 120 volts. The heaters are suspended from the ceiling near their 20-ampere, 120-volt receptacles. (The circuits for the heater receptacles are covered in unit 16.)

The branch-circuit conductors and the overcurrent devices must be at least 125 percent of the heater's rating, *Section 424-3(b)*

The electric heating load used to determine the size of the service-entrance feeder is computed at 100% of the total connected fixed space heating load, *Section 220-15*.

The specifications indicate that the total connected load for the residence is 15 500 watts, figure 23-8. The method of calculating this value is covered in unit 28, "Service-entrance Calculations."

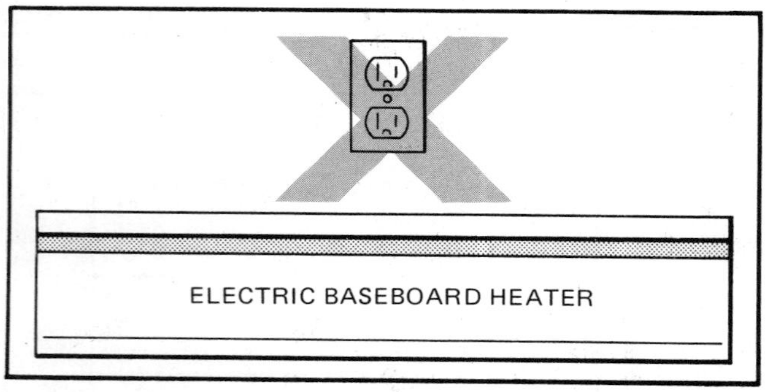

Fig. 23-6 CODE VIOLATION. Baseboard electric heaters shall not be installed below receptacle outlets. This Underwriters' Laboratories Standards requirement is part of the instructions furnished by the manufacturer with all electric baseboard heaters. See NEC *Section 110-3(b)*. In this type of installation, cords attached to the receptacle outlet would hang over the heater, creating a fire hazard.

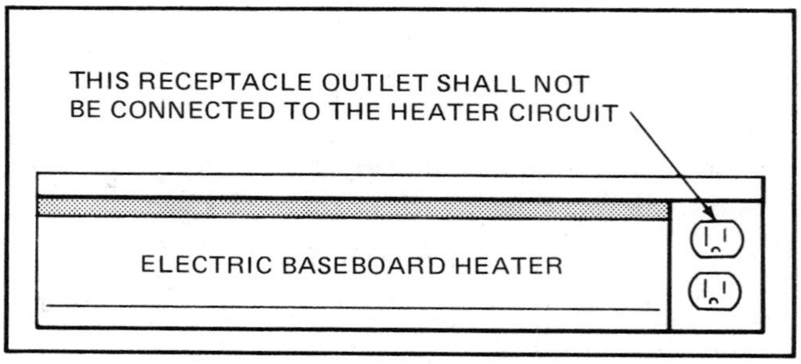

Fig. 23-7 Factory-mounted receptacle outlets on permanently installed electric baseboard heaters may be counted as the required outlets for the space occupied by the baseboard unit. See NEC *Section 210-52(a)*.

Room	Wattage	Volts	Circuit No.	Ampere Rating	Poles	Wire Size	Thermostat
Living Room	2500	240	A5-7	20	2	No. 12	Yes
Dining Room	750	240	A5-7	20	2	No. 12	LR thermostat
Kitchen	1250	240	A10-12	15	2	No. 12	Yes
Front Hall	500	240	A9-11	15	2	No. 12	Yes
Bedroom No. 1	1500	240	A2-4	15	2	No. 12	Yes
Bedroom No. 2	1500	240	A1-3	15	2	No. 12	Yes
Bathroom	500	240	A9-11	15	2	No. 12	Yes
Recreation Room	3000	240	A6-8	20	2	No. 12	Yes
Lavatory	500	240	A13-15	15	2	No. 12	Yes
Workshop	1750	120	A14	20	1	No. 12	Built into heater
Utility Room	1750	120	A16	20	1	No. 12	Built into heater
TOTAL	15 500						

Fig. 23-8 Summary of electric heating circuits.

Marking the Identified Conductor

Two-wire cable contains one white and one black conductor. Since 240-volt circuits supply the electric baseboard heaters, it would appear that the use of two-wire cable is in violation of the Code. However, *Section 200-7* shows that this is not a Code violation.

According to *Section 200-7*, two-wire cable may be used for the 240-volt heaters if the white identified conductor is made unidentified by paint, colored tape, or other effective means. This step is necessary since it may be assumed that an unmarked white conductor is a grounded conductor having no voltage to ground. Actually, the white wire is connected to a "hot" phase and has 120 volts to ground. **Thus, a person can be subject to a harmful shock by touching this wire and the grounded baseboard heater (or any other grounded object) at the same time.** Most electrical inspectors accept black paint or tape as a means of changing the color of the white conductor. The white identified conductor must be made permanently unidentified at the electric heater terminals and at the panels where these cables originate.

Unit 21 shows that Circuits A9-11, A10-12, and A13-15 serve both the heating units and the air-conditioning units. NEC *Section 220-21* permits this arrangement because it is unlikely that the air-conditioning units and the heating units will be operating at the same time.

REVIEW

Note: Refer to the Code or the plans where necessary.

1. a. What is the allowance in watts made for electric heat in this dwelling? ____________
   b. What is the value in amperes of this load? ____________
2. What are some of the advantages of electric heating? ____________
3. List the different types of electric heating system installations. ____________
4. There are two basic voltage classifications for thermostats. What are these classifications? ____________

5. What device is required when the total connected load exceeds the maximum rating of a thermostat? ______

6. The electric heat in this residence is provided by what type of units (with the exception of the workshop and utility room)? ______

7. At what voltage do the first floor electric heating elements operate? ______

8. a. How many circuits supply the baseboard electric heating units? ______
   b. Are these circuits double pole or single pole? ______

9. The electric heating units are supplied by what size conductors? ______

10. What type of electric heating units are provided in the workshop and utility room?
    ______

11. a. How much current passes through the recreation room thermostat when it is in the ON position? Show all calculations. ______

    b. How much current passes through the living room and dining room thermostats? Show all calculations. ______

12. A certain type of control connects electric heating units to a 120-volt supply or a 240-volt supply, depending upon the amount of the temperature drop in a room. These controls are supplied from a 120/240-volt, three-wire, single-phase source. Assuming that this type of device controls a 240-volt, 2000-watt heating unit, what is the wattage produced when the control supplies 120 volts to the heating unit? Show all calculations. (Note: The theory involved in this problem is covered in unit 19.)

13. What advantages does a 240-volt heating unit have over a 120-volt heating unit? Use the recreation room heating unit as an example.

14. a. How many wall-mounted thermostats are provided for this residence? ____________
b. At what height are these thermostats mounted? ______________________________
c. What type of thermostats are used? ______________________________________

15. The white conductor of a cable may be used to connect to a hot circuit conductor only if ____

____.

16. How many feet (meters) of cable are required to complete all of the baseboard electric heating circuits? Run the circuits first to the thermostat and then to the baseboard heating unit. ____

17. Electric baseboard heating units (shall) (shall not) be installed beneath receptacle outlets. Underline the correct answer.

18. The branch circuit supplying a heater must be sized to at least _________% of the heater's rating according to *Section* _______________.

unit 24

Oil Burner Hot Water Heating System

OBJECTIVES

After studying this unit, the student will be able to

- interpret schematic diagrams provided by manufacturers of heating system controls.
- list the components of a typical oil burner heating system.
- describe the functions of the control devices provided in a typical system.

An electric heating system is used in the residence described in this text. However, hot water heating systems using oil burners are very common and the student should be familiar with the operation of such a system.

INSTALLATION

The heating contractor usually furnishes and installs all of the equipment, including controls, required for an oil burner system. The controls are then wired by the electrician.

It is a good practice to provide a separate branch circuit for the oil burner and its controls. Many local electrical codes require such a separate circuit. In addition, these codes also require that conduit (EMT) be used to install exposed wiring in basements. Thus, the oil burner, circulating pump, stack switch, and any other line voltage equipment must be wired in conduit. Some controls require flexible wiring and so flexible conduit may be used. Low-voltage thermostat cable may be installed as the required low-voltage conductors between the control relay and the thermostat.

Section 430-102 requires that a disconnect switch be provided within sight of the controller. To meet this requirement, a single-pole switch or box-cover unit may be mounted on the oil burner or furnace. To service or repair the unit, the circuit controlling the boiler or furnace can be shut off quickly and safely using this switch or the box-cover unit.

WIRING DIAGRAM

The specifications for a dwelling usually indicate the type of heating system to be installed and the controls required for the system. The listing of materials and components often includes the phrase *or equivalent*. This means that the heating contractor may substitute similar equipment for specified equipment. However, substitutions can be made only if the replacement is of equal quality and can perform like the equipment specified.

The plans and specifications do not show the actual hookup of the electrical devices that are part of the oil burner circuit. However, equipment manufacturers provide wiring diagrams to insure correct installation, figure 24-1. Many types of building construction are possible, so the actual installation is a matter of personal judgment by the electrician, combined with local code requirements.

OIL BURNER CONTROLS

The oil burner controls shown in figure 24-1 are only a few of the many types manufactured. The rest of this unit describes a hot water heating system using a pump to circulate the hot water

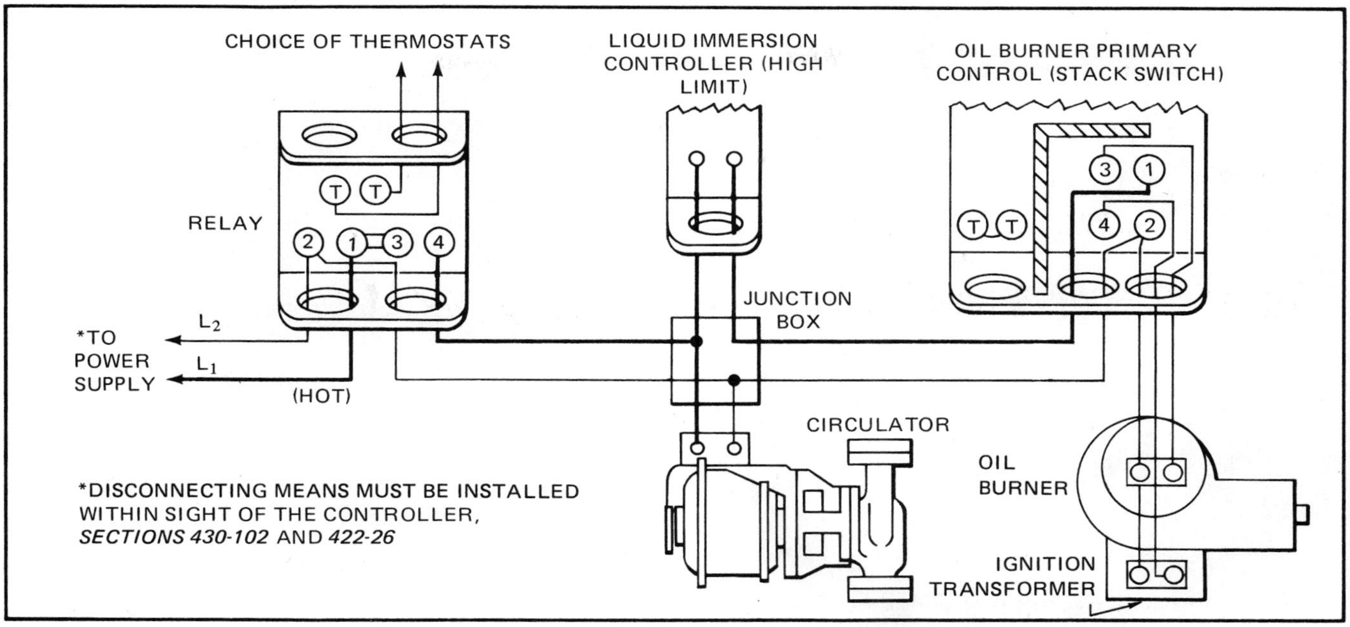

Fig. 24-1 Wiring diagram of an oil burner heating installation.

through radiators or convectors to deliver heat to the various rooms. The controls of the heating system are as follows:

- a thermostat controls both the burner and the circulating pump through a relay.
- a primary control, also known as a stack switch, provides switching or on-off control for the burner and the ignition transformer. This control guards against flame failure. The stack switch is mounted in the stack between the boiler and the chimney.
- a high-limit control prevents very high boiler temperatures.

Features of the Oil Burner Heating System

1. The burner and circulator are operated by the thermostat in response to temperature conditions in the living areas.
2. High-limit temperature protection prevents excessive boiler water temperatures and overheating of the living area.
3. Ignition and flame failure are monitored to insure the safe operation of the burner.

Principles of Operation

1. When the thermostat calls for heat, the switching relay is actuated and starts the circulator pump. Hot water begins to move through the heating system.
2. At the same time, a parallel circuit is made with the primary control which starts the oil burner. The liquid immersion control (high-limit control) is connected in series with the oil burner.
3. The circulator and burner continue to operate until the thermostat no longer calls for heat.
4. If the boiler water temperature reaches the value set on the high-limit control and the thermostat is still calling for heat, the burner shuts down. The burner cannot be restarted until the boiler water cools to the *on* point of the limit control. The circulator continues to operate as long as the thermostat calls for heat.

Multizone control can be achieved by providing a circulating pump and thermostat for each desired zone. In zone control, the high-limit control is set to maintain the boiler water temperature at a certain value. Each zone thermostat controls its own circulating pump.

The thermostat is usually the brain of a control system. It senses the conditions in a heated area. The thermostat operates the heating plant to maintain the area at a preset, comfortable heat level. Thermostats generally are accurate to 1/2 degree.

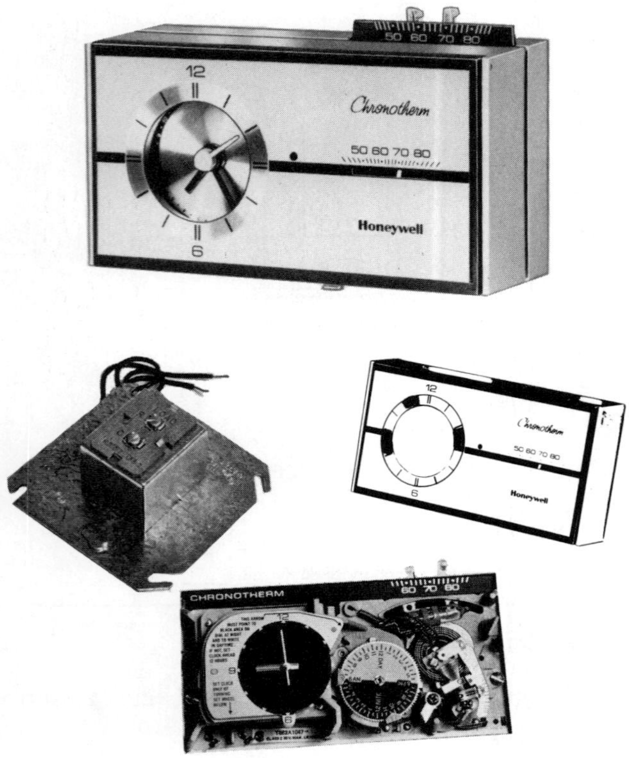

Fig. 24-2 Thermostat.

Thermostat

The thermostat shown in figure 24-2 makes the heating operation completely automatic. It maintains a reduced temperature at night. During the day, the thermostat allows the temperature to return to a set value. This thermostat generally is used with two-wire control circuits. The thermostat features the heat anticipation principle and operates the burner according to specific comfort requirements.

Primary Control

A primary control is shown in figure 24-3. This component starts and stops the oil burner motor in response to commands from the thermostat. Primary controllers are available for both constant and intermittent ignition burners. Both types of controllers will recycle once if the flame goes out for any reason. If the flame fails to ignite, the controller trips to safety. The control shown in figure 24-3 is designed for stack mounting. It operates quickly and reliably in low stack temperatures.

Fig. 24-3 Oil burner primary control.

Fig. 24-4 Switching relay.

Switching Relay

A switching relay, figure 24-4, starts and stops the burner and circulating pump at the command of the thermostat. A built-in transformer provides low voltage for the thermostat circuit. The switching relay is mounted vertically on a firm partition or wall close to the boiler.

Liquid Immersion Controller

The liquid immersion controller, figure 24-5, is the high-limit temperature control recommended for an installation of this type. This control must be mounted in the boiler and the sensing tube must be in direct contact with the water in the boiler. This setting of the controller is usually in the range

Fig. 24-5 Liquid immersion controller.

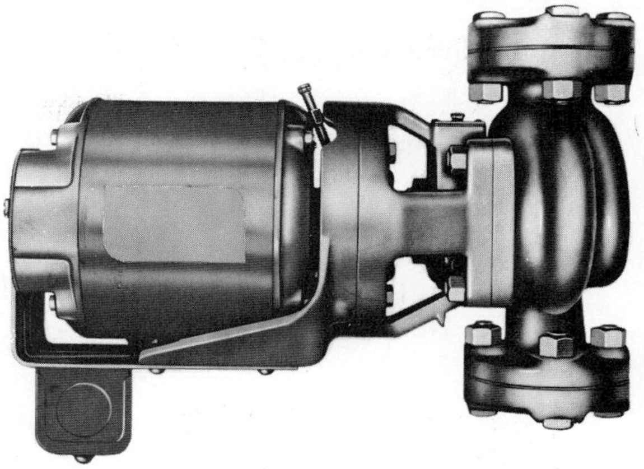

Fig. 24-6 Water circulator.

Fig. 24-7 Flow valve.

between 180°F and 200°F. The setting can be changed by turning the adjusting screw on the face of the control. Turn the screw clockwise for a lower temperature and counterclockwise for a higher temperature.

Water Circulator

The water circulator, figure 24-6, features quiet operation, leakproof construction, and freedom from lubrication problems. It is designed to speed up circulation in sluggish gravity systems. A circulator is required on forced hot water systems. The circulator should be installed in the return from the radiator.

Flow Valve

The flow valve shown in figure 24-7 prevents the gravity circulation of hot water through the heating system. In general, flow valves are installed in the main riser just above the boiler. The expansion tank connection must be made on the boiler side of the flow valve.

REVIEW

Note: Refer to the Code or the plans where necessary.

1. How is the residence in this text heated? ____________________
2. Can the wiring diagram in figure 24-1 be applied to all types of installations? Explain.

____

____

____

3. What contractor is responsible for the oil burner installation? ____________

4. What authority determines many of the methods used in oil burner installations?
____

5. Is conduit required for the installation of an oil burner? ____________

6. The Code states that a disconnecting means must be located within ____________ of the ____________.

7. What wires or devices connect to the following terminals on the stack switch?
   a. No. 1 terminal ____________
   b. No. 2 terminal ____________
   c. No. 3 terminal ____________
   d. No. 4 terminal ____________

8. What piece of equipment is controlled by the thermostat? ____________

9. What equipment furnishes the spark for ignition? ____________

10. Is the ignition transformer action continuous or of short duration? ____________
____

11. Name two actions controlled by the stack switch.
   a. ____________ b. ____________

12. What is the function of the circulator? ____________
____

13. How is the continuous operation of the circulator affected by the thermostat?
____
____

14. Name all of the control devices used in this installation. ____________
____
____

15. a. Where is the stack switch located? ____________
   b. Where is the switching relay located? ____________
   c. Where is the liquid immersion controller located? ____________

16. What is the function of the liquid immersion controller? ____________
____

17. What is the function of the flow valve? ____________
____

unit 25

Gas Burner Heating System

OBJECTIVES

After studying this unit, the student will be able to

- list the similarities and differences between oil-fired hot water heating systems and gas-fired hot water heating systems.
- describe three types of gas heating systems.
- list the wiring requirements of each of the three systems.
- explain the principles on which the thermocouple and the thermopile are based.

When it is economical to use natural or manufactured gas, a gas burner may be installed rather than an electric heating system or an oil burner.

A gas burner system is similar to an oil burner system (unit 24). Both systems require hot and cold water piping, a circulatory pump, radiators or convectors, and control equipment. The wiring for these systems is also similar. A thermostat, a high-limit control switch, and a relay are required in each system. However, the stack switch used in an oil burner system is not used in a gas burner system. A safety shutoff valve is used in its place. This valve blocks the flow of gas to the main burner if the pilot flame goes out or is too small to ignite the main burner.

TYPES OF GAS BURNER SYSTEMS

Forced System

Figure 25-1 is a wiring diagram for a typical forced, gas-fired hot water heating system. The gas valve is a line voltage solenoid. Note that the high-limit control, the safety shutoff valve, and the gas valve are connected in series. This means that the flow of gas can be shut off quickly at the gas valve if there is a very high boiler temperature or a flame-out. The gas valve is held open electrically. It is spring loaded to remain closed when there is no current in the circuit.

A pump or circulator is required in a forced system to circulate the water within the system. Some controllers have the following features combined in one unit: high-limit protection, switching for the gas valve, and switching for the circulator. All wiring is line voltage.

Gravity System

A gravity system does not need a pump to force the water through the system. Only a transformer is connected to the 120-volt supply. Thus, this system is much easier to wire than a forced system. A 120-volt power supply is connected to a transformer rated to handle the gas valve, figure 25-2. The thermostat, safety shutoff valve, limit control, and gas valve are connected in series as part of a low-voltage circuit.

Self-generating System

Several manufacturers of gas burners provide *self-generating systems.* These systems do not require an outside power supply. Figure 25-3 shows

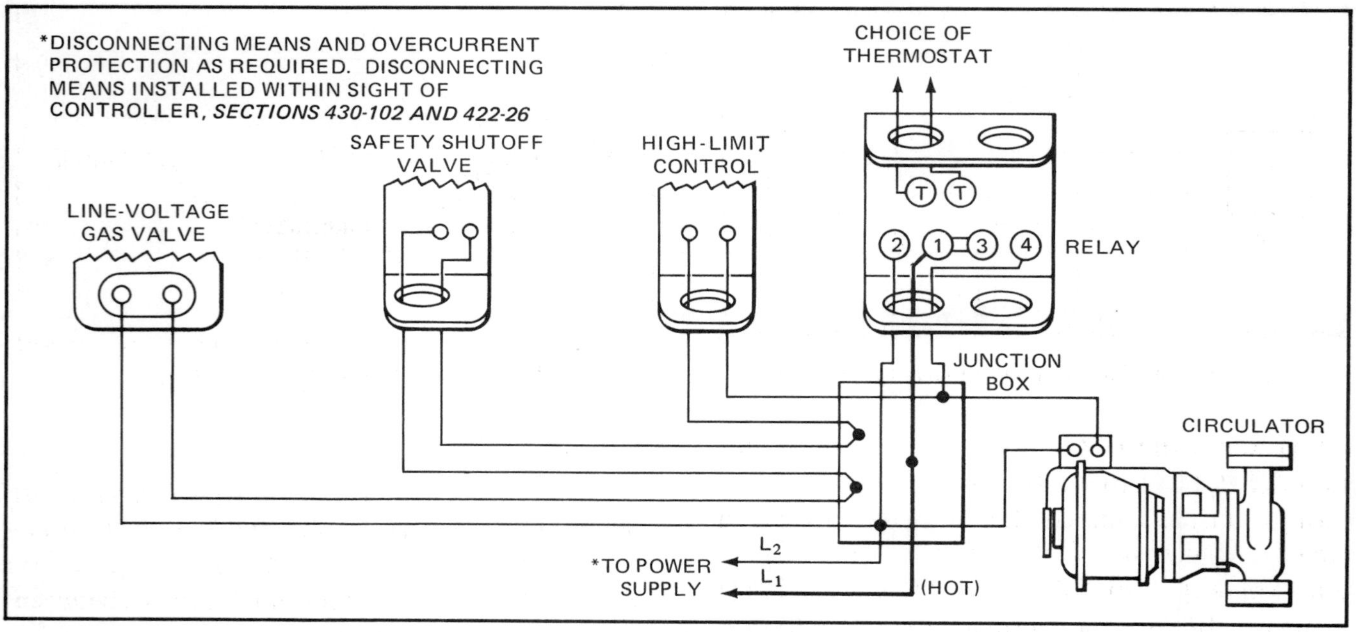

Fig. 25-1 Typical wiring for a gas burner installation using a circulator.

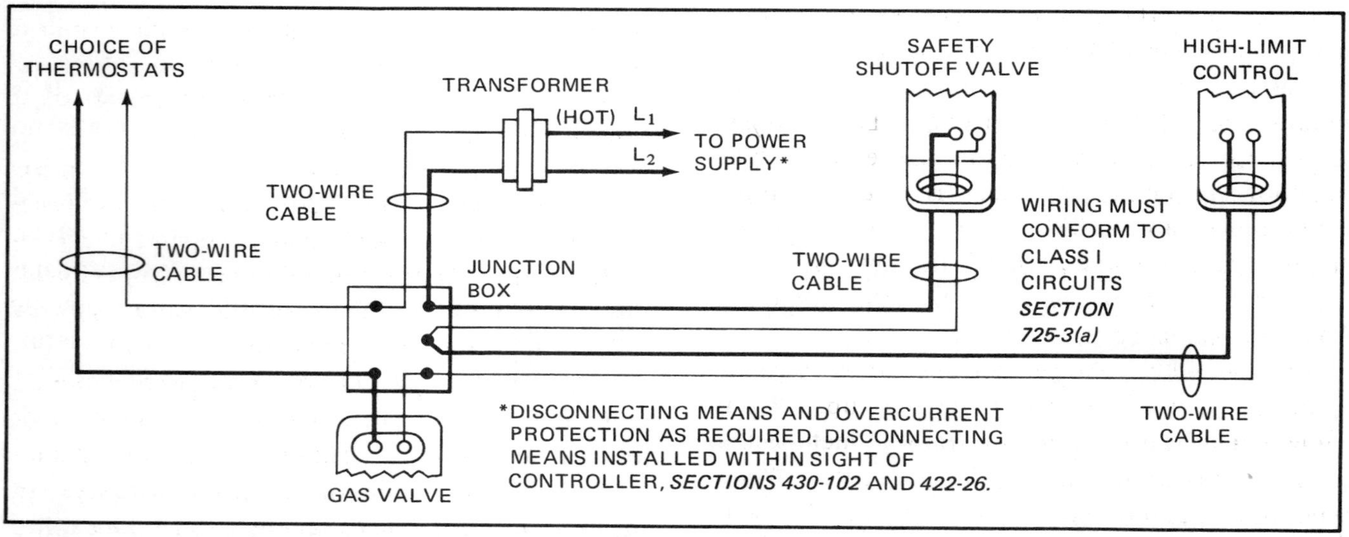

Fig. 25-2 Wiring for a gas burner heating installation using a gravity system.

the wiring diagram of a typical self-generating system. Note that the components are connected in series. The small amount of energy required by the gas valve is supplied by a thermopile. (A thermopile is defined and described later in this unit.) Many of these burners are prewired at the factory. Thus, the only wiring at the site is done by the electrician who runs a low-voltage cable to the thermostat. This system is not affected by a power failure because there is no outside power source.

The self-generating system operates in the millivolt range. A special thermostat is provided to function at this low voltage level. All other components of the system are matched to operate at the low voltage.

CIRCUIT PROTECTION

NEC *Section 725-4* states that **remote-control circuits to safety-control equipment shall be Class 1 if the failure of the equipment to operate poses a**

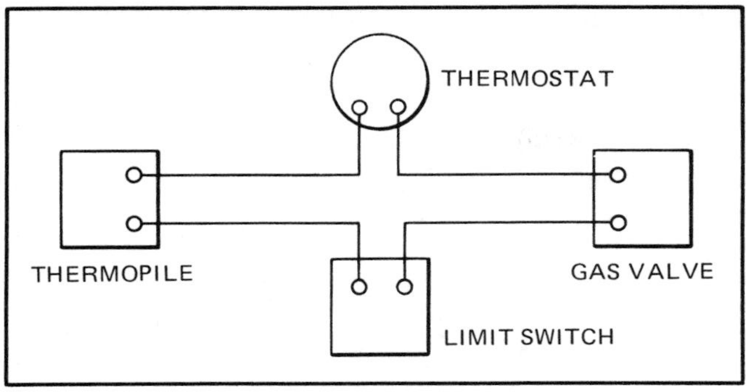

Fig. 25-3 Wiring diagram of a self-generating system.

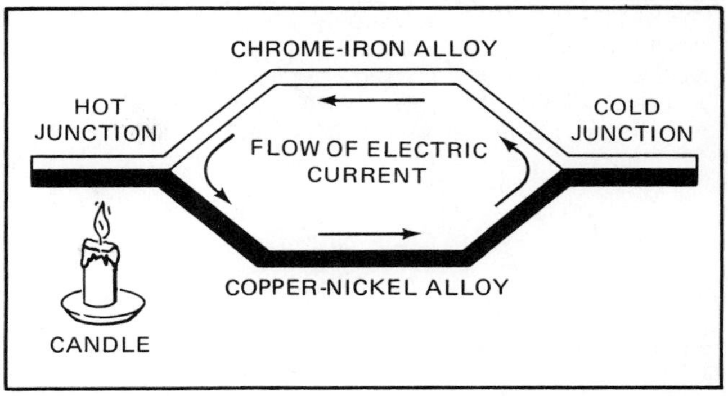

Fig. 25-4 Principle of a thermocouple.

life or fire hazard. The use of a Class 1 circuit means that the part of the circuit which can fail must be installed in conduit, intermediate metal conduit, electrical metallic tubing, rigid nonmetallic conduit, Type MI cable, or other suitable material. In this way, protection from physical damage is provided, *Section 725-18*.

A short circuit can occur in the wiring from the limit control of the heating system to the gas burner valve. This is especially true for low-voltage wiring. To prevent short circuits, a special type of low-voltage cable is listed by Underwriters' Laboratories. *Low-energy safety control wire*, marked as such at two-foot (610-mm) intervals, is a tough, high-temperature cable (105°C). It is available in two or three parallel conductors. This cable meets the requirements of *Section 725-18* for protection from physical damage.

Color-coding choices for the wires depends upon the method used for the hookup. When nonmetallic-sheathed cable or armored metal cable are used, the colors are black and white. When conduit containing individual conductors is used, the conductor can be black, red, white, blue or yellow, or almost any color except green. Low-voltage conductors come in different color combinations such as red-white, black-white, blue-white, red-blue-black, as well as others.

THERMOPILE FOR A SELF-GENERATING SYSTEM

A *thermopile* is a series of thermocouples connected together. The following paragraphs describe what a thermocouple is and what it does.

Thermocouple

Whenever two unlike metals are connected to form a circuit, an electric current flows if the junctions of the metals are at different temperatures, figure 25-4. As the temperature difference between the two junctions increases, the electric current increases. Thus, a thermocouple is formed by combining metals such as iron and copper, copper and constantan, copper-nickel alloy and chrome-iron alloy, platinum and platinum-rhodium alloy, or chromel and alumel. The types of metals used depend upon the temperatures involved.

In a gas burner system, the source of heat for the thermocouple is the pilot light. The cold junction of the thermocouple remains open and is connected to the electric solenoid (gas valve). (Recall that the thermostat, limit switch, safety shutoff valve, and gas valve are connected in series.)

A thermocouple develops a voltage of about 25 to 30 dc millivolts. Thus, the circuit resistance must be kept low. The electrician should follow the manufacturer's recommendations regarding wire size. In addition, it is suggested that any connections other than those at screw terminals be soldered.

Thermopile

When thermocouples are connected in series, a thermopile is formed, figure 25-5. The power output of the thermopile is greater than that of a single thermocouple; typically either 250 or 750 dc millivolts. For both a thermocouple and a thermopile, there must be a temperature difference between the metal junctions. If the junctions marked A in figure 25-5 are heated and the junctions marked B remain cold, the resultant flow of current is shown by the arrows.

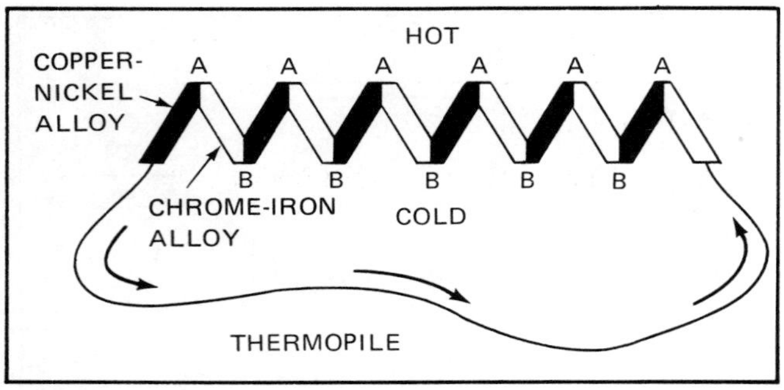

Fig. 25-5 Principle of a thermopile.

REVIEW

Note: Refer to the Code or the plans where necessary.

1. Does a gas burner system require a stack switch? ________________

2. What is the function of a safety shutoff valve? ________________

3. All protective devices such as a limit switch, safety shutoff valve, and low-water cutout, if used, are connected in (series) (parallel). Underline one.

4. On the diagram of the gas-fired, forced hot water system, figure 25-1, indicate whether the wiring is line voltage or low voltage for the following:
   a. Gas valve __________ d. Relay __________
   b. Safety shutoff valve __________ e. Thermostat __________
   c. Limit switch __________ f. Circulating pump __________

5. Does a gravity system require a pump to circulate the hot water through the system? Explain. ________________

6. What is meant by the term *self-generating*? ________________

7. Because a self-generating unit is not connected to an outside source of power, the electrician need not use care in the selection of the conductor size or in making electrical connections. Is this statement true or false? Explain. ________________

8. In each of the following groups (a, b, and c), which of the choices is easier to wire? (Underline the correct answer.)
   a. A forced hot water oil burner or a forced hot water gas burner.
   b. A forced hot water system or a gravity system.
   c. A gravity, self-generating gas system or a gravity gas system connected to a separate line voltage circuit.

9. A limit switch cuts off the circuit supplying the solenoid gas valve or the oil burner motor, depending upon the type of system installed. **If the wires connecting the limit switch are shorted together, even though the limit switch contacts open, the heat can increase to a dangerous level in the system.** To protect the system, the Code classifies

it as Class ________ , which must be wired with ________________________________,
________________________, ________________________, ________________________,
________________________ , or other suitable material.

10. What is a thermocouple? Explain how it operates. Use a diagram to illustrate the operation.

____

____

____

____

____

____

____

____

____

____

____

____

11. Does a thermocouple generate current when there is no difference in temperature between the junctions? Explain. ____

____

12. What is a thermopile? Use a diagram to illustrate the principle of operation of the thermopile.

____

____

____

____

____

____

____

____

____

____

unit 26

Heat and Smoke Detectors

OBJECTIVES

After studying this unit, the student will be able to

- identify the National Fire Protection Association (NFPA) standard which refers to household fire warning equipment.
- name the two basic types of smoke detectors.
- discuss the location requirements for the installation of heat and smoke detectors.
- list the locations where heat and smoke detectors may not be installed.
- describe the wiring requirements for the installation of heat and smoke detectors.

THE IMPORTANCE OF HEAT AND SMOKE DETECTORS

Fire is the third leading cause of accidental death. Home fires account for the biggest share of these fatalities, most of which occur at night during sleeping hours.

Heat and smoke detectors are installed in a residence to give the occupants an early warning of the presence of a fire. Fires produce smoke and toxic gases that can overcome the occupants while they sleep. Most fatalities result from the inhalation of smoke and toxic gases, rather than from burns.

Fig. 26-1 Heat detector.

NATIONAL FIRE PROTECTION ASSOCIATION (NFPA) *STANDARD NO. 74*

The National Fire Protection Association (NFPA) Standard No. 74, entitled *Household Fire Warning Equipment*, discusses the proper selection, installation, operation, and maintenance of fire warning equipment commonly used in residential occupancies.

Fire warning devices commonly used in a residence are a heat detector, figure 26-1, and a smoke detector, figure 26-2.

Fig. 26-2 Smoke detector.

Heat and smoke detectors should be installed in critical areas of a home. A sufficient number of detectors should be installed so as to provide full coverage.

Section 2-4.1.1 of NFPA *Standard No. 74* states that smoke detectors shall be installed outside of each separate sleeping area in the immediate vicinity of the bedrooms, and on each additional story of the dwelling, including basements, but excluding crawl spaces and unfinished attics.

Minimum requirements are stated in the standard. It is recommended that smoke and heat detectors be installed strategically so as to provide protection for living rooms, dining rooms, bedrooms, kitchens, hallways, attics, furnace and utility rooms, basements, and heated attached garages. Kitchen installations must be not directly above the stove.

TYPES OF SMOKE DETECTORS

Two types of smoke detectors commonly used are the photoelectric type and the ionization type.

Photoelectric Type

The photoelectric type of smoke detector has a light sensor that measures the amount of light in a chamber. When smoke is present, an alarm is sounded which indicates that there is a reduction in light due to the obstruction of the smoke. This type of sensor detects smoke from burning materials that produce great quantities of smoke, such as furniture, mattresses, and rags. This type of detector is less effective for gasoline and alcohol fires which do not produce heavy smoke.

Ionization Type

The ionization type of detector contains a low-level radioactive source which supplies particles that ionize the air in the smoke chamber of the detector. Plates in this chamber are oppositely charged. Because the air is ionized, an extremely small amount of current (millionths of an ampere) flows between the plates. Smoke entering the chamber impedes the movement of the ions, reducing the current flow. The detector senses and monitors this reduction in current flow, which causes an alarm to go off.

The ionization type of detector is effective for detecting small amounts of smoke, as is the case with gasoline and alcohol fires.

INSTALLATION REQUIREMENTS

The basic installation requirement is to locate heat and smoke detectors *between* the sleeping areas and the rest of the house, figure 26-3.

The following information concerns specific recommendations for installing a smoke or heat detector. (Refer to figures 26-4 and 26-5.)

- Install the detector as close as possible to the center of a room or hallway.
- Install the detector at the top of an open stairway, because heat and smoke travel upward. Do *not* install the detector in the *dead* air space at the top of a stairway that can be closed off by a door.
- The edge of the detector must be not closer than 4 inches (102 mm) from the wall.
- The top of the detector must be not closer than 4 inches (102 mm) from the ceiling, and not farther than 12 inches (305 mm) from the ceiling.
- Do not install a detector in *dead* air space where the ceiling meets the walls.

The 4-inch (102 mm) and 12-inch (305 mm) requirements apply because smoke and heat may not reach into the corners where the walls and

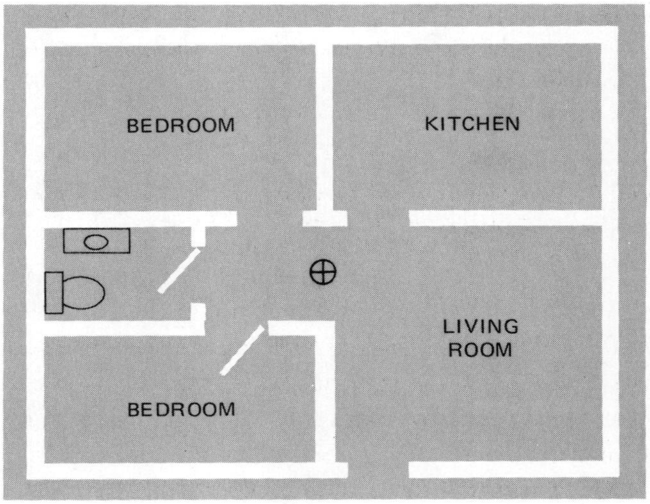

Fig. 26-3 Recommended location of heat or smoke detector *between* sleeping areas and rest of house.

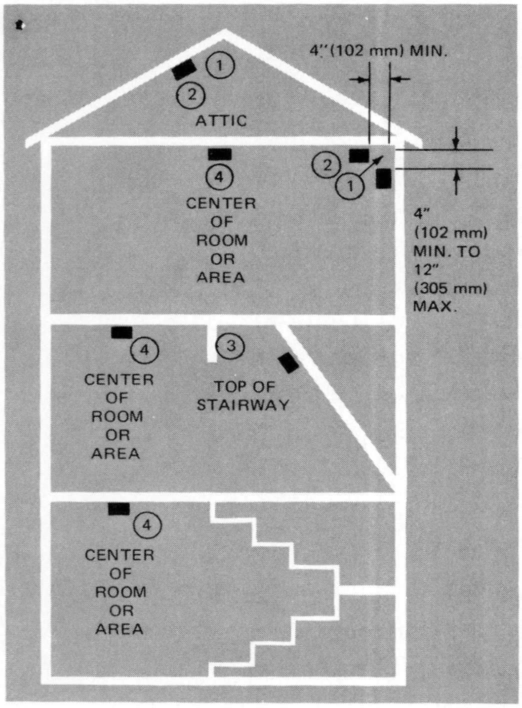

Notes: (1) Do *not* install detectors in *dead* air spaces.

(2) Mount detectors on the bottom edge of joists or beams. The space between these joists and beams is considered to be *dead* air space.

(3) Do *not* mount detectors in *dead* air space at the top of a stairway if there is a door at the top of the stairway that can be closed. Detectors *should* be mounted at the top of an open stairway because heat and smoke travel upward.

(4) Mount detectors in the center of a room or area.

Fig. 26-4 Recommendations for the installation of heat and smoke detectors.

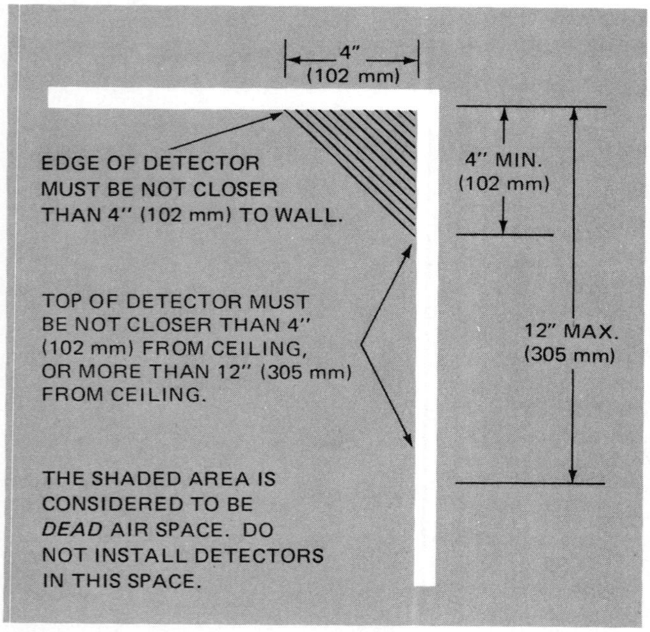

Fig. 26-5 Do not mount detectors in the *dead* air space where the ceiling meets the wall.

ceiling meet. This area can be considered as a *dead* air pocket which entraps air in such a manner as to prevent the rapid entry of smoke and heat into the detector.

INSTALLATION RESTRICTIONS

Heat and smoke detectors must *not* be installed

- where normal ambient temperatures exceed 100°F or drop below 40°F.
- where relative humidity exceeds 85 percent.
- in bathrooms, laundry areas, or other areas where large amounts of visible water vapor exist.
- in front of air registers, air conditioners, or any high-draft areas where the moving air will keep the smoke or heat from entering the detector.
- in *dead* air spaces.
- in kitchens where the accumulation of household smoke can result in false alarms with certain types of detectors. The photoelectric type of detector *may* be installed in kitchens, but must *not* be installed directly above the range or cooking appliances.

FEATURES OF SMOKE DETECTORS

Smoke detectors may contain an indicating light to show that the unit is functioning properly. They also may have a test button which actually simulates smoke. When the button is pushed, it is possible to test the detector's smoke-detecting abillity, as well as its circuitry and alarm.

Heat detectors are available that sense a specific *fixed* temperature, such as 135°F or 200°F. Available also are *rate-of-rise* heat detectors that sense rapid changes in temperature (12 degrees to 15 degrees per minute) such as those caused by flash fires.

Fixed and *rate-of-rise* temperature detectors are available as a combination unit. Combination *smoke, fixed,* and *rate-of-rise* temperature detectors are also available in one unit.

The spacing of heat detectors shall be as recommended by the manufacturer, because each

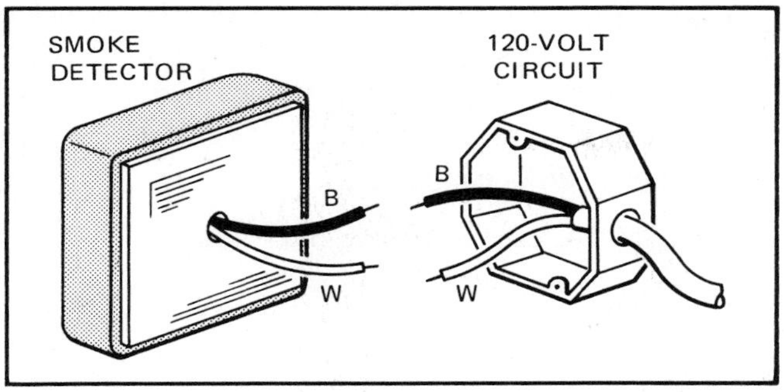

Fig. 26-6 Wiring of direct-connected detector units.

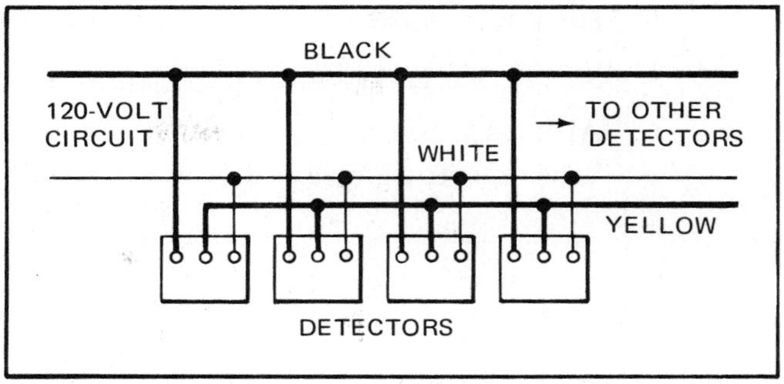

Fig. 26-7 Wiring of feedthrough (tandem) detector units.

detector is capable of sensing heat within a given space in a given time limit. NFPA *Standard No. 74* discusses this subject.

WIRING REQUIREMENTS

Direct-connected units. These units are connected to a 120-volt circuit. The black and white wires on the unit are spliced to the black and white wires of the 120-volt circuit, figure 26-6.

Feedthrough (tandem) units. These units may be connected in tandem up to 10 units. Any one of these detectors can sense smoke, then send a signal to the remaining detectors, setting them off so that all detectors sound an alarm. These units contain three wires: one black, one white, and one yellow, figure 26-7. The yellow wire is the interconnecting wire.

Cord-connected units. These units operate in the same way as direct-connected units, except that they are plugged into a wall outlet (one that is not controlled by a switch). These units are mounted to the wall on a bracket or screws. A clip on the cord is attached to the wall outlet with the same screw that fastens the faceplate to the outlet, figure 26-8. This prevents the cord from being pulled out, rendering the unit inoperative.

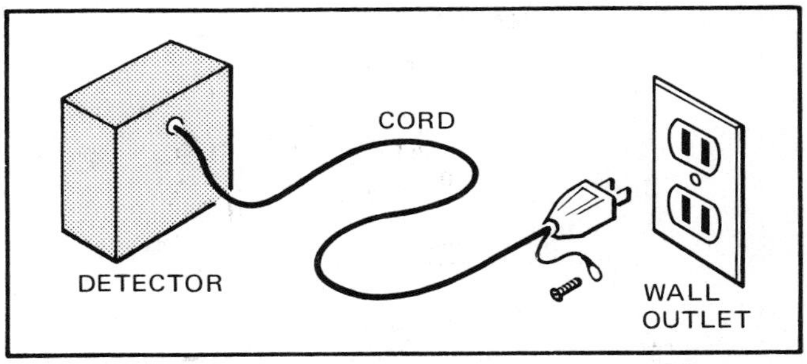

Fig. 26-8 Cord-connected detector units are plugged into wall outlet.

Battery-operated units. These units require no wiring; they are simply mounted in the desired locations. Batteries last for about a year. Battery-operated units should be tested periodically, as recommended by the manufacturer.

Always follow the installation requirements and recommendations of the manufacturer of the detecting unit.

REVIEW

Note: Refer to the Code or diagrams where necessary.

1. What is the name and number of the standard written about smoke and heat detectors?
   ____
2. Name the two basic types of smoke detectors. ______________________________
3. Why is it important to mount a smoke or heat detector not closer than 4 inches (102 mm) from a wall? ______________________________
4. A basic rule is to install smoke detectors (select one) ______________________________
   a. between the sleeping area and the rest of the house.
   b. at the top of a basement stairway that has a closed door at the top of the stairs.
   c. in a garage that is subject to subzero temperatures.

5. Heat detectors are available in two temperature types. Name them. ______________________

____

6. Although NFPA *Standard No. 74* gives many rules for the installation of smoke and heat detectors, always follow the installation recommendations of ______________________

____.

unit 27

Service Entrances and Equipment

OBJECTIVES

After studying this unit, the student will be able to

- define electrical service, overhead service, service drop, and underground service.
- list the various Code sections covering the installation of a mast-type overhead service and an underground service.
- discuss the Code requirements for disconnecting the electrical service using a main panel and load centers.
- discuss the grounding of interior alternating-current systems and the bonding of all service-entrance equipment.
- describe the various types of fuses.
- select the proper fuse for a particular installation.
- explain the operation of fuses and circuit breakers.
- explain the term "interrupting capacity."
- determine available short-circuit current using a simple formula.

An electric service is required for all buildings containing an electrical system and receiving electrical energy from a utility company. The National Electrical Code describes the term *service* as the conductors and equipment required to deliver energy from the electrical supply source to the wiring system of the premises.

OVERHEAD SERVICE

The Code terms a *service drop* as the overhead service conductors, including splices, if any, which are connected from an outdoor support to the service-entrance conductors at the structure.

The overhead service includes all of the service equipment and installation means from the attachment of the service-drop wires on the outside of the building to the point where the circuits or feeders are tapped to supply specific loads or load centers. In general, watthour meters are located on the exterior of a building. Local codes may permit the watthour meter to be mounted inside the building. In some cases, the entire service-entrance equipment may be mounted outside the building. This includes the watthour meter and the disconnecting means.

Mast-type Service Entrance

The *mast* service, figure 27-1, is a commonly used method of installing a service entrance. The mast service is often used on buildings with low roofs, such as ranch-style dwellings, to insure adequate clearance between the ground and the lowest service conductor.

The service conduit must be run through the roof as shown in figure 27-1. The conduit is fastened to comply with local code requirements. Methods of fastening this type of service are not covered by the National Electrical Code. Figure 27-1 shows the installation of the service entrance for the residence in the plans.

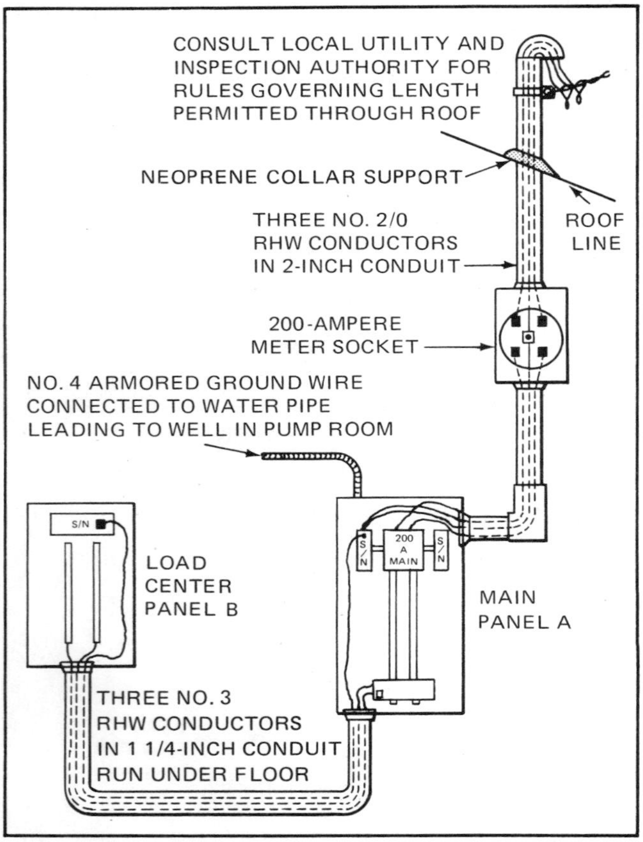

Fig. 27-1 Mast-type service entrance for the residence.

Clearance Requirements for Mast Installations

Several factors determine the maximum length of conduit that can be installed between the roof support and the point where the service-drop conductors are attached. These factors are the service-drop length, the system voltage, and the roof pitch, figure 27-2.

The bending force on the conduit increases with an increase in the distance between the roof support and the point where the service-drop conductors are attached. The pulling force of a service drop on a mast service conduit increases as the length of the service drop increases. As the length of the service drop decreases, the pulling force on the mast service conduit decreases.

If extra support is not to be provided, the mast service conduit must be at least two inches in diameter. This size prevents the conduit from bending due to the strain of the service-drop conductors. If extra support is provided, it is usually in the form of a guy wire attached to the roof rafters by approved fittings. On extremely long service drops, *Section 230-28* recommends that conduit of a larger diameter be installed and a guy wire attached.

NEC *Table 3A, Chapter 9,* shows that 1 1/2-inch conduit can be used for this type of service. The specifications for the residence state that 2-inch conduit must be installed.

The NEC rules for insulation and clearances apply to the service-drop and service-entrance conductors. For example, the service conductors must be insulated, except where the voltage to ground does

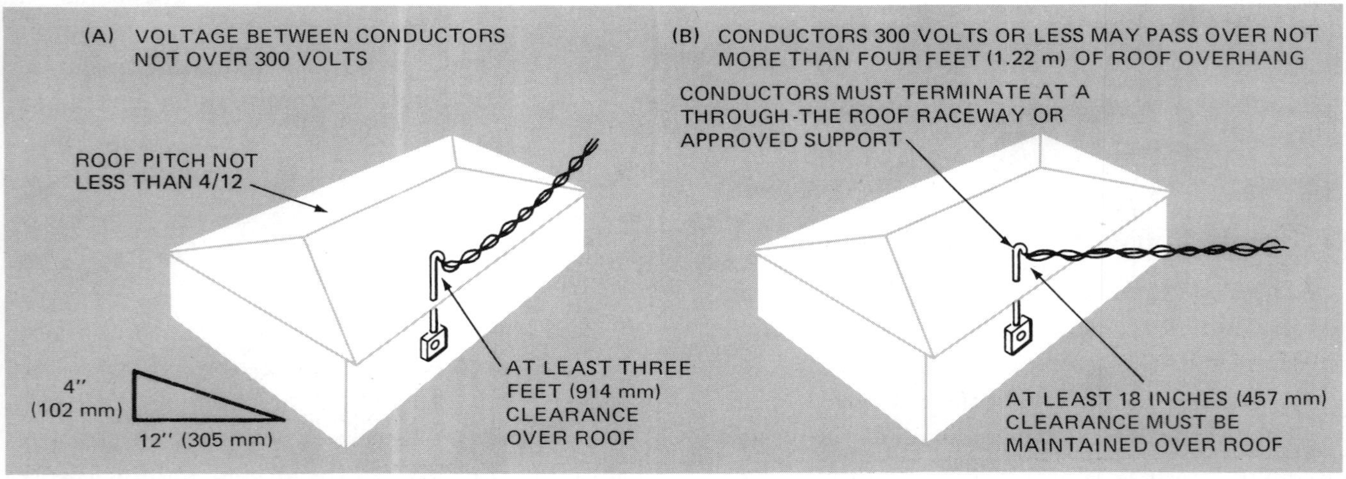

Fig. 27-2 Clearance requirements for service-drop conductors passing over roofs, *Section 230-24(a)*.

not exceed 300 volts. In this case, the grounded neutral conductors are not required to have insulation.

Section 230-24(a) gives clearance allowances for the service drop passing over the roof of a dwelling, figure 27-2.

The installation requirements for a typical service entrance are shown in figures 27-3 and 27-4. Figure 27-3 shows the required clearances above the ground. The wiring connections and Code references are given in figure 27-4.

MAIN SERVICE DISCONNECT LOCATION

The main service disconnect means shall be installed at a readily accessible location, so that the service-entrance conductors within the building are as short as possible, figure 27-5.

The reason for this rule is that the service-entrance conductors do not have overcurrent protection other than that provided by the utility's transformer fuses. **Should a fault occur on these service-entrance conductors within the building (at the bushing where the conduit enters the main switch, for instance), the arcing could result in a fire.**

The electrical inspector must make a judgment as to what is considered to be a readily accessible location nearest the point of entrance of the service-entrance conductors, NEC *Section 230-72(c)*.

UNDERGROUND SERVICE

The *underground service* means the cable installed underground from the point of connection to the system provided by the utility company.

New residential developments often include underground installations of the high-voltage electrical systems. The conductors in these distribution systems end in the bases of pad-mounted transformers or in *pedestals.* These pedestals are placed at the rear lot line or in other inconspicuous locations in the development.

The conductors installed between the pad-mounted transformers and the meter are called *service lateral conductors.* Normally, the electric utility supplies and installs the service laterals. Figure 27-6 shows a typical underground installation.

The wiring from the external meter to the main service equipment is the same as the wiring for a service connected from overhead lines, figure 27-4. Some local codes may require conduit to be installed from the pole to the service-entrance equipment. NEC requirements for underground service are given in *Part D* of *Article 230, Sections 230-30*, and *230-31*. The underground conductors must be suitable for direct burial in the earth.

If the electric utility installs the underground service conductors, the work must comply with the

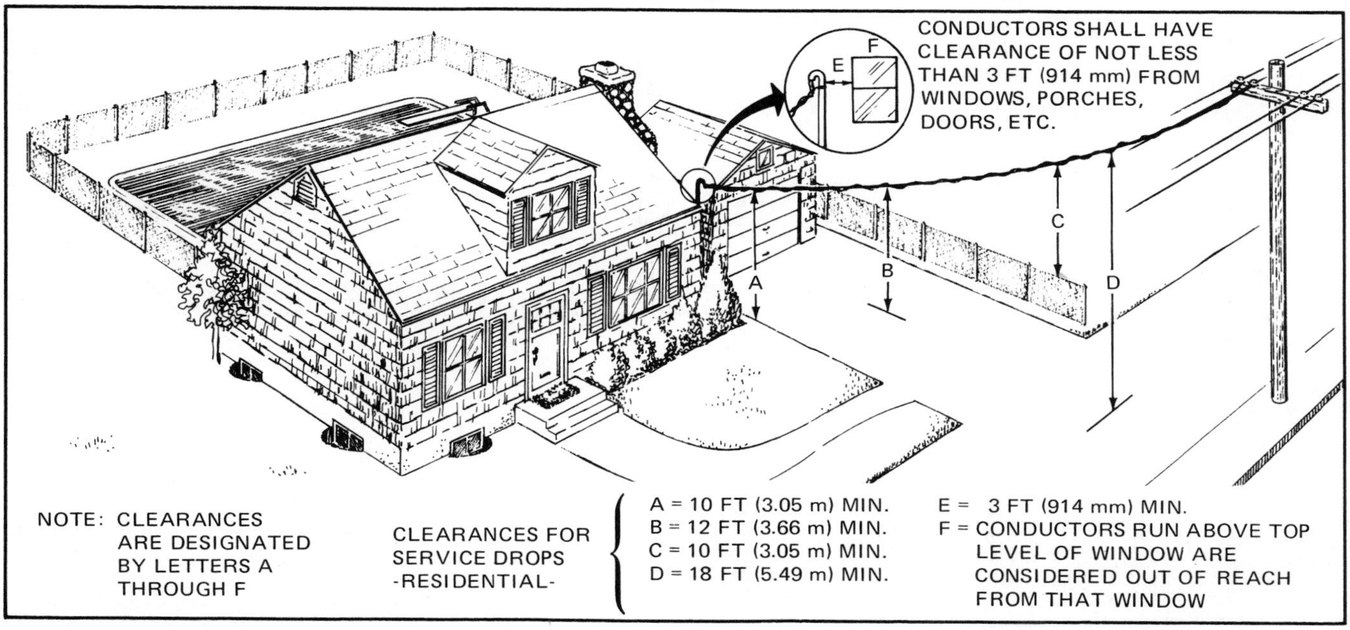

Fig. 27-3 Clearances for a typical service-entrance installation, *Sections 230-24(b)* and *(c)*. (For clearances above swimming pool, see figure 30-6.)

rules established by the utility. These rules may not be the same as those given in NEC *Section 90-2*.

When the underground conductors are installed by the electrician, *Sections 230-48* and *230-49* apply. These sections deal with the protection of conductors against damage and the sealing of underground conduits where they enter a building. *Section 230-49* refers to *Section 300-5* which covers all situations involving underground wiring.

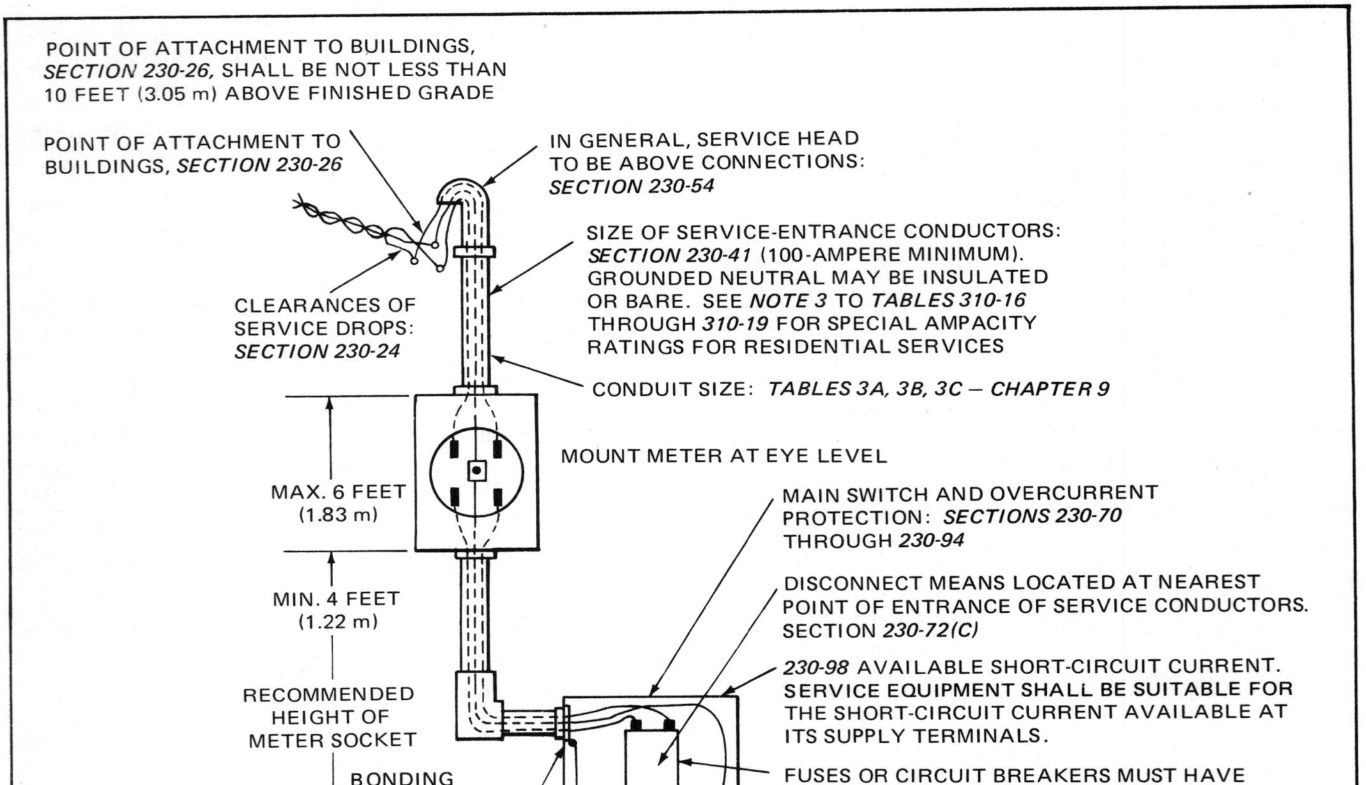

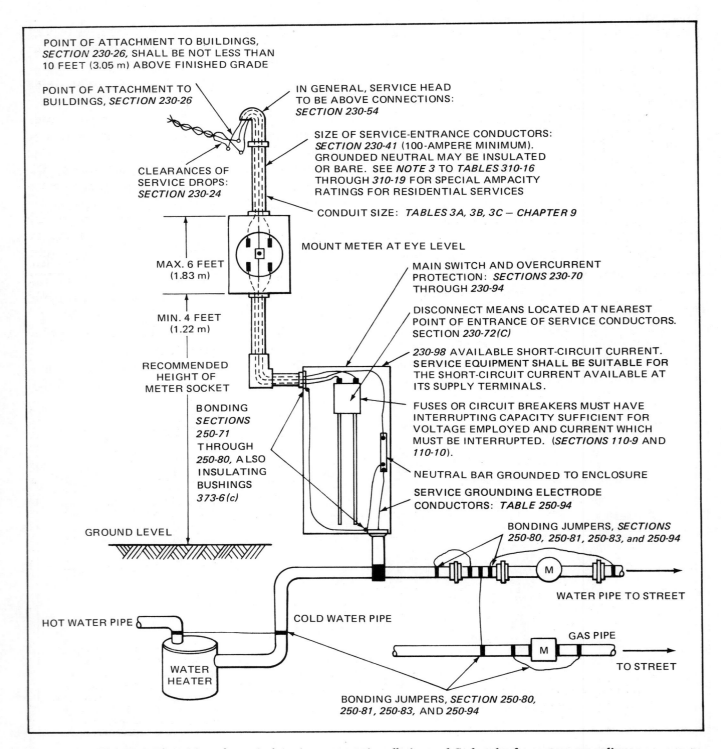

Fig. 27-4 The wiring of a typical service-entrance installation and Code rules for system grounding.

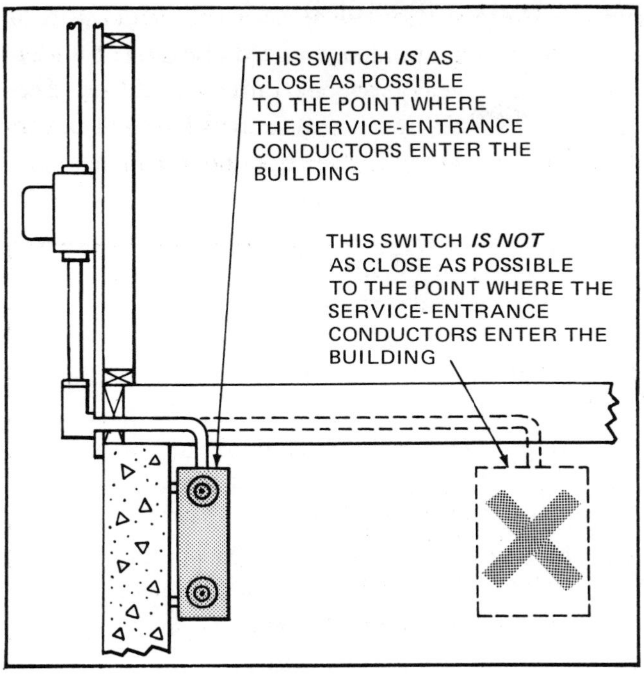

Fig. 27-5 Main service disconnect location.

DISCONNECT MEANS (PANEL A)

The requirements for disconnecting the electrical service are covered in *Sections 230-70* through *230-84*. *Section 230-71(a)* requires that the service disconnect means consist of not more than six switches or six circuit breakers mounted in a single enclosure, in a group of separate enclosures, or in or on a switchboard. By complying with *Section 230-71(a)*, all electrical equipment in a building can be disconnected, if necessary, with no more than six hand operations. Some local codes take exception to this rule and state that each service must have a single main disconnect. A panelboard may not contain more than 42 overcurrent devices, *Section 384-15*.

For the residence, Panel A provides a 200-ampere, pullout-type fuse block for the main disconnect. Panel A also has a number of branch-circuit overcurrent devices. These devices are provided for the electric heating system, the air conditioning system, and general lighting circuits,

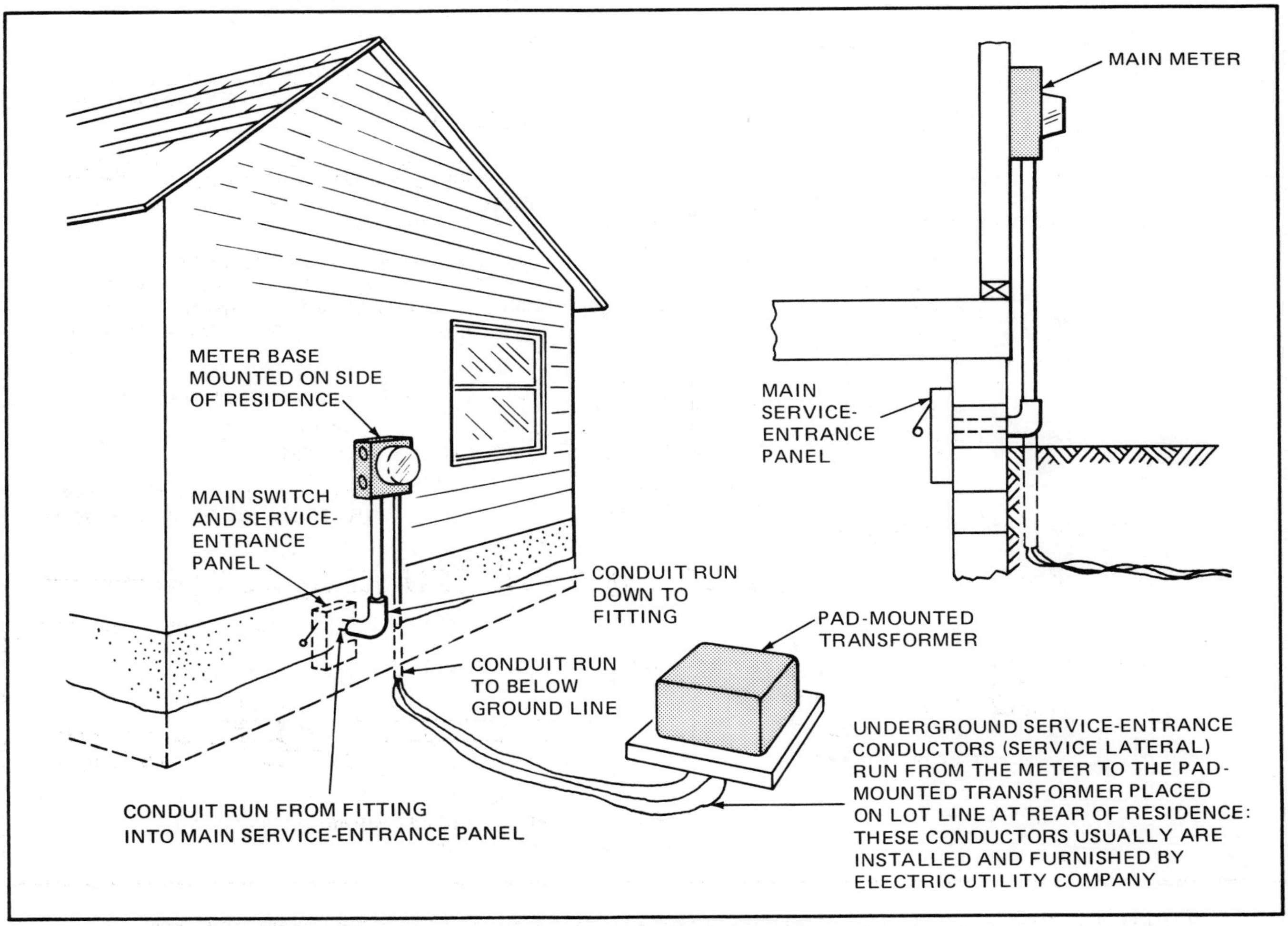

Fig. 27-6 Underground service.

figures 27-7 and 27-8. This type of panel is approved by Underwriters' Laboratories as a load center and as service equipment. The panel meets code requirements in most localities.

Panel A is located in a corner in the south end of the workshop. The placement of the main disconnect is determined by the location of the meter. The disconnect is mounted at an accessible point as close as possible to the place where the service conductors enter. The location of the meter depends upon the service drop. The local utility company generally decides where the service drop is to be located.

Section 250-61(b) states that a grounded circuit conductor shall not be used for grounding noncurrent-carrying equipment on the load side of the service disconnecting means. Exceptions to this rule are the frames of ranges, wall-mounted ovens, counter-mounted cooking units, and clothes dryers, under the conditions specified by *Section 250-60.*

The panelboard must have an approved means for attaching the grounding conductors when nonmetallic-sheathed cable is used, figure 27-9.

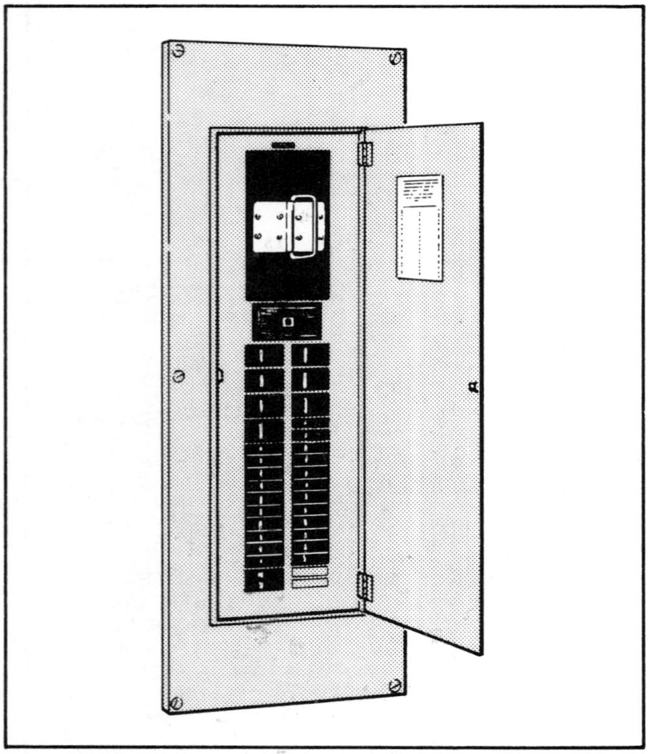

Fig. 27-7 Typical main panel with 200-ampere pullout.

LOAD CENTER (PANEL B)

It is recommended that load centers be installed near a heavy concentration of loads. The individual branch-circuit conductors in the area can be run to the load center, not back to the main panel. Thus, the circuit conductors are short and line losses are reduced.

A typical load center is shown in figure 27-10. The circuit schedule of Panel B, the load center in the residence, is shown in figure 27-11. Panel B is mounted in the utility room. It is fed by three No. 3 RHW conductors run in a 1 1/4-inch conduit originating from Panel A. The conductors are protected by a 100-ampere, 240-volt, two-pole overcurrent device in Panel A.

The conduit runs under the basement concrete floor. Thus, Type RHW conductors are installed. *Article 100* states that an installation of this type is considered to be in a wet location.

METER

Many electric utility companies offer lower rates for water heaters and electric heating units connected to separate meters. The residence in the plans has only one meter. Thus, all lighting, heating, cooking, and water heater loads are registered on one meter.

The meter socket, figure 27-12, is mounted at eye level on the outside of the southeast wall of the house. The service conduit is connected to the socket at a threaded boss on the top of the socket. A second length of conduit is attached to the bottom of the meter socket. This conduit is carried down into the basement workshop where it is attached to Panel A. Since this conduit must pass through the basement wall, figure 27-13, some form of fitting must be used. Some electricians prefer to use a conduit fitting and others prefer to use a bent elbow.

COST OF USING ELECTRICAL ENERGY

A watthour meter is always connected into some part of the service-entrance equipment. In the case of residential metering, a watthour meter would normally be installed as part of the service entrance. (See figure 27-4.) Some utilities furnish and install watthour meters on a *pedestal* located on the lot line between residential properties.

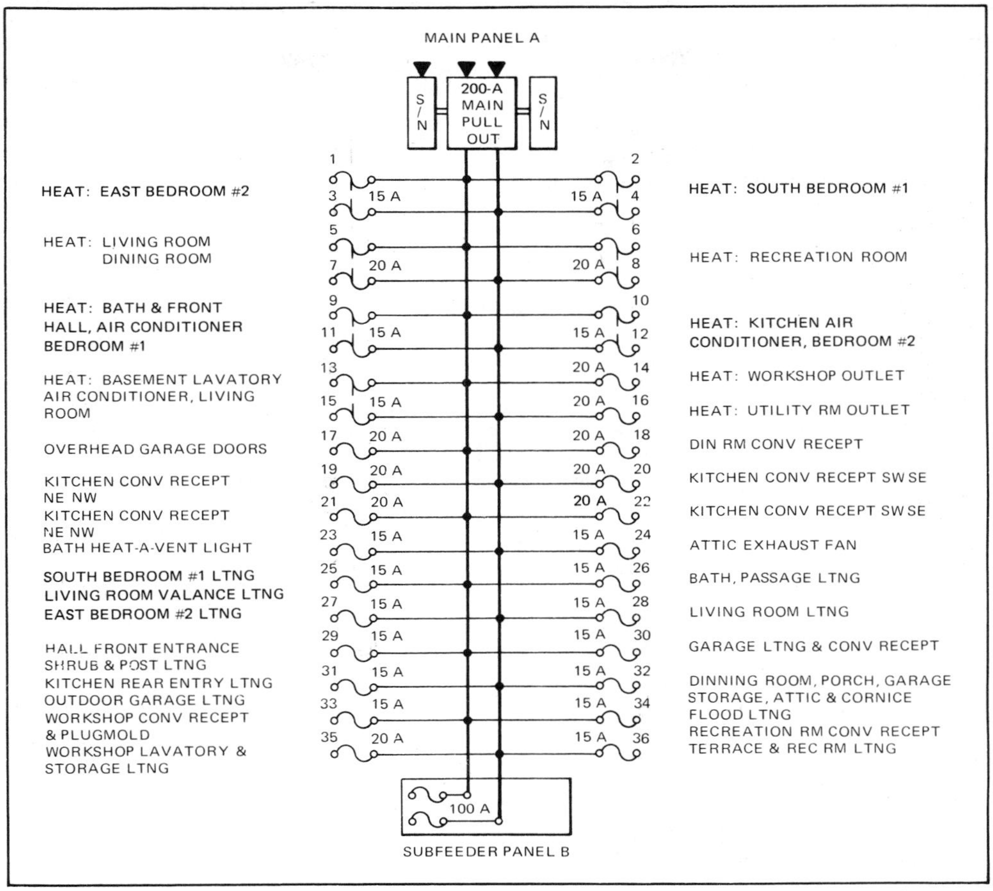

Fig. 27-8 Circuit schedule of main Panel A for the residence.

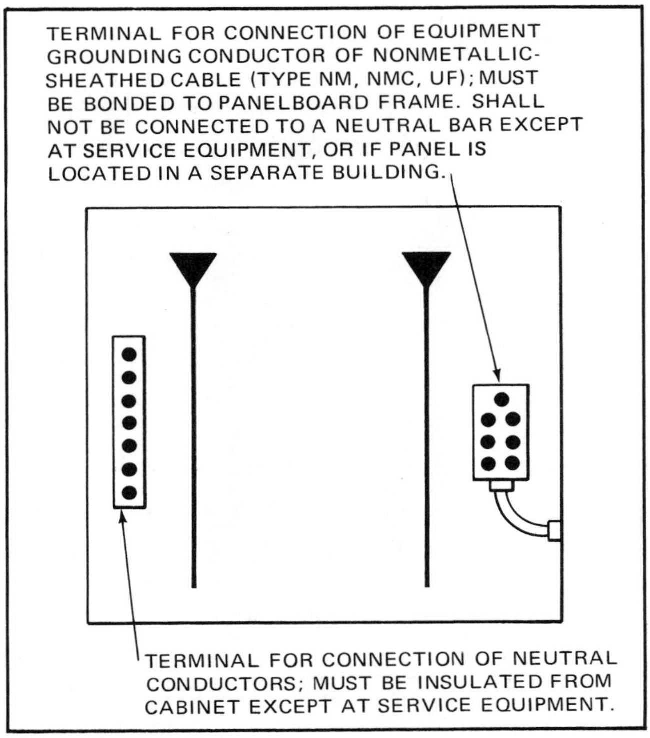

TERMINAL FOR CONNECTION OF EQUIPMENT GROUNDING CONDUCTOR OF NONMETALLIC-SHEATHED CABLE (TYPE NM, NMC, UF); MUST BE BONDED TO PANELBOARD FRAME. SHALL NOT BE CONNECTED TO A NEUTRAL BAR EXCEPT AT SERVICE EQUIPMENT, OR IF PANEL IS LOCATED IN A SEPARATE BUILDING.
TERMINAL FOR CONNECTION OF NEUTRAL CONDUCTORS; MUST BE INSULATED FROM CABINET EXCEPT AT SERVICE EQUIPMENT.

Fig. 27-9 Grounding of panelboards, *Section 384-27.*

Fig. 27-10 Typical load center.

All electrical energy consumed is metered so that the utility can bill the customer on a monthly or bimonthly basis.

The kilowatt (kW) is a convenient unit of electrical power. One thousand watts (W) is equal to one kilowatt (kW). The watthour meter measures both wattage and time. As the dials of the meter turn, the kilowatt-hour (kWh) consumption is continually recorded.

Utility rates are based upon "so many cents per kilowatt hour."

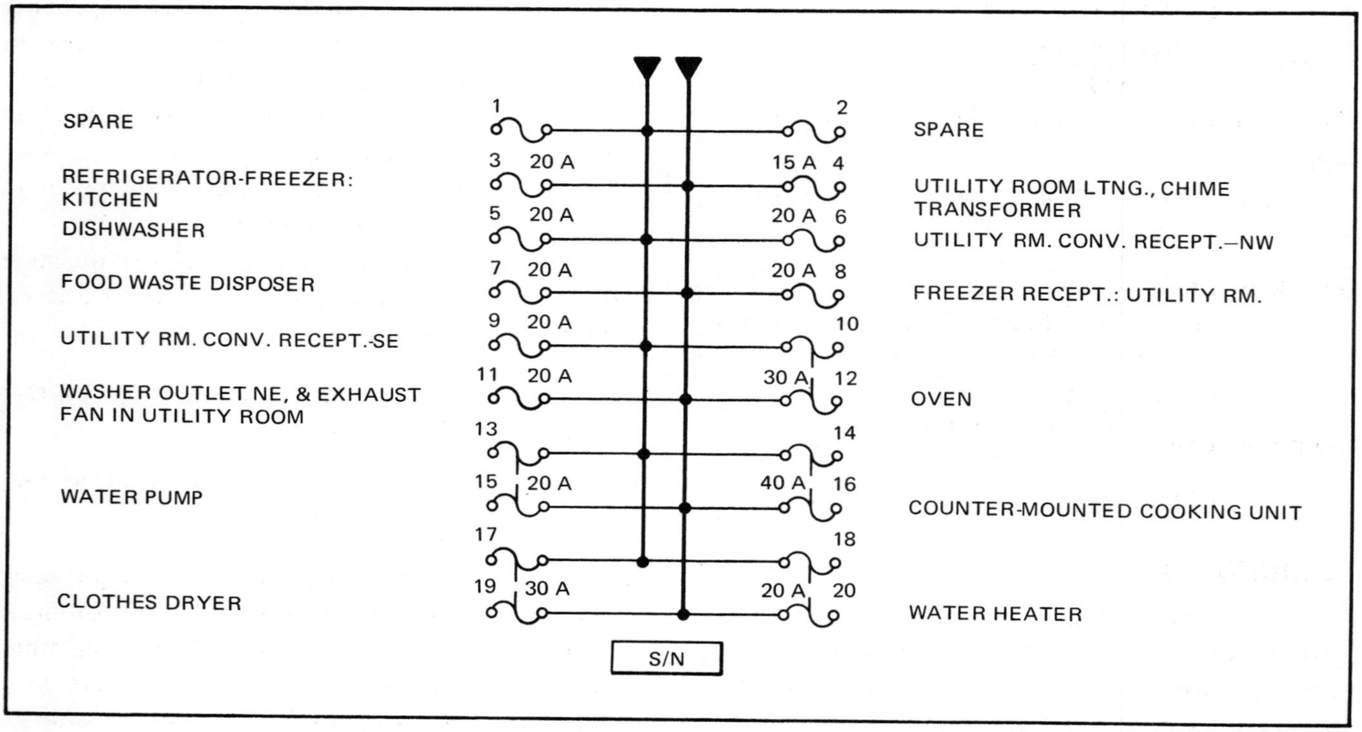

Fig. 27-11 Circuit schedule of Panel B.

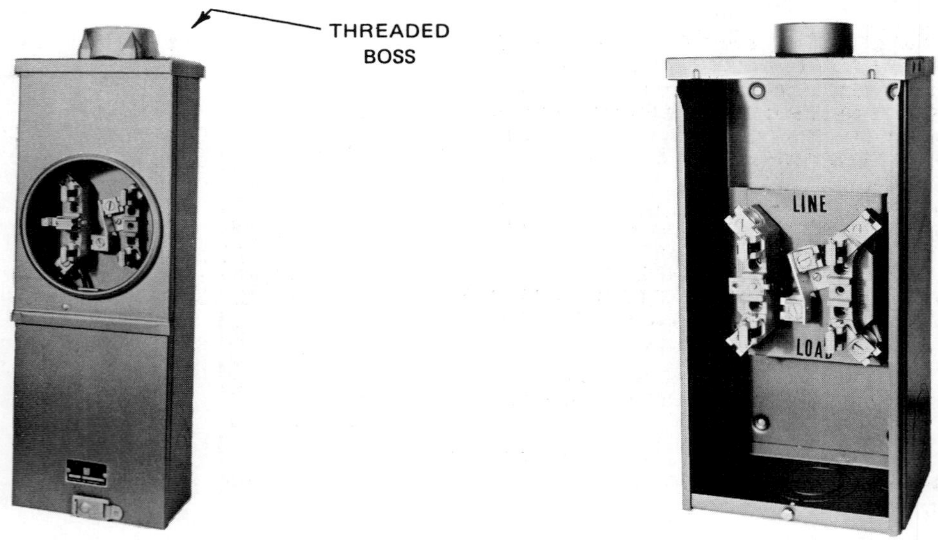

Fig. 27-12 Meter socket.

Example: One kWh will light a 100-watt light bulb for 10 hours. One kWh will operate a 1000-watt electric heater for 1 hour. Therefore, if the electric rate is 4 cents per kWh, the use of a 100-watt bulb for 10 hours would cost 4 cents. Or, a 1000-watt electric heater could be used for 1 hour at a cost of 4 cents.

$$\text{kWh} = \frac{\text{watts} \times \text{hours}}{1000} = \frac{100 \times 10}{1000} = 1 \text{ kWh}$$

To determine the cost of the energy used by an appliance:

$$\text{cost} = \frac{\text{watts} \times \text{hours used} \times \text{cost per kWh}}{1000}$$

Problem: Find the cost of operating a color television for 8 hours. The set is rated at 175 watts. The electric rate is 4.2 cents per kilowatt-hour.

Solution: $\text{cost} = \frac{175 \times 8 \times 0.042}{1000} = 0.0588$

(5.88 cents)

GROUNDING

Electrical systems and their conductors are grounded to minimize voltage spikes when lightning strikes, or when other line surges occur. Grounding stabilizes the normal voltage to ground.

Electrical metallic conduits and equipment are grounded so that the equipment voltage to ground is kept to a low value. In this way, the shock hazard is reduced.

Proper grounding means that overcurrent devices can operate faster to respond to ground faults. A low-impedance (ac resistance) ground path results with proper grounding. A low-impedance ground path means that there is a high value of ground current. As the ground current increases, there is an increase in the speed with which a fuse will blow or a circuit breaker will trip. Thus, as the overcurrent device senses and opens the circuit faster, less equipment damage results.

The arcing damage to electrical equipment is closely related to the value of ampere-squared-seconds (I^2t), where

I = current flowing from phase-to-ground, or from phase-to-phase, in amperes.

t = time needed by the overcurrent device to open the circuit, in seconds.

This expression shows that there is less equipment damage if the smallest possible value of current is permitted to flow for the shortest possible time. The quantity ampere-squared-seconds is covered in detail in the text *Electrical Wiring–Commercial.*

SECTION 300-7 STATES THAT WHEN RACEWAYS PASS THROUGH AREAS HAVING GREAT TEMPERATURE DIFFERENCES, SOME MEANS MUST BE PROVIDED TO PREVENT PASSAGE OF AIR BACK AND FORTH THROUGH THE RACEWAY. NOTE THAT OUTSIDE AIR IS DRAWN IN THROUGH THE CONDUIT WHENEVER A DOOR OPENS. COLD OUTSIDE AIR MEETING WARM INSIDE AIR CAUSES THE CONDENSATION OF MOISTURE. THIS CAN RESULT IN RUSTING AND CORROSION OF VITAL ELECTRICAL COMPONENTS. EQUIPMENT HAVING MOVING PARTS, SUCH AS CIRCUIT BREAKERS, SWITCHES, AND CONTROLLERS, IS ESPECIALLY AFFECTED BY MOISTURE. THE SLUGGISH ACTION OF THE MOVING PARTS IN THIS EQUIPMENT IS UNDESIRABLE.

INSULATION OR OTHER TYPE OF SEALING COMPOUND CAN BE INSERTED AS SHOWN TO PREVENT THE PASSAGE OF AIR.

SECTION 230-48 REQUIRES SEALING WHERE THE SERVICE RACEWAY ENTERS FROM UNDERGROUND

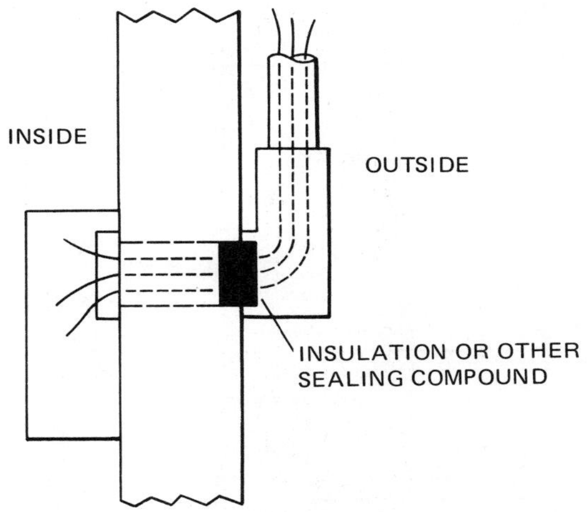

Fig. 27-13 Installation of conduit through a basement wall.

System Grounding

In the *system grounding* concept, rather than grounding a single item, the electrician must be concerned with grounding an entire system. The term "system" means the service neutral conductor, hot and cold water pipes, gas pipes, service-entrance equipment, and jumpers installed around meters. If any of these system parts become disconnected or open, the integrity of the system is maintained through other paths. This means that all parts of the system must be tied (bonded) together.

Figure 27-14 and the following steps illustrate what can happen if an entire system is not grounded.

1. A live wire contacts the gas pipe. The bonding jumper Ⓐ is not installed originally.
2. The gas pipe now has 120 volts on it. The pipe is "hot."
3. The insulating joint in the gas pipe results in a poor path to ground; assume the resistance is 8 ohms.
4. The 20-ampere fuse does not blow.
   $I = \frac{E}{R} = \frac{120}{8} = 15$ amperes
5. If a person touches the hot gas pipe and the water pipe at the same time, current flows through the person's body. If the body resistance is 12 000 ohms, the current is:
   $I = \frac{E}{R} = \frac{120}{12\,000} = 0.01$ ampere
   This value of current passing through a human body can cause death.
6. The fuse is now "seeing" 15 + 0.01 = 15.01 amperes; however, it still does not blow.
7. If the *system grounding* concept had been used, bonding jumper Ⓐ would have kept the voltage difference between the water pipe and the gas pipe at zero. Thus, the fuse would blow. If 10 feet (3.05 m) of No. 4 AWG copper wire were used as the jumper, then the resistance of the jumper is 0.002 59 ohm. The current is:
   $I = \frac{E}{R} = \frac{120}{0.002\,59} = 46\,332$ amperes.
   (In an actual system, the impedance of all of the parts of the circuit would be much higher. Thus, a much lower current would result. The value of current, however, would be enough to cause the fuse to blow.)

Advantages of System Grounding

The system grounding concept has several advantages.

- The potential voltage differences between the parts of the system are minimized, reducing the shock hazard.
- The impedance of the ground path is minimized. This results in a higher current flow in the event of a ground fault. The lower the impedance, the higher the current flow. This means that the overcurrent device will open sooner.

Reviewing figure 27-4 shows that both the metal hot and cold water pipes, the service raceways,

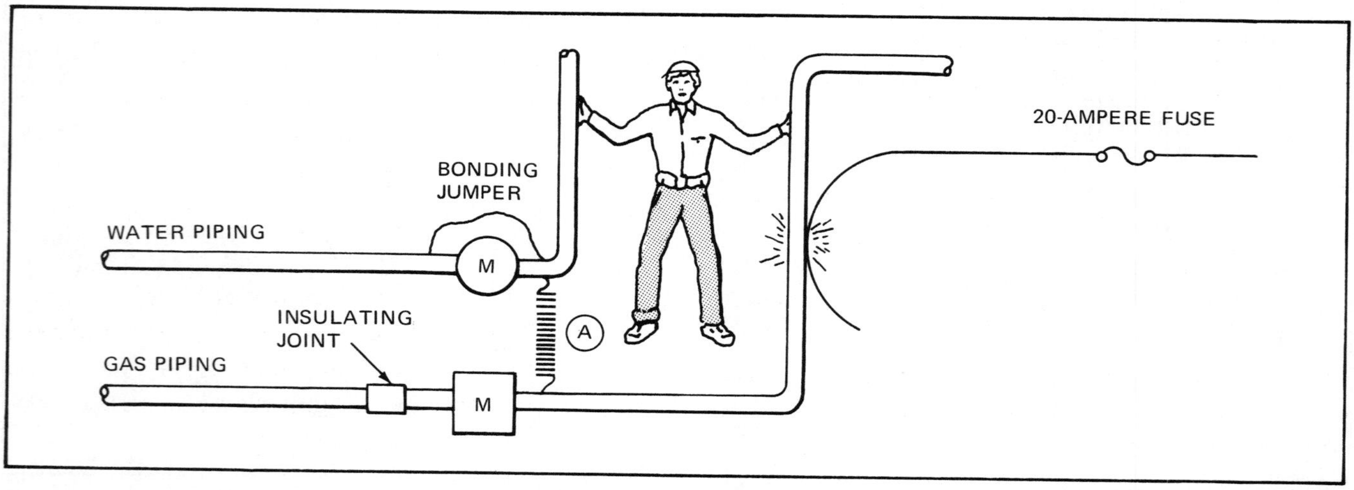

Fig. 27-14 System grounding.

the metal enclosures, the service switch, and the neutral conductor are bonded together to form a *grounding electrode system, Section 250-81*.

For discussion purposes regarding this residence, the gas piping is selected as the additional electrode as required by *Section 250-81(a)*. The gas pipe supplements the water pipe ground. This arrangement is permitted only if it is authorized by the gas utility and approved by the electrical inspector. The gas piping must not contain insulating sections or insulating joints.

If the gas piping is not selected as the required supplemental grounding electrode, then any one of the following items may be used:

- the metal frame of a building (the residence is constructed of wood).
- at least 20 feet (6.1 m) of steel reinforcing bars, 1/2-inch (12.7 mm) minimum diameter, or at least 20 feet (6.1 m) of bare copper conductor not smaller than No. 4 AWG. Both of these must be encased in concrete at least two-inches (50.8-mm) thick and in direct contact with the earth, such as near the bottom of a foundation or footing.
- at least 20 feet (6.1 m) of bare copper wire encircling the building, having a minimum size of No. 2 AWG, buried directly in the earth at least 2-1/2 feet (762-mm) deep.
- ground rods – at least one rod*, not less than 8 feet (2.44 m) in length. See *Section 250-83(c)*.
- ground plates – at least 2 square feet (0.186 m^2).* See *Section 250-83(d)*.

  *More than one rod or plate must be installed if the resistance to ground exceeds 25 ohms. See *Section 250-84*.

The requirements for a grounding electrode system are covered in NEC *Article 250*, parts G, H, J, and K.

There is little doubt that this concept of *grounding electrode systems* gives rise to many interpretations of the Code. The electrician must check with the local code authority to determine the local interpretation. For example, some electrical inspectors may not require a bonding jumper to be installed between the cold and hot water pipes, as shown in figure 27-4. They are of the opinion that an adequate bond is made through the water heater itself.

Other electrical inspectors will require that the cold and hot water pipes be bonded together. One reason for this requirement is that some water heaters contain insulating fittings that reduce corrosion caused by electrolysis inside the tank. Another reason given to justify their stand is that while the water heater installed first may contain no insulating fittings, a replacement heater may have insulating fittings. Thus, there will be no bond between the hot and cold water pipes. When there is any doubt as to the bonding requirements, bond the pipes together as shown in figure 27-4.

When grounding service-entrance equipment, figure 27-15, the following Code rules must be observed.

- The system must be grounded when the maximum voltage to ground does not exceed 150 volts, *Section 250-5(b)(1)*.
- All grounding schemes shall be installed so that no objectionable currents will flow over the grounding conductors and other grounding paths, *Section 250-21(a)*.
- The ground electrode conductor must be connected to the supply side of the service disconnecting means. It must not be connected to any grounded circuit conductor on the load side of the service disconnect *Section 250-23(a)*.
- The neutral conductor must be grounded, *Section 250-25*.
- Tie (bond) everything together. See the following section of this unit on bonding, *Sections 250-80(a)* and *(b)*, *250-81*, and *250-71*.
- The grounding electrode conductor used to connect the grounded neutral conductor to the grounding electrode must not be spliced, *Section 250-53*.
- The grounding electrode conductor is to be sized acccording to *Table 250-94*.
- The metal hot and cold water piping system shall be bonded to the service equipment enclosure, to the grounded conductor at the service, and to the grounding electrode conductor, *Section 250-80(a)*.
- The grounding electrode conductor must be connected to the metal underground water

pipe when 10 feet (3.05 m) long or more, including the well casing, *Section 250-81.*

- In addition to grounding the service equipment to the underground water pipe, an additional electrode must be used, such as a bare conductor in the footing, *Section 250-81(c)*, a grounding ring, *Section 250-81(d)*, a metal underground gas piping system, *Section 250-83(a)*, rod or pipe electrodes, *Section 250-83(c)*, or plate electrodes, *Section 250-83(d).*
- The grounding electrode conductor shall be copper, aluminum, or copper-clad aluminum *Section 250-91.*
- The grounding electrode conductor may be solid or stranded, uninsulated, covered or bare, and must not be spliced, *Section 250-91.*
- Bonding shall be provided around all insulating joints or sections of the metal piping system that may be disconnected, *Section 250-112.*
- The connection to the grounding electrode must be accessible, *Section 250-112.*
- The grounding conductor must be connected tightly using the proper lugs, connectors, clamps, or other approved means, *Section 250-115.*

Section 230-98 should be reviewed at this point (this section is described in detail in *Electrical Wiring—Commercial*). This section requires adequate interrupting capacity for services (discussed later in this unit).

This residence is supplied by three No. 2/0 RHW service-entrance conductors. According to *Table 250-94*, a No. 4 AWG grounding conductor is required. This conductor may be run in conduit or cable armor, or it may be run exposed if it is not to be subjected to severe physical damage. Thus, No. 4 AWG armored ground cable is run from the top of Panel A. The cable is concealed above the lavatory, passage, and recreation room ceilings. It is stapled beside the utility room joists and runs into the pump room. Using a grounding clamp, the ground cable is then connected to the water pipe where it enters the basement wall.

BONDING

Section 250-71 lists the parts of the service-entrance equipment that must be bonded. *Section 250-72* lists the methods approved for bonding this

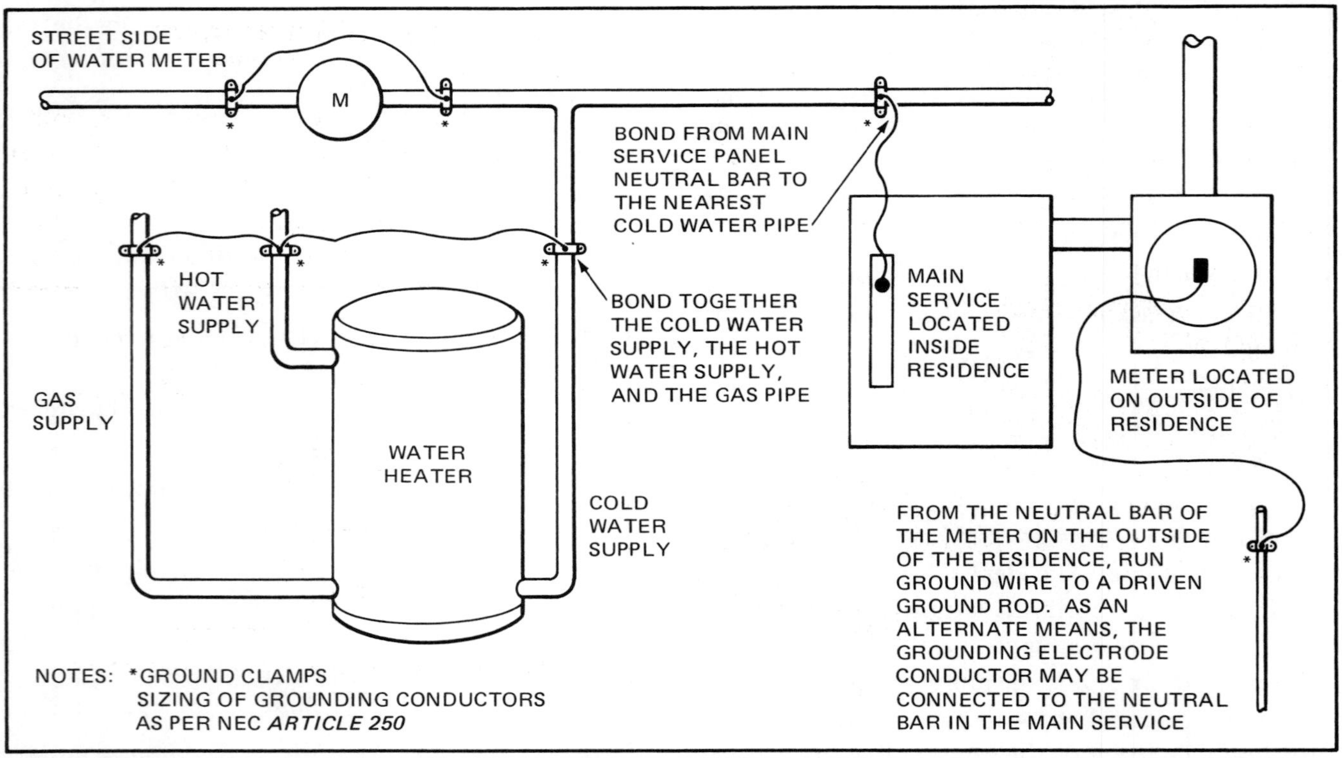

Fig. 27-15 Diagram shows one method that may be used to provide proper grounding and bonding of service-entrance equipment for a typical one-family residence.

service equipment. Bonding bushings, figure 27-16, and bonding jumpers are installed on service-entrance equipment to insure a low-impedance path to ground if a fault occurs on any of the service-entrance conductors. Service-entrance conductors are not fused at the service head. Thus, the short-circuit current on these conductors is limited only by the following: (1) the capacity of the transformer or transformers supplying the service equipment and (2) the distance between the service equipment and the transformers. The short-circuit current can easily reach 20 000 amperes or more in dwellings. Fault currents can easily reach 40 000 to 50 000 amperes or more in apartments, condominiums, and similar dwellings. These installations are served by a large-capacity transformer located close to the service-entrance equipment and the metering. **This extremely high fault current produces severe arcing which is a fire hazard.** The use of proper bonding reduces this hazard to some extent.

Fig. 27-16 Insulated bonding bushing with grounding lug.

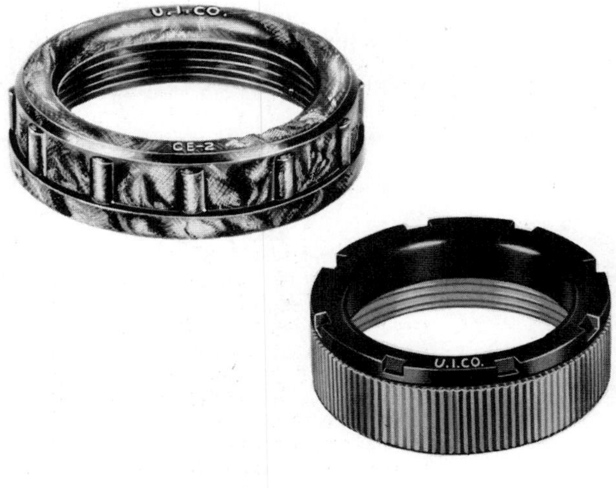

Fig. 27-17 Insulated bushings.

Fault current calculations are presented later in this unit. *Electrical Wiring–Commercial* covers these calculations in much greater detail.

Section 250-79 states that the main bonding jumpers must be not smaller than the grounding electrode conductor. The grounding lugs on bonding bushings are sized by the trade size of the bushing. The lugs become larger as the size of the bushing increases.

Section 373-6(c) states that if No. 4 AWG or larger conductors are installed in a raceway, an insulating bushing or equivalent must be used, figure 27-17. This bushing protects the wire from shorting or grounding itself as it passes through the metal bushing. Combination bushings can be used. These bushings are metallic (for mechanical strength) and have plastic insulation. When conductors are installed in electrical metallic tubing, they can be protected at the fittings by the use of connectors with insulated throats.

If the conduit bushing is made of insulating material only, as in figure 27-17, then two locknuts must be used, figure 27-18.

BRANCH-CIRCUIT OVERCURRENT PROTECTION

The overcurrent devices commonly used to protect branch circuits in dwellings are fuses and circuit breakers. NEC *Article 240* is concerned with overcurrent protection.

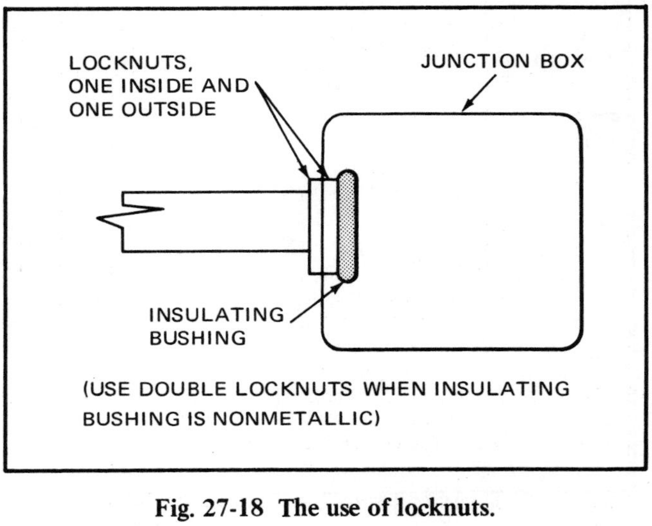

Fig. 27-18 The use of locknuts.

Plug Fuses, Fuseholders, and Adapters (*Article 240, Part E*)

Fuses are a reliable and economical form of overcurrent protection. *Sections 240-50* through *240-54* give the requirements for plug fuses, fuseholders and adapters. These requirements include the following:

- the protective devices shall not be used in circuits exceeding 125 volts between conductors. An exception to this rule is for a system having a grounded neutral where no conductor is more than 150 volts to ground. (This is the case for the 120/240-volt system used in the residence in the plans.)
- the fuses shall have ampere ratings of 0 to 30 amperes.
- plug fuses shall have a hexagonal configuration somewhere on the fuse when rated at 15 amperes or less.
- the screw shell of the fuseholder must be connected to the load side of the circuit.
- Edison-base plug fuses may be used only to replace fuses in existing installations where there is no sign of overfusing or tampering.
- all new installations shall be in type S fuses.
- Type S fuses are classified at 0 to 15 amperes, 16 to 20 amperes, and 21 to 30 amperes. The reason for this classification is given in the following paragraph.

When the electrician installs fusible equipment, the ampere rating of the various circuits must be determined. Based on this rating, an adapter of the proper size is inserted into the Edison-base fuseholder. The proper Type S fuse is then placed in the adapter. Because of the adapter, the fuseholder is nontamperable and noninterchangeable. For example, assume that a 15-ampere adapter is inserted for the 15-ampere branch circuits in the residence. It is impossible to substitute a Type S fuse with a larger rating without removing the 15-ampere adapter. Another type of fuseholder that can be used is molded into the shape required for the various sizes of Type S fuses.

Type S fuses and adapters, figure 27-19, are available with ratings in the range from 3/10 of an ampere to 30 amperes. The Type S fuse may have a time-delay feature. If it does, it is known as a dual-element fuse. Momentary overloads do not cause a dual-element fuse to blow. An example of such an overload is the current surge as an electric motor is started. However, a dual-element fuse opens rapidly if there is a heavy overload or a short circuit.

Dual-element Fuse

A dual-element fuse has two fusible elements connected in series. These elements are known as the *thermal cutout element* and the *fuse link element*. When an excessive current flows, one of the elements opens. The amount of excess current determines which element opens.

The electrical characteristics of these two elements are very different. Thus, a dual-element fuse has a greater range of protection than the single-element fuse. The thermal cutout element opens circuits on currents in the low overload range (up to about 500 percent of the fuse rating). The fuse

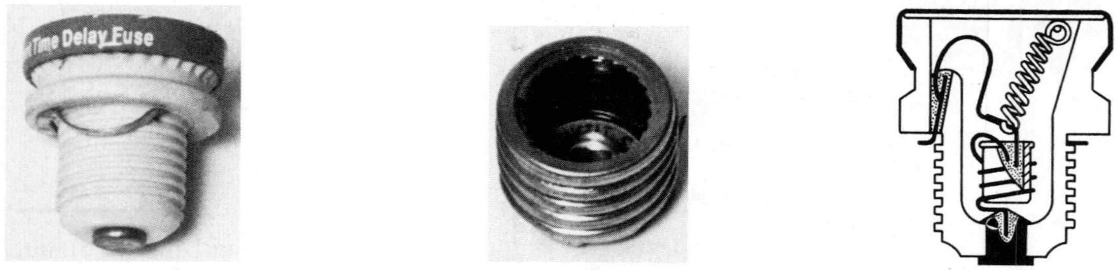

TYPE S FUSE (0-30 AMPERES)

ADAPTER

TYPE DELAY (DUAL-ELEMENT) TYPE S FUSE

Fig. 27-19 Type S fuses and adapter.

link element handles only short circuits and heavy overload currents (about 500 percent of the fuse rating).

Thermal Cutout Element. The thermal cutout opens when excessive heat is developed in the element. This heat may be the result of a loose connection or poor contact in the fuseholder or excessive current. When the temperature reaches 280°F, a fusible alloy melts and the element opens. Any excessive current produces heat in the element. However, the mass of the element absorbs a great deal of heat before the cutout opens. A small excess of current will cause this element to open if it continues for a long period of time. This characteristic gives the thermal cutout element a large time lag on low overloads (up to 10 seconds at a current of 500 percent of the fuse rating). In addition, it provides very accurate protection for prolonged overloads.

Fuse Link Element. The capacity of the fuse link element is high enough to prevent it from opening on low overloads. This element is designed to clear the circuit quickly of short circuits or heavy overloads above 500 percent of the fuse ratings.

Dual-element fuses are used on motor and appliance circuits where the time-lag characteristic of the fuse is required. Single-element fuses do not have this time lag. Such fuses blow as soon as an overcurrent condition occurs. The homeowner may be tempted to install a fuse with a higher rating. The fuseholder adapter, however, prevents the use of such a fuse. Thus, the dual-element fuse is recommended for this type of situation.

Three basic types of dual-element fuses are available. The standard plug fuse can be used only to replace blown fuses on existing installations. The type S fuse is required on all new installations. The cartridge fuse is a dual-element fuse which is available in both ferrule and knife blade styles.

Cartridge Fuses (*Article 240, Part F*)

The most common type of cartridge fuse is shown in figure 27-20. These fuses are available in 250-volt and 600-volt sizes with ratings from 0 to 600 amperes.

Section 240-60(a) recognizes the use of a type of cartridge fuse that is rated at not over 60 amperes for circuits of 300 volts or less to ground. The physical size of this fuse is smaller than that of standard cartridge fuses. The time-delay characteristics of this fuse mean that it can handle harmless current surges or momentary overloads without blowing. However, these fuses open very rapidly under short-circuit conditions.

The small cartridge fuses are Type S fuses in cartridge form. They are called Type SC fuses, figure 27-21. These fuses prevent the practice of overfusing since they are size limiting for their ampere ratings. For example, a fuseholder designed to accept a 15-ampere, Type SC fuse will not accept a 20-ampere, Type SC fuse. In the same manner, a fuseholder designed to accept a 20-ampere, Type SC fuse will not accept a 30-ampere, Type SC fuse.

The standard ratings for fuses are given in *Section 240-6*. These ratings range from 1 ampere to 6000 amperes.

Another type of physically small fuse is the Class T fuse, figure 27-22. This type of fuse has a high-interrupting rating in sizes from 0 to 600 amperes in both 300-volt and 600-volt ratings. Such fuses can be used as the main fuses in a panel

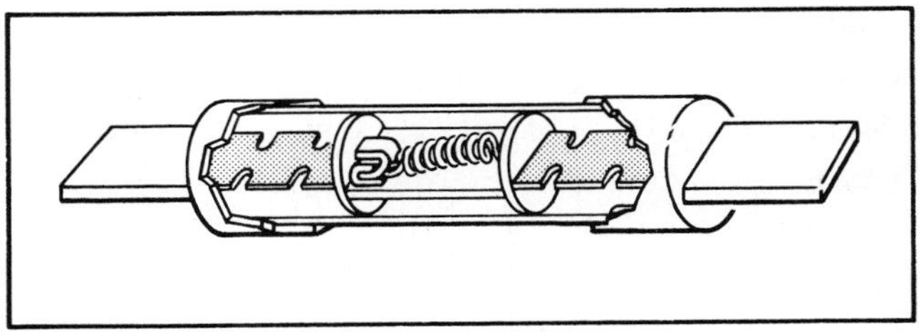

Fig. 27-20 Cartridge-type dual-element fuse (0–600 amperes).

Fig. 27-21 Type SC fuses.

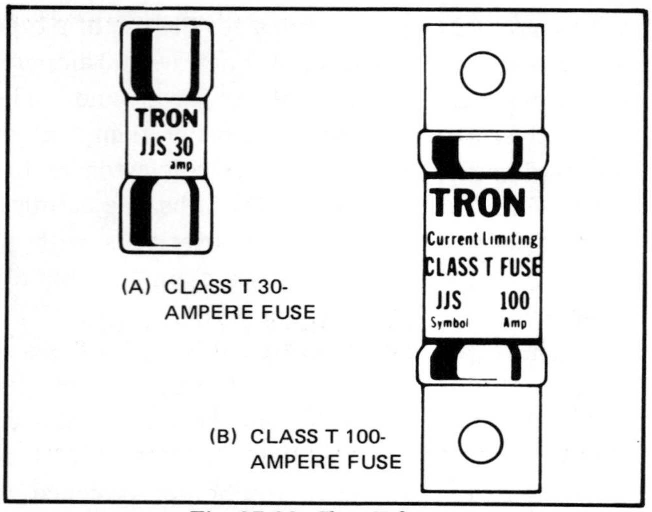

Fig. 27-22 Class T fuses.

having circuit breaker branches, figure 27-23. In this case, the Class T fuses protect the low-interrupting capacity breakers against high-level short-circuit currents.

Circuit Breakers (*Article 240, Part G*)

Installations in dwellings normally use thermal-magnetic circuit breakers. On a continuous overload, a bimetallic element in such a breaker moves until it unlatches the inner tripping mechanism of the breaker. Momentary small overloads do not cause the element to trip the breaker. If the overload is heavy or if there is a short circuit, a magnetic coil in the breaker causes it to interrupt the branch circuit instantly. (Unit 11 covers the effect of tungsten lamp loads on circuit breakers.)

Sections 240-80 through *240-83* give the requirements for circuit breakers. The following points are taken from these sections.

- Circuit breakers shall be trip free so that even if the handle is held in the on position, the internal mechanism will trip to the off position.
- Breakers shall indicate clearly whether they are on or off.
- A breaker shall be nontamperable so that it cannot be readjusted (trip point changed) without dismantling the breaker or breaking the seal.
- The rating shall be durably marked on the breaker. For small breakers rated at 100 amperes or less and 600 volts or less, the rating must be molded, stamped, or etched on the handle (or on another part of the breaker that will be visible after the cover of the panel is installed).
- Every breaker with an interrupting rating other than 5000 amperes shall have this rating marked on the breaker.
- Circuit breakers rated at 120 volts and used for fluorescent loads, shall not be used as switches unless marked "SWD."

Most circuit breakers are ambient temperature compensated. This means that the tripping point of the breaker is not affected by an increase in the surrounding temperature. An ambient-compensated breaker has two elements. One element heats up due to the current passing through it and the heat in the surrounding area. The other element heats up because of the surrounding air only. The actions of these elements oppose each other. Thus, as the tripping element tends to lower its tripping point because of external heat, the second element opposes the tripping element and stabilizes the tripping point. As a result, the current through the tripping element is the only factor that causes the element to open the circuit. It is a good practice to turn the breaker on and off periodically to "exercise" its moving parts.

Section 240-6 gives the standard ampere ratings of circuit breakers.

INTERRUPTING RATINGS FOR FUSES AND CIRCUIT BREAKERS

Section 110-9 states that all fuses, circuit breakers, and all other electrical devices that break current shall have an interrupting capacity sufficient for the voltage employed and for the current which must be interrupted.

According to *Section 110-10*, all overcurrent devices, the total circuit impedance, and the withstand capability of all circuit components (wires, contactors, and so on), must be selected so that

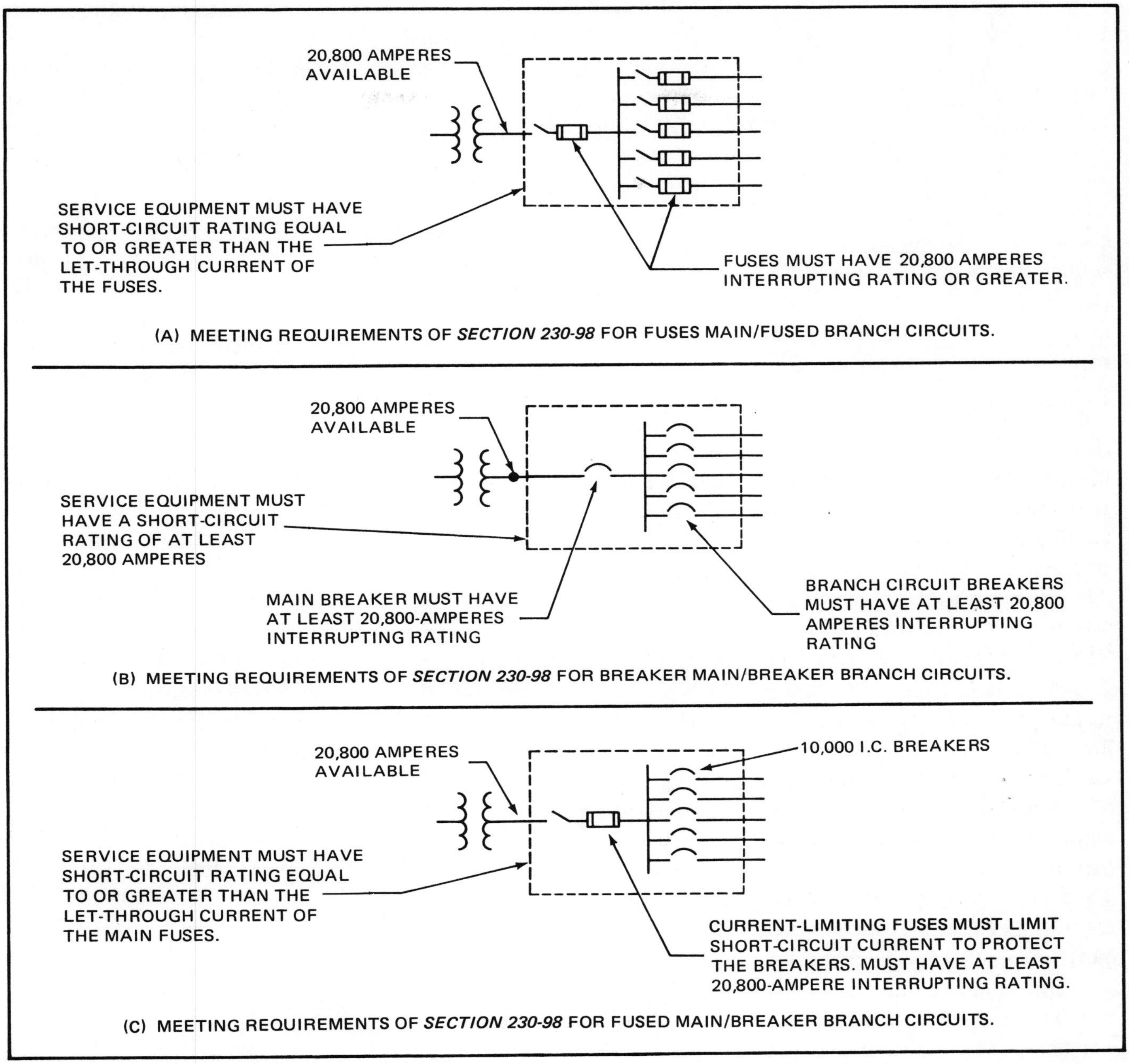
20,800 AMPERES
AVAILABLE
SERVICE EQUIPMENT MUST HAVE SHORT-CIRCUIT RATING EQUAL TO OR GREATER THAN THE LET-THROUGH CURRENT OF THE FUSES.
FUSES MUST HAVE 20,800 AMPERES INTERRUPTING RATING OR GREATER.
(A) MEETING REQUIREMENTS OF SECTION 230-98 FOR FUSES MAIN/FUSED BRANCH CIRCUITS.
20,800 AMPERES
AVAILABLE
SERVICE EQUIPMENT MUST HAVE A SHORT-CIRCUIT RATING OF AT LEAST 20,800 AMPERES
MAIN BREAKER MUST HAVE AT LEAST 20,800-AMPERES INTERRUPTING RATING
BRANCH CIRCUIT BREAKERS MUST HAVE AT LEAST 20,800 AMPERES INTERRUPTING RATING
(B) MEETING REQUIREMENTS OF SECTION 230-98 FOR BREAKER MAIN/BREAKER BRANCH CIRCUITS.
20,800 AMPERES
AVAILABLE
10,000 I.C. BREAKERS
SERVICE EQUIPMENT MUST HAVE SHORT-CIRCUIT RATING EQUAL TO OR GREATER THAN THE LET-THROUGH CURRENT OF THE MAIN FUSES.
CURRENT-LIMITING FUSES MUST LIMIT SHORT-CIRCUIT CURRENT TO PROTECT THE BREAKERS. MUST HAVE AT LEAST 20,800-AMPERE INTERRUPTING RATING.
(C) MEETING REQUIREMENTS OF SECTION 230-98 FOR FUSED MAIN/BREAKER BRANCH CIRCUITS.

Fig. 27-23 Applications of fuses and breakers to meet the requirements of the National Electrical Code, *Section 230-98*.

minimal damage will result in the event of a fault, either line-to-line or line-to-ground.

The overcurrent protective device must be able to interrupt the current that may flow under any condition (overload or short circuit). Such interruption must be made with complete safety to personnel and without damage to the panel or switch in which the overcurrent device is installed. **Overcurrent devices with inadequate interrupting ratings are, in effect, bombs waiting for a short circuit to trigger them into an explosion. Personal injury may result and serious damage will be done to the electrical equipment.**

Type SC fuses have an interrupting rating of 100 000 amperes root mean square (rms) symmetrical. Plug fuses can interrupt no more than 10 000 amperes rms symmetrical. Cartridge dual-element fuses (figure 27-20) and Class T fuses (figure 27-22) have interrupting ratings of 200 000 amperes rms symmetrical. These interrupting ratings are listed by Underwriters' Laboratories. The interrupting rating of a fuse is marked on its label when the rating is other than 10 000 amperes.

Short-circuit Currents

This text does not cover in detail the methods of calculating short-circuit currents. (See *Electrical Wiring–Commercial.*) The ratings required to determine the maximum available short-circuit current delivered by a transformer are the kVA and impedance values of the transformer. The size and length of wire installed between the transformer and the overcurrent device must be considered as well.

The transformers used in modern electrical installations are efficient and have very low impedance values. A low-impedance transformer having a given kVA rating delivers more short-circuit current than a transformer with the same kVA rating and a higher impedance. When an electrical service is connected to a low-impedance transformer, the problem of available short-circuit current is very serious. *Section 230-98* gives the interrupting rating requirements for services. The possible situations are shown in figure 27-23.

The examples given in figure 27-23 show that the available short-circuit current at the transformer secondary terminals is 20 800 amperes.

Three types of installations meet the requirements of NEC *Section 230-98.*

a. Fusible panels are used where the mains and branches all have fuses with adequate interrupting capacities: 10 000, 50 000, or 100 000 amperes, or whatever fault current is available.
b. Panels are used with main breakers and branch-circuit breakers having adequate interrupting capacities: 10 000, 25 000, 50 000, or 100 000 amperes, or whatever fault current is available.
c. A fusible main disconnect switch or pullout is used with current-limiting fuses and 10 000-ampere interrupting capacity branch-circuit breakers. The current-limiting fuses protect the breakers by virtue of their current-limiting characteristic.

Determining Short-circuit Current

The local power utility and the electrical inspector are good sources of information when short-circuit current is to be determined.

A simplified method is given as follows to determine the approximate available short-circuit current at the terminals of a transformer.

1. Determine the normal full-load secondary current delivered by the transformer.
   For single-phase transformers:
   $$I = \frac{kVA \times 1000}{E}$$
   For three-phase transformers:
   $$I = \frac{kVA \times 1000}{E \times 1.73}$$
   Where I = current, in amperes
   kVA = kilovolt-amperes (transformer nameplate rating)
   E = secondary line-to-line voltage transformer nameplate rating)
2. Using the impedance value given on the transformer nameplate, find the multiplier to determine the short-circuit current.
   $$\text{Multiplier} = \frac{100}{\text{percent impedance}}$$
3. The short-circuit current = normal full-load secondary current × multiplier.

Example: A transformer is rated at 100 kVA and 120/240 volts. It is a single-phase transformer with an impedance of one percent (from the transformer nameplate). Find the short-circuit current.

For a single-phase transformer:

$$I = \frac{\text{kVA} \times 1000}{E}$$

$$I = \frac{100 \times 1000}{240}$$

$$= 416 \text{ amperes, full-load current}$$

The multiplier for a transformer impedance of one percent is:

$$\text{multiplier} = \frac{100}{\text{percent impedance}}$$

$$= \frac{100}{1}$$

$$\text{multiplier} = 100$$

The short-circuit current = 1 × multiplier
= 416 amperes × 100
= 41 600 amperes

Thus, the available short-circuit current at the terminals of the transformer is 41 600 amperes.

This value decreases as the distance from the transformer increases. If the transformer impedance is 1.5 percent, the multiplier is:

$$\text{multiplier} = \frac{100}{1.5} = 66.6$$

The short-circuit current = 416 × 66.6
= 27 706 amperes

If the transformer impedance is two percent:

$$\text{multiplier} = \frac{100}{2} = 50$$

The short-circuit current = 416 × 50
= 20 800 amperes

(*Note:* A short-circuit current of 20 800 amperes is used in figure 27-23).

REVIEW

Note: Refer to the Code or the plans where necessary.

1. Where does an overhead service start and end? ____________________

2. What are service-drop conductors? ____________________

3. Who is responsible for determining the service location? ____________________

4. a. The service head must be located (above) (below) the point where the service-drop conductors are spliced to the service-entrance conductors (underline one).
   b. What Code section provides the answer to part a? ____________________

5. What is a mast-type service entrance? ____________________

6. a. What size and type of conductors are installed for this service? ________
   b. What size conduit is installed? ________
   c. What size ground wire is installed? (Not neutral) ________
   d. Is the ground wire insulated, armored, or bare? ________

7. How and where is the ground wire attached to the water pipe? ________

   ____

8. a. When a conduit is extended through a roof, must it be "guyed"? ________
   b. How far through the roof is this service conduit extended? ________

9. What are the minimum distances or clearances for the following?
   a. Service-drop clearance over private driveway ________
   b. Service-drop clearance over private sidewalks ________
   c. Service-drop clearance over alleys ________
   d. Service-drop clearance over a roof having a roof pitch of not less than 4/12. (Voltage between conductors does not exceed 300 volts.) ________

   ____

   e. Service-drop horizontal clearance from a porch ________
   f. Service-drop clearance from a fence that can be climbed ________

10. What size ungrounded conductors are installed for each of the following residential services (use Type THW copper conductors). See *Note 3* to *Table 310-16.*
    a. 60-ampere service? No. ________ THW Copper
    b. 100-ampere service? No. ________ THW Copper
    c. 200-ampere service? No. ________ THW Copper

11. What size grounding conductors are installed for the services listed in question 10?
    a. 60-ampere service? No. ________ AWG grounding conductor
    b. 100-ampere service? No. ________ AWG grounding conductor
    c. 200-ampere service? No. ________ AWG grounding conductor

12. What is the recommended height of a meter socket from the ground? ________

13. a. May the bare grounded neutral conductors of a service be buried directly in the ground? ________
    b. What section of the Code covers this? ________

14. What exceptions are made regarding the use of bare neutral conductors installed underground? ________

15. How far must mechanical protection be provided when underground service conductors are carried up a pole? ________

16. a. A method of service disconnect may consist of how many switches or circuit breakers?

    ____

    b. Must these devices be in one enclosure? ________
    c. What type of main disconnect is provided for this residence? ________

17. Complete the following table by filling in the columns with the appropriate information.

	Circuit Number	Ampere Rating	Poles	Volts	Wire Size
a. Dining Room Convenience Receptacle					
b. Workshop Convenience Receptacle					
c. Water Pump					
d. Attic Fan					
e. Kitchen Lighting					
f. Utility Room Heating					
g. Attic Light					
h. Counter-mounted Cooking Unit					
i. Utility Room Lighting					

18. a. What size conductors supply Panel B? ______
    b. What size conduit? ______
    c. Is this conduit run in the form of electrical metallic tubing or rigid conduit? ______

    ______

    d. What size overcurrent device protects the feeders to Panel B? ______

19. What is the ampere rating of the main copper buses (or bus bars) in Panel B? (See Specifications) ______

20. How many electric meters are provided for this residence? ______

21. a. According to the Code, is it permissible to ground rural service-entrance systems and equipment to driven ground rods only when a water system is available? ______
    b. What section of the Code applies? ______

22. What table of the Code lists the sizes of grounding conductors to be used for service entrances of various sizes similar to the type found on this residence? ______

23. Do the following conductors require mechanical protection?
    a. No. 8 grounding conductor ______
    b. No. 6 grounding conductor ______
    c. No. 4 grounding conductor ______

24. Why is bonding of service-entrance equipment necessary? ______

    ______

    ______

25. What special types of bushings are required on service entrances? ______

26. When No. 4 AWG conductors or larger are installed in conduit, what additional provision is required on the conduit ends? ______

    ______

    ______

27. What size bonding conductors must be used for this residence? ______

28. a. What is a Type S fuse? ______
    b. Where must Type S fuses be installed? ______

29. a. What is the maximum voltage permitted between conductors when using plug fuses? ______
    b. May plug fuses (Type S) be installed in a switch that disconnects a 120/240-volt clothes dryer? ______
    c. Give a reason for the answer to b. ______

30. Will a 20-ampere, Type S fuse fit properly into a 15-ampere adapter? ______

31. What part of a circuit breaker causes the breaker to trip:
    a. on an overload? ______ b. on a short circuit? ______

32. What is meant by an ambient-compensated circuit breaker? ______

33. List the standard sizes of circuit breakers up to and including 100 amperes. ______

34. Using the method shown in this unit, what is the approximate short-circuit current available at the terminals of a 50-kVA single-phase transformer rated at 120/240 volts? The transformer impedance is one percent. ______

35. Where is the service for this residence located? ______

36. a. On what type of wall is Panel A fastened? ______
    b. On what type of wall is Panel B fastened? ______

37. a. How many feet (meters) of conduit are required for the service as measured from the service head to Main Panel A? ______
    b. How many feet (meters) of wire are required? Allow three feet (9.14 mm) to extend beyond the service head. ______
    c. How many feet (meters) of ground wire are required? (To system ground) ______

38. a. How many feet (meters) of conduit are required to install the feeders from Panel A to Panel B? ______
    b. How many feet (meters) of wire are required for this installation? ______

39. If the top edge of the service conduit through the basement wall is 12 inches (305 mm) below the bottom edge of the joists in the workshop, approximately how far above the finish grade is the conduit fitting? ______

40. State three possible combinations of service equipment which meet the requirements of *Section 230-98*.

a. ______

b. ______

c. ______

41. When conduits pass through the wall from outside to inside, the conduit must be ______ to prevent air circulation through the conduit.

42. Briefly explain why electrical systems and equipment are grounded? ______

43. What Code section states that all overcurrent devices must have adequate interrupting ratings for the current to be interrupted? ______

44. All electrical components have some sort of "withstand rating." This rating indicates the ability of the component to withstand fault currents for the time required by the overcurrent device to open the circuit. What Code section refers to withstand ratings with reference to overcurrent protection? ______

45. Arcing fault damage is closely related to the value of ______

46. In general, systems are grounded so that the maximum voltage to ground does not exceed: (underline one)

a. 120 volts b. 150 volts c. 300 volts

47. To insure a complete grounding electrode system, (underline one)

a. everything must be tied together.
b. all metal pipes and conduits must be isolated from one another.
c. the service neutral is grounded to the water pipe only.

48. An electric clothes dryer is rated at 5700 watts. The electric rate is 5 cents per kWh. The dryer is used continuously for 3 hours. Find the cost of operation, assuming that the heating element is on continuously. ______

49. A heating cable rated at 750 watts is used continuously for 72 hours to prevent snow from freezing in the gutters of the house. The electric rate is 4.7 cents per kWh. Find the cost of operation. ______

unit 28

Service-Entrance Calculations

OBJECTIVES

After studying this unit, the student will be able to

- determine the total calculated load of the residence using the methods of *Article 220*.
- calculate the size of the service-entrance conductors, including the size of the neutral conductor.

The branch-circuit load values determined in earlier units of this text are now used to illustrate the proper method of determining the size of the service-entrance conductors for the residence. The calculations are based on National Electrical Code requirements. The student must check local and state electrical codes for any variations in requirements that may take precedence over the National Electrical Code.

SIZE OF SERVICE-ENTRANCE CONDUCTORS AND SERVICE DISCONNECTING MEANS

Service-entrance conductors, *Section 230-41 (b)(1)(2)*, and the disconnecting means, *Section 230-79(c)*, shall be not smaller than:

(1) for a single-family dwelling with six or more two-wire circuits – 100 ampere, three wire.

(2) for a single-family dwelling with an initial computed load of 10 kW or more – 100 ampere, three wire.

The electric range used in a dwelling generally has a rating of at least 8 kW, *Table 220-19*. Thus, only 2 kW are available for the remaining appliance, lighting, and special-purpose circuits. There are few, if any, single-family dwellings for which a service smaller than 100 amperes can be installed.

Two methods are permitted by the Code to determine the size of the service-entrance conductors for a dwelling. Method No. 1 is outlined in *Article 220, Parts A and B*; the second method is given in *Section 220-30, Article 220, Part C.*

METHOD NO. 1 – *Article 220, Parts A and B*

(*Note:* The values used in this procedure are taken from calculations in previous units of this text.)

General lighting load:		
2 582 square feet at 3 watts per square foot		7 746 watts
Appliance circuits:		
Dining Room	1	
Kitchen	4	
Laundry	4	
Workshop	1	
Total	10 circuits	
Ten circuits at 1 500 watts per circuit		15 000 watts
Section 220-16		22 746 watts
Application of demand factor, *Table 220-11:*		
3 000 watts at 100%		3 000 watts
22 746 - 3 000 = 19 746 at 35%		6 911 watts
Net computed load, less range:		9 911 watts

Wall-mounted oven and counter-mounted cooking unit, *Table 220-19, Note 4*

Wall-mounted oven load	7 450 watts	
Cooking unit load	6 600 watts	
	14 050 watts or 14 kW	
14 kW exceeds 12 kW by 2 kW		
2 kW × 5% = 10% increase		
8 kW × 0.10 = 0.8 kW (from *Column A, Table 220-19*)		
8 + 0.8 = 8.8 kW		8 800 watts
Net computed load *with* range:		18 711 watts
Electric space heat, *Section 220-15*		15 500 watts
Dryer, *Table 220-18*		5 000 watts
Subtotal		39 211 watts

Air conditioners, three at 8 amperes each:

3 × 8 = 24 amperes

24 × 240 volts = 5 760 watts; this value is less than 15 500 watts of space heating. Therefore, the air-conditioning load need not be included in the service calculation, *Section 220-21.*

Water heater		3 000 watts
Dishwasher		
Motor: 7.2 × 120 =	864 watts	
Heater:	1 000 watts	
Therefore, maximum demand is the heater only		1 000 watts
Food Waste Disposer 7.2 × 120		864 watts
Water pump 8 × 240 × 1.25 *Section 430-22*		2 400 watts
Garage door openers 2 × 5.8 × 120		1 392 watts
Built-in refrigerator-freezer 5.8 × 120		696 watts
Combination light, heater, and exhaust fan		1 475 watts
Exhaust fan 5.8 × 120		696 watts
Total		11 523 watts

Since there are more than four appliances, in addition to the cooking units, air conditioning, space heating, and clothes dryer, a demand factor of 75 percent may be applied to the fixed load, *Section 220-17.*

11 523 × 0.75	8 642 watts
Total calculated load 8 642 + 39 211	47 853 watts

$$\text{Amperes} = \frac{\text{watts}}{\text{volts}} = \frac{47\,853}{240} = 199.4 \text{ amperes}$$

Table 310-16 and *Note 3* to this table indicate that No. 2/0 RHW wires or equivalent are required for the service-entrance conductors.

Size of Neutral Conductor

Section 220-22 states that the neutral feeder load shall be the maximum unbalance of the load determined by the method of *Article 220*. The maximum unbalanced load shall be the maximum connected load between the neutral and any one ungrounded conductor.

When these calculations are checked, it is evident that much of the load is due to 240-volt equipment. Thus, when only this equipment is operating, no current flows in the neutral conductor. The size of the neutral conductor may be reduced in this case as follows:

General lighting load and small appliance load after applying demand factors	9 911 watts
Range load = 8 800 watts at 70 percent, *Section 220-22*	6 160 watts
Dryer motor = 5.8 × 120	696 watts
Total	16 767 watts
Electric space heat, Water heater, Water pump, Air conditioners, Dryer heating element } 240 volts	
Dishwasher (maximum demand, heater only)	1 000 watts
Food waste disposer 7.2 × 120 × 1.25, *Section 220-14*	1 080 watts
Garage door openers 2 × 5.8 × 120	1 392 watts
Built-in refrigerator-freezer 5.8 × 120	696 watts
Combination light, heater, and exhaust fan	1 475 watts

Exhaust fan		
5.8 X 120		696 watts
	Total	6 339 watts

Apply a 75-percent demand factor, *Section 220-17*

6 339 X 0.75 = 4 754 watts

Then, 16 767 + 4 754 = 21 521 watts

$$\text{Amperes} = \frac{\text{watts}}{\text{volts}} = \frac{21\,521}{240} = 89.67 \text{ amperes}$$

The neutral to be installed must be able to carry 89.67 amperes. Referring to *Table 310-16, Note 3*, the neutral can be a No. 4 RHW (100-ampere) conductor or equivalent. The grounded neutral conductor may be bare as outlined in *Sections 230-30* and *230-40*. Many electricians prefer to reduce the neutral conductor one or two sizes below the size of the hot conductors. This procedure does not always result in a correctly sized neutral conductor. However, for most installations and especially for dwellings equipped with electric ranges, dryers, and water heaters, a neutral conductor one size smaller than the ungrounded conductors is more than adequate. The correct neutral size should be determined first. The electrician must check local and state codes which may have more stringent requirements than the National Electrical Code.

METHOD No. 2 – *Section 220-30 (Article 220, Part C)*

A second method of determining the load for a one-family dwelling is given in *Section 220-30*. This method simplifies the calculations. In most cases, the service-entrance conductors are smaller than those permitted by *Article 220, Parts A and B*.

General lighting load, 2 582 square feet at 3 watts per square foot	7 746 watts
Ten appliance circuits at 1 500 watts per circuit	15 000 watts
Wall-mounted oven (nameplate rating)	6 600 watts
Counter-mounted cooking unit (nameplate rating)	7 450 watts
Electric space heat (each room separately controlled)	15 500 watts
Water heater	3 000 watts
Dryer	5 000 watts
Dishwasher (maximum demand, heater only)	1 000 watts
Food waste disposer 7.2 X 120	864 watts
Water pump, 8 X 240 X 1.25, *Section 430-22*	2 400 watts
Garage door openers 2 X 5.8 X 120	1 392 watts
Built-in refrigerator-freezer, 5.8 X 120	696 watts
Combination light, heater, and exhaust fan	1 475 watts
Exhaust fan	696 watts
Total	68 819 watts

The three air conditioners draw 8 amperes each:
3 X 8 = 24 amperes
Therefore, 24 X 240 = 5 760 watts

This value is less than the 15 500-watt load of the electric heat. Thus, the air-conditioner load need not be included in the service calculation, *Section 220-21*.

First 10 kW at 100%	10 000 watts
*Remaining load at 40%: 58 819 X 0.40	23 528 watts
Total calcualted load	33 528 watts

$$\text{Amperes} = \frac{\text{watts}}{\text{volts}} = \frac{33\,528}{240} = 139.7 \text{ amperes}$$

* (*Note:* The remaining load includes the electric space heating load. This is permissible only if the electric heating load is controlled by four or more separately controlled heating units. The baseboard heating units in this residence are separately controlled by eight thermostats.)

No. 1 RHW conductors or equivalent could be installed for this service, *Table 310-16, Note 3.* However, the residence in the plans is to have a full 200-ampere, 240-volt service consisting of three No. 2/0 RHW conductors.

The main disconnect in the residence is a 200-ampere pullout-type panel. Many electric utilities state that the conductors feeding this type of disconnect must be the same size as the ampere rating of the disconnect. Certain utilities also state that the neutral conductor cannot be reduced in size because it is not always possible to foresee the type of load that may be connected to the panel. As a result, it is sometimes difficult to conform to all the rules of the electric utility as well as the local electrical code. It is believed, however, that by installing three No. 2/0 RHW conductors, most local and state electrical code requirements will be satisfied.

REVIEW

Note: Refer to the Code or the plans where necessary.

1. When a service-entrance calculation results in a value of 10 kW or more, what is the minimum size service permitted by the Code? ____________________

2. a. What is the unit load per square foot for the general lighting load of a residence?

   ____

   b. What are the demand factors for the general lighting load in dwellings? __________

   ____

3. a. What is the ampere rating of the circuits that are provided for the small appliance loads? ____________________

   b. What is the minimum number of small appliance circuits permitted by the Code?

   ____

   c. How many small appliance circuits are included in this residence? __________

4. Why is the air-conditioning load for this residence omitted in the service calculations?

   ____

   ____

5. What demand factor may be applied when four or more fixed appliances are connected to a service, in addition to an electric range, air conditioner, dryer, or space heating equipment? *Section 220-17* ____________________

   ____

6. What load may be used for an electric range rated at not over 12kW? *Table 220-19*

   ____

7. What is the load for an electric range rated at 16 kW? *Table 220-19*. Show calculations.

   ____

   ____

8. What is the computed load when fixed electric heating is used in a residence? *Section 220-15* ____________________

   ____

9. On what basis is the neutral conductor of a service entrance determined? __________

   ____

   ____

10. Why is it permissible to omit the electric space heater, water heater, and certain other 240-volt equipment when calculating the neutral service-entrance conductor for this residence? ____________________

    ____

    ____

11. a. What section of the Code contains an optional method for determining service-entrance loads? ______
    b. Is this section applicable to a two-family residence? ______

12. Calculate the minimum size of service-entrance conductors required for a residence containing the following: floor area 24′ x 38′ (7.3 m x 11.6 m); 12-kW electric range; 5-kW dryer with a 240-volt heating element, 4-kW, 120-volt motor, and 1-kW light; 1800-W, 120-volt bath heater; 12-kW, 240-volt electric heat (six units); 12-ampere, 240-volt air conditioner; 3-kW, 240-volt water heater. Determine the sizes of the ungrounded conductors and the neutral conductor. Use Type RHW conductors. Calculate the required sizes using both methods outlined in this unit. Be sure to include the small appliance load.

Using Method 1: Two No. ____ RHW ungrounded conductors
One No. ____ RHW (or bare, if permitted) neutral
One No. ____ grounding conductor (to water meter)

Using Method 2: Two No. ____ RHW ungrounded conductors
One No. ____ RHW (or bare, if permitted) neutral
One No. ____ grounding conductor (to water meter)

(STUDENT CALCULATIONS)

a complete sweep of the 25 positions. Thus, the switch controls the 25 relays connected to these positions.

LOW-VOLTAGE RELAY

Figure 29-5 shows the wiring of a relay used in low-voltage remote-control systems. This relay operates contacts in the 120-volt lighting circuit. The relay has a split coil design. One coil is used to close the 120-volt circuit. The other coil opens the contactors in the 120-volt circuit. This relay operates on 24 volts. There are two No. 12 AWG black leads for the 120-volt circuits and three No. 20 AWG colored leads for the 24-volt connections.

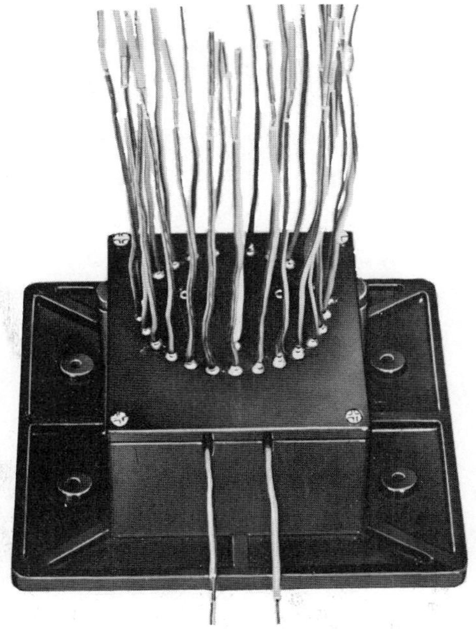

Fig. 29-4 Motorized master control.

Basically, the low-voltage relay is a double solenoid. By energizing the coil connected to the red and blue leads, the iron core tries to center itself within the coil. The core moves to the left until it closes the switch contacts. When the power supply to the coil is removed, the switch contacts remain closed. They are held in the on position by a mechanical latching device in the relay. When the coil connected to the black and blue leads is energized, the iron core tries to center itself in this coil. The core now moves to the right to open the switch contacts. The relay is mechanically latched in the off position.

Relay Mounting

The low-voltage relay is small and can be mounted from the inside of a standard outlet box through a 1/2-inch knockout opening. For quiet operation, the relay may be mounted in a 3/4-inch knockout opening using a rubber grommet. When mounted from the inside of the box, the two high-voltage leads of the relay remain inside the outlet box. The low-voltage wires are kept outside the box. The wall of the box serves as a partition between the high and low voltages. The two high-voltage leads inside the box are connected like a standard single-pole switch. The hot or black wire from the 120-volt source goes to one of these leads. The other lead wire from the relay is connected to the lamp terminal. The blue lead of the three low-voltage wires is common to both the off

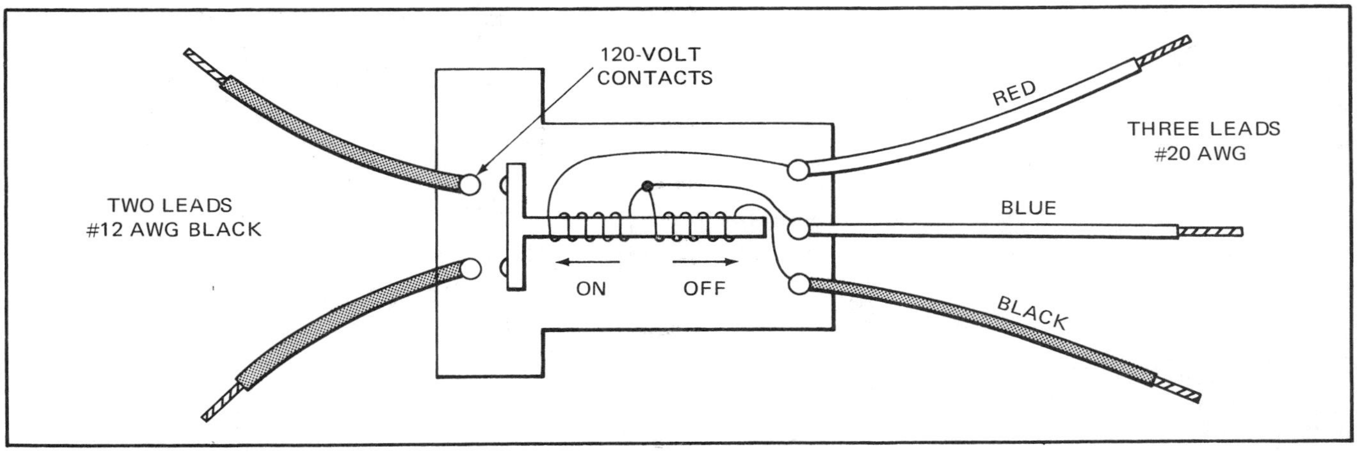

Fig. 29-5 Wiring for a low-voltage relay.

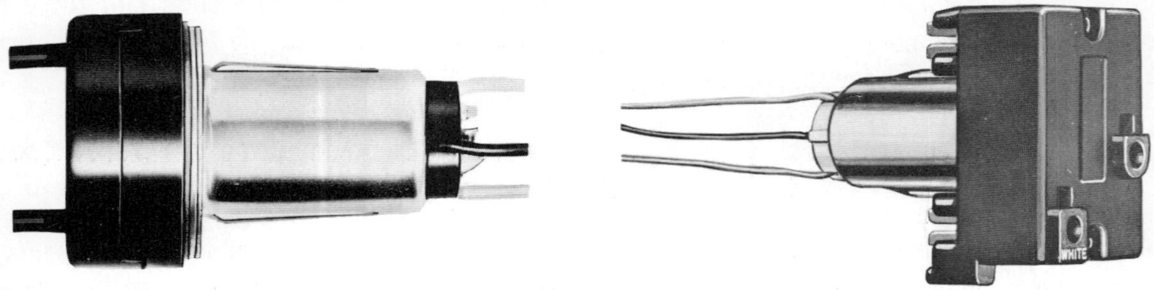

Fig. 29-6 Plug-in relays.

coil and the on coil. The red lead connects internally to the on coil and the black lead connects internally to the off coil.

Plug-in Relay

Plug-in relays are shown in figure 29-6. It is not necessary to splice the line voltage leads for this type of relay. Plug-in relays may be used when a number of relays are to be mounted in one enclosure designed for this purpose.

The relays shown in figure 29-6 are rated for 1 hp, 20 amperes, 125 volts ac/20 amperes, 277 volts ac. The 20-ampere ac rating for both 125 volts and 277 volts means that the relay can be used to control alternating-current loads, including tungsten filament loads and fluorescent loads, up to the full rating of the relay.

TRANSFORMERS

A low-voltage, remote-control system requires just one transformer, figure 29-7. The transformer steps down the lighting circuit voltage from 120 volts to 24 volts for the low-voltage control circuit. The transformer rating is no larger than 100 volt-amperes. When the transformer is overloaded, its output voltage decreases and there is less current output. In other words, the transformer has energy-limiting characteristics which counteract any overload. A specially designed core limits the amount of electrical energy that can be delivered at the secondary or output terminals. Some transformers have a thermal breaker which opens the primary circuit to protect the transformer from overheating. As soon as it cools down, the thermal breaker resets itself and the transformer is automatically reconnected to the line.

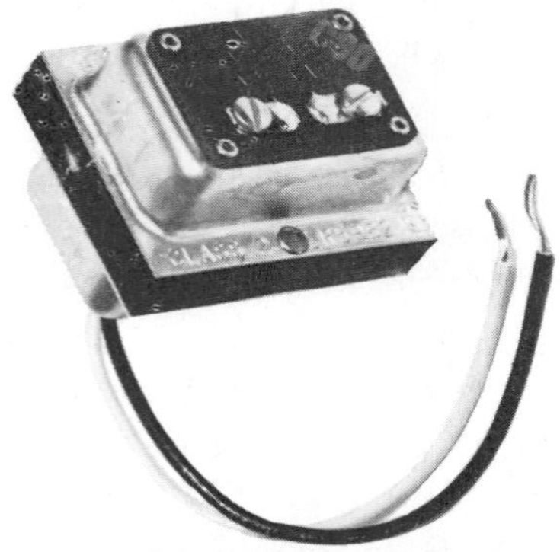

Fig. 29-7 Transformer for remote-control system.

Rarely is more than one switch pressed at the same time in a low-voltage, remote control system. Although 25 or more relays may be connected to the transformer, it is the same as if there were only one. The transformer seems to be underrated for the connected load. Actually, it can control many relays. The transformer manufacturer usually recommends the maximum number of relays that may be operated at one time.

Some manufacturers of remote-control systems suggest that a rectifier be added to the secondary circuit. This device changes the alternating-current supply to direct or pulsating direct current (depending upon whether a half-wave or a full-wave rectifier is used). If a relay is energized for a long time, the alternating-current supply will produce eddy currents inside the laminations. As a result, there is a temperature buildup that may damage the relay. (Recall that this type of relay is meant to provide momentary contact operation only.) Direct current or pulsating direct current does not produce as much heat in this type of relay as does alternating current.

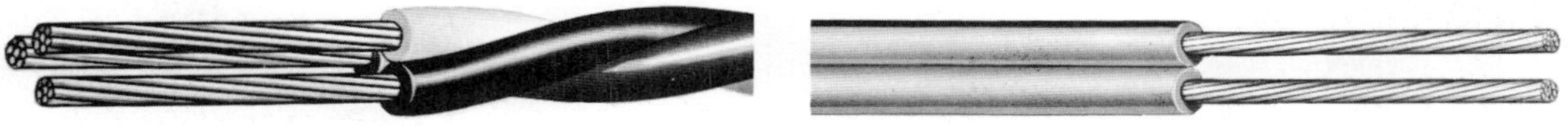

Fig. 29-8 Cable for remote-control systems.

CONDUCTORS

No. 18 AWG conductors are generally used for low-voltage, remote-control systems, figure 29-8. Larger conductors may be required for long runs. A two- or three-conductor cable is used for the installation. Insulation on the individual conductors may be a double-cotton wrapped covering or a plastic material. Regardless of the type of insulation used, the individual conductors in the cable can be identified easily.

The installation of a remote-control system is simplified by using color-coded low-voltage cables. Cables are available in color combinations such as blue-white, red-black, or black-white-red. When these cables are installed correctly in a dwelling, the wires are connected like color to like color. A cable suitable for either overhead or underground installations is available for use in a low-voltage, remote-control system installed outdoors.

INSTALLATION PROCEDURE

The installation of a remote-control system in new construction begins with the roughing-in of the 120-volt system. All of the switch legs or switch loops are omitted for the system. In other words, the 120-volt supply conductors are brought directly to each outlet or switch box that will have a switched lighting fixture or switched outlet attached to it. Although the various devices used may be numbered or color coded in different ways, the electrician must remember that only three low-voltage conductors are required between the relay and the switch. In addition, two low-voltage conductors must be run from the transformer to either the relay or the switch to carry the low-voltage supply. These wires are spliced like color to like color. Another method is to assign certain colors to correspond to the numbers on the relays and switches. The wiring is then completed by slipping the relay through the 1/2-inch knockout in the outlet or switch box. The 120-volt connections are made in the standard way.

A relay with a built-in transformer can be used. No. 18/3 cable is run between the relay and the switch. The connections are made like color to like color. The black and white line voltage relay leads are connected to the supply circuit. The red line voltage relay lead is connected to the lamp terminal. (This red lead is actually the switch leg.)

The low-voltage cables are stapled in place in new construction. When existing buildings are rewired, the low-voltage cables may be fished through the walls and ceiling or run behind moldings and baseboards.

Under certain conditions, many or all of the relays should be placed at one location. For example, this is done when a motorized master control is used. Special relay ganging boxes can be used. In this type of box, all of the relays are inserted through a metal barrier. The line voltage connections are made on one side of this barrier and the low-voltage connections are made on the other side.

TYPICAL WIRING INSTALLATION

One method of installing remote-control, low-voltage switches is shown in figure 29-9. Multipoint control of the light results when the low-voltage, remote-control switches are connected in parallel.

Figure 29-10 shows one method of connecting a master selector control with individual low-voltage switches. The lighting outlets can be placed in different parts of the building and can be connected to operate 120-volt circuits. These outlets are then controlled by their own individual switches or the master selector control. Three lighting circuits are shown but the master selector switch can control up to twelve different relays.

If a master selector switch were installed in the residence, it could control the shrub lights, front entrance light, post light (front), garage lights (front), garage light and post light (rear), garage light (inside), cornice lights, terrace entrance light,

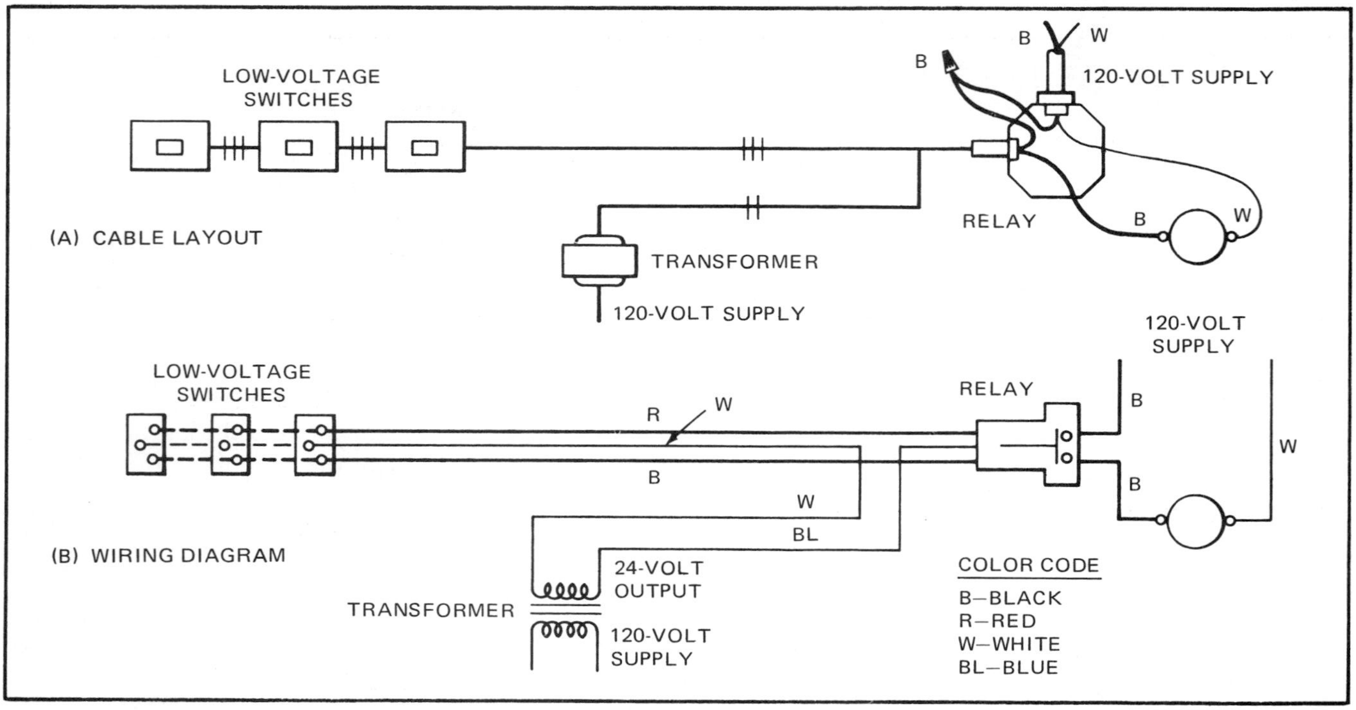

Fig. 29-9 Typical cable layout and wiring diagram for low-voltage switches.

and rear porch light. Thus, the entire outside of the residence could be flooded with light instantly, if necessary.

NATIONAL ELECTRICAL CODE

The low-voltage, remote-control system is not subject to the same Code restrictions as the standard 120-volt system. The low-voltage (24-volt) portion of the remote-control system is a Class 2 system and is regulated by NEC *Sections 725-31* through *725-42.*

For a remote-control system, all wiring on the supply side of the transformer (120 volts) must conform to the wiring methods given in *Chapter 3* of the National Electrical Code. All wiring on the load side of the transformer (24 volts) must conform to *Section 725-38.*

The transformer for a remote-control, low-voltage system is designed so that the power output is limited to meet the requirements of Underwriters' Laboratories, *Table 725-31(a).* The power output can also be limited by a combination of the power source and the overcurrent protection, *Table 725-31(b).* The maximum nameplate rating of the transformer is 100 volt-amperes.

Conductors with thermoplastic insulation are normally used for a low-voltage system. However, the Code does not require thermoplastic insulation, *Section 725-40.*

Conductors for Class 2 systems may not be run with regular light and power wiring, *Section 725-38.* However, power circuit conductors used solely to supply power to the equipment connected to the Class 2 circuit may be run with conductors for Class 2 systems.

The insulation used on remote-control conductors is rather thin. The electrician must be careful not to damage this insulation. Any damage may result in a short circuit or a relay malfunction. The electrician usually bores holes through the studs and joists of the residence for the small remote-control conductors. They are not pulled through the holes used for the cables for the regular wiring because of the possibility of mechanical damage to the insulation. The electrician must use care when stapling these low-voltage conductors. If the insulation is pinched, a short may occur between the conductors.

The low-voltage system conductors and installation methods are similar to those used to install the chimes in this residence.

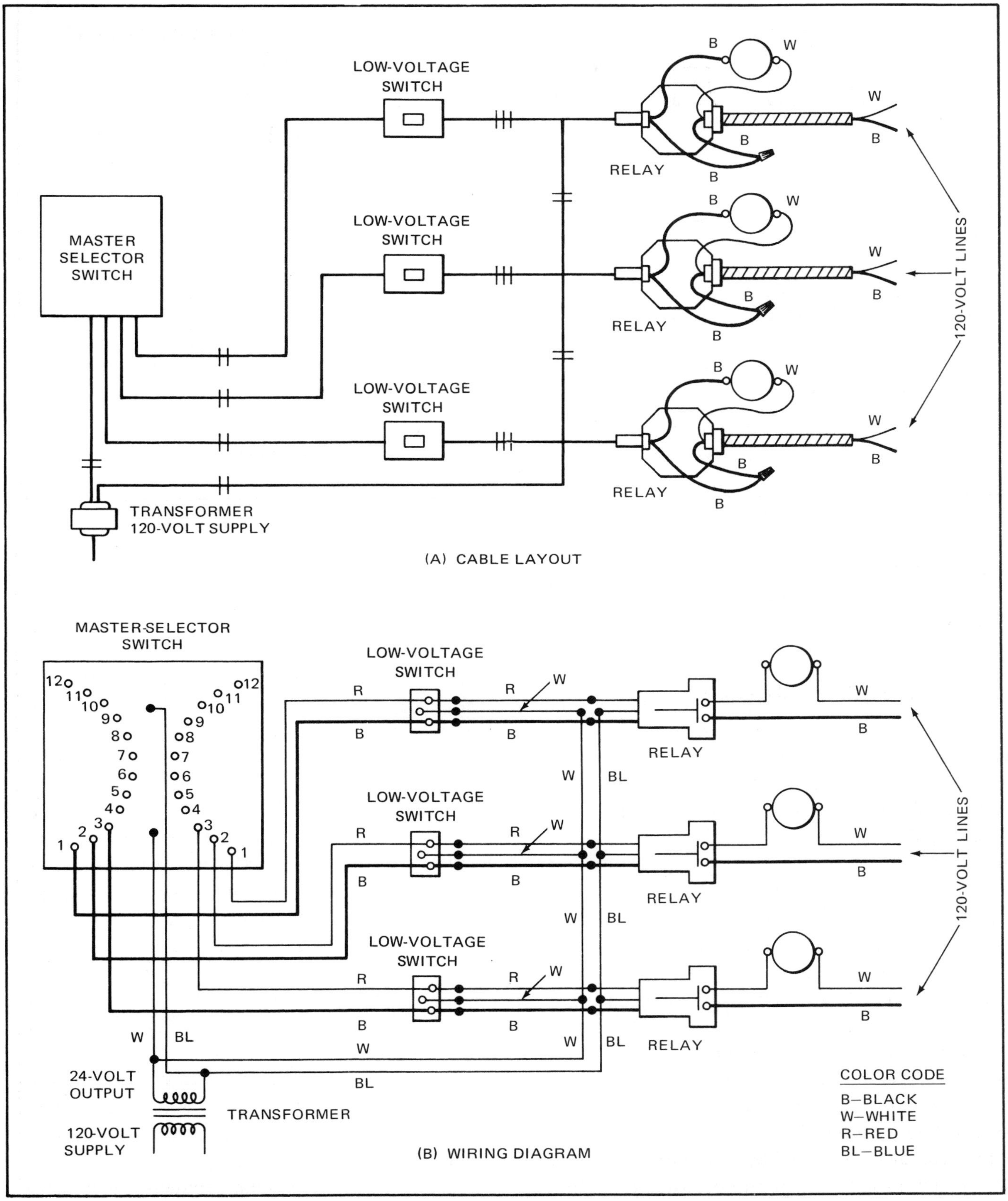

Fig. 29-10 Cable layout and wiring diagram for a remote-control system.

REVIEW

Note: Refer to the Code or the plans where necessary.

1. What is the approximate voltage used on low-voltage, remote-control systems? ______
2. What are the advantages of a low-voltage system? ______
3. What type of switch is used to control the relays of a low-voltage system? ______
4. Are switch boxes recommended at all low-voltage switch locations? Explain your answer. ______
5. Are three-way and four-way switches used in remote-control circuits? Explain your answer. ______
6. What is a master selector switch? ______
7. What is a motorized master control? ______
8. a. Do the relays connected to a motorized master control operate at the same time? Explain your answer. ______

   b. Do the relays connected to a master selector switch operate at the same time? Explain your answer. ______
9. Explain briefly the operating principle of a low-voltage relay. ______
10. What is the electrical rating of the type of relays discussed in this unit? ______
11. What will happen if the common lead from the transformer is connected to the red lead of a relay instead of the blue lead, and the blue and black leads are connected to the switch leads? ______
12. Is it permitted to run low-voltage conductors and line voltage (120 volt) conductors into the same box? Explain your answer. ______
13. The maximum nameplate rating of Class 2 transformers is ______ volt-amperes.
14. Since the insulation on low-voltage wire is much thinner than the insulation on regular building wire, the electrician should ______ when installing the conductors to avoid ______ the insulation.

15. a. What is the purpose of a rectifier? ______________________
    b. Must rectifiers be used on low-voltage systems? ______________________

16. What type of conductors are generally used for a low-voltage, remote-control system?
____

17. Why are color-coded wires and cable used in low-voltage control systems? __________
____

18. The type of low-voltage wiring discussed in this unit is classified as (Class 1) (Class 2). Underline the correct answer.

19. When overcurrent protection is not provided in the secondary circuits of the transformers described in this unit, the transformers must ______________________ their power output.

20. Conductors supplying transformers are governed by what part of the Code? __________

21. Complete all line voltage and low-voltage connections in the following diagram. Use colored pencils to indicate conductors. Show low-voltage connections with a dot. Switch No. 1 controls lamps A and C, switch No. 2 controls lamp B, and switch No. 3 controls the bottom of each convenience receptacle. The top of each receptacle is to be hot at all times. The line voltage wiring is in armored cable.

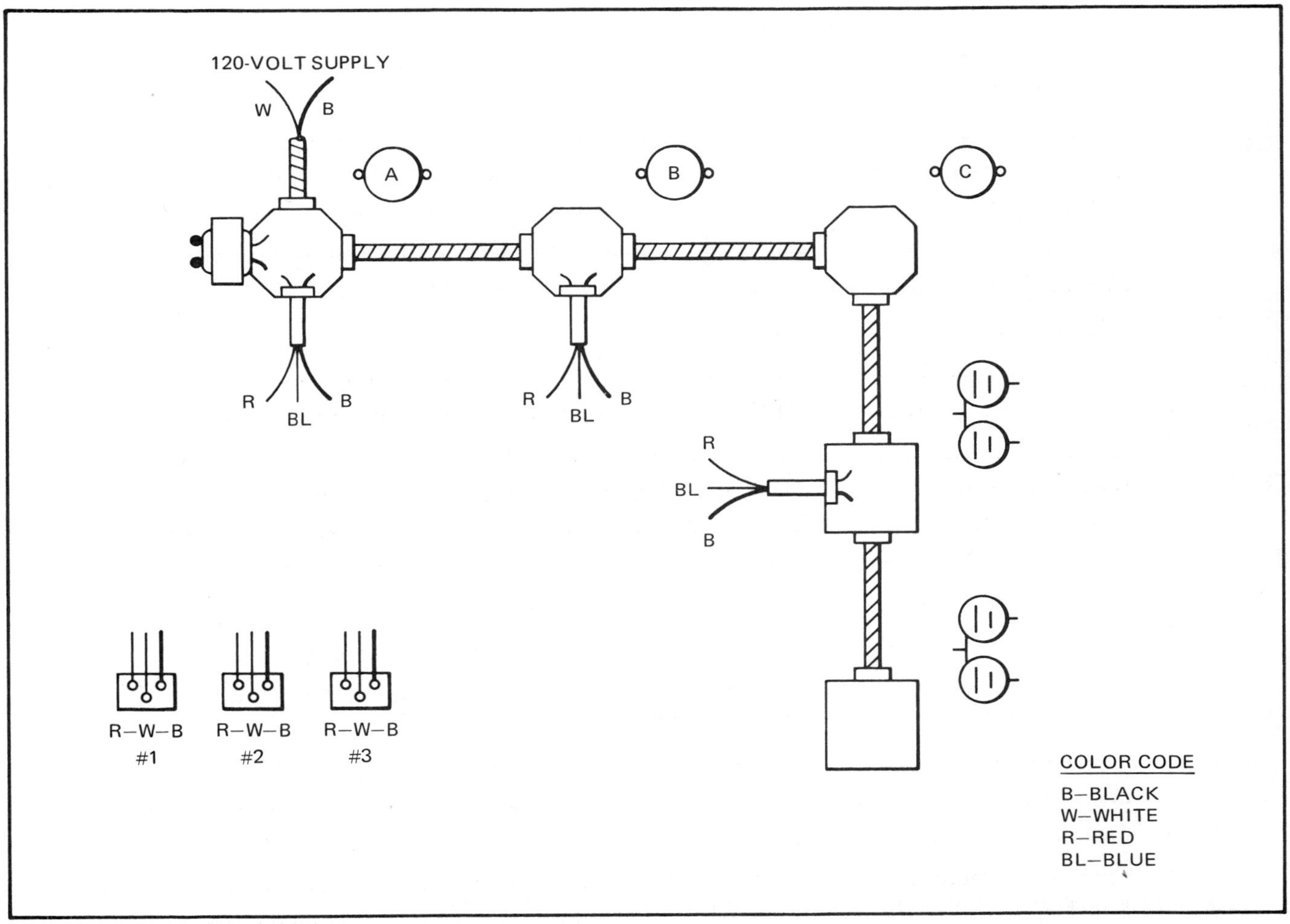

unit 30

Swimming Pools, Spas, and Hot Tubs

OBJECTIVES

After studying this unit, the student will be able to

- recognize the importance of proper pool wiring with regard to human safety.
- discuss the hazards of electrical shock associated with faulty wiring in, on, or near pools.
- describe the differences between permanently installed pools and pools which are portable or storable.
- understand and apply the basic Code requirements for the wiring of swimming pools, spas, and hot tubs.

POOL WIRING (*ARTICLE 680*)

Swimming pools, wading pools, therapeutic pools, decorative pools, hot tubs, and spas must be wired according to the National Electrical Code, *Article 680*. To protect people using such pools, specific rules governing pool wiring have been enacted in recent years. Extreme care is required when wiring, grounding, and bonding pools and the switches, outlets, and fixtures located near pools.

ELECTRICAL HAZARDS

A person can suffer an immobilizing or lethal shock in a residential-type pool in either of two ways.

1. An electrical shock can be transmitted to someone in a pool who touches a "live" wire, or the "live" casing or enclosure of an appliance, such as a hair dryer, radio, or extension cord, among others, figure 30-1.
2. In the event that an appliance falls into the pool, an electrical shock can be transmitted to a person in the water, by means of voltage gradients in the water. Refer to figure 30-2 for an illustration of this life-threatening hazard.

As shown in the figure, "rings" of voltage radiate outwardly from the radio to the pool walls. These rings can be likened to the rings that result when a rock is thrown into the water. The voltage rings or gradients range from 120 volts at the radio to zero volts at the pool walls. The pool walls are assumed to be at ground or zero potential. The gradients, in varying degrees, are found in the entire body of water.

Figure 30-2 shows voltage gradients in the pool of 90 volts and 60 volts. (This figure is a simplification of the actual situation in which there are many voltage gradients.) In this case, the voltage differential, 30 volts, is an extremely hazardous value. **The person in the pool, who is surrounded by these voltage gradients, is subject to severe shock, immobilization (which can result in drowning), or actual electrocution.**

The figures which follow show the Code rules relating to safety procedures in pool wiring. See *Article 680* of the National Electrical Code for more detailed information and exceptions to the Code rules.

CODE-DEFINED POOLS

The Code describes a *permanently installed swimming pool* as one that is located in the ground,

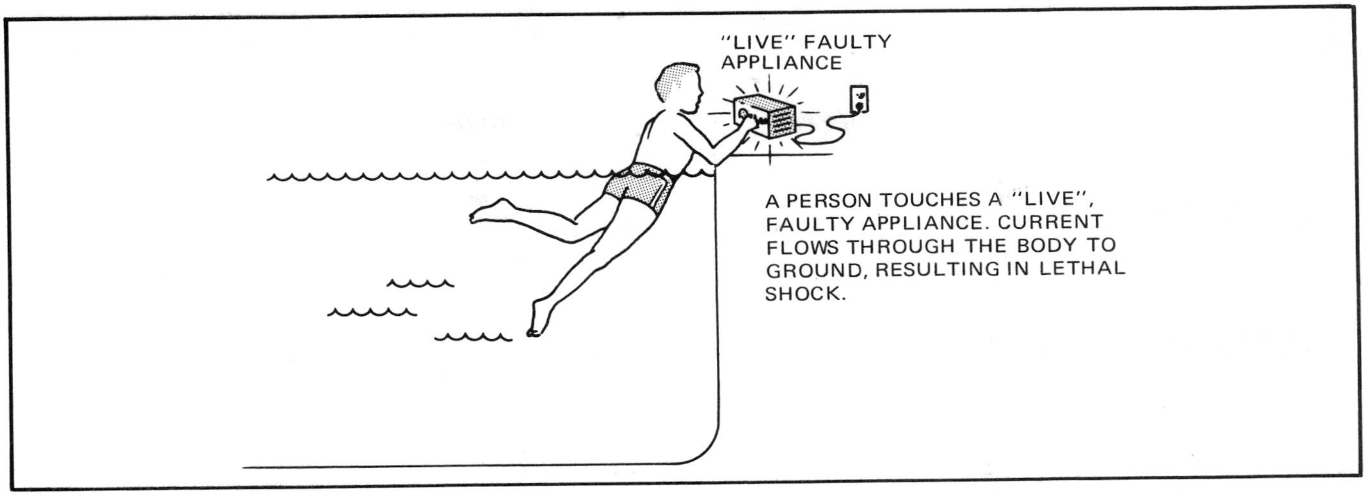

Fig. 30-1 Touching a "live," faulty appliance can cause lethal shock.

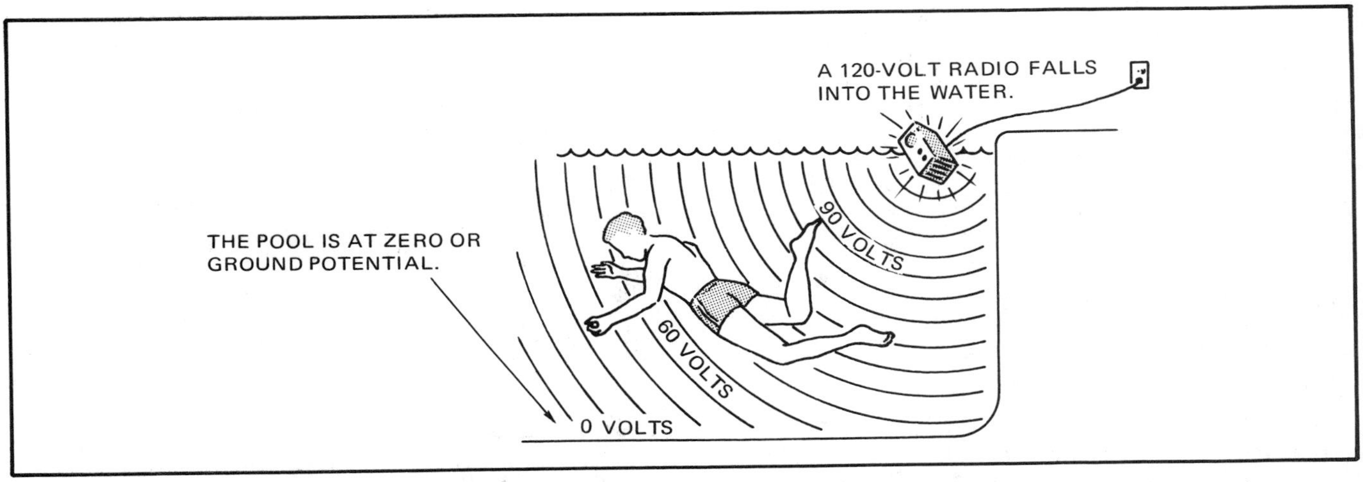

Fig. 30-2 Voltage gradients surrounding a person in the pool can cause severe shock, drowning, or electrocution.

on the ground, or in a building. The manner of construction is such that the pool cannot be disassembled readily for storage, whether or not it is served by any electrical circuits.

The Code describes a *storable swimming pool* or *wading pool* as one that has a maximum lengthwise dimension of 18 feet (5.49 m), and a maximum wall height of 42 inches (1.07 m). The construction of this type of pool is such that it can be disassembled for storage and reassembled to its original form.

GROUNDING AND BONDING OF SWIMMING POOLS

GROUNDING (*SECTION 680-24*)

Section 680-24 of the Code requires that all of the following items *must* be grounded, as illustrated in figure 30-3:

- wet- and dry-niche lighting fixtures.
- all electrical equipment within 5 feet (1.52 m) of the inside wall of the pool.
- all electrical equipment associated with the recirculating system of the pool.
- junction boxes.
- transformer enclosures.
- ground-fault circuit interrupters.
- panelboards that supply the electrical equipment for the pool.

Proper grounding and bonding ensures that all of the metal parts in and around the pool area are at the same ground potential, thus reducing the shock hazard. Proper grounding and bonding practices also facilitate the opening of the overcurrent pro-

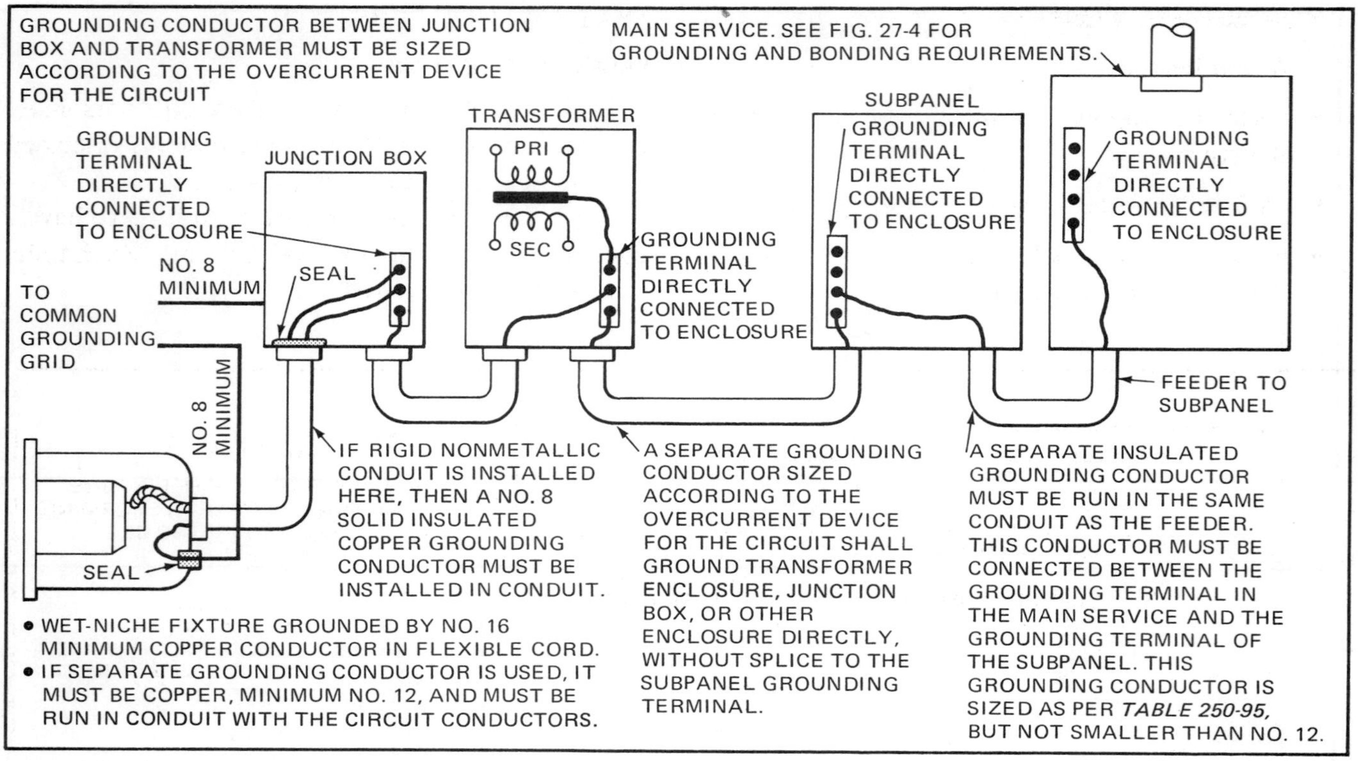

GROUNDING CONDUCTOR BETWEEN JUNCTION BOX AND TRANSFORMER MUST BE SIZED ACCORDING TO THE OVERCURRENT DEVICE FOR THE CIRCUIT
MAIN SERVICE. SEE FIG. 27-4 FOR GROUNDING AND BONDING REQUIREMENTS.
TRANSFORMER
SUBPANEL
GROUNDING TERMINAL DIRECTLY CONNECTED TO ENCLOSURE
JUNCTION BOX
PRI
SEC
GROUNDING TERMINAL DIRECTLY CONNECTED TO ENCLOSURE
GROUNDING TERMINAL DIRECTLY CONNECTED TO ENCLOSURE
GROUNDING TERMINAL DIRECTLY CONNECTED TO ENCLOSURE
NO. 8 MINIMUM
SEAL
TO COMMON GROUNDING GRID
NO. 8 MINIMUM
FEEDER TO SUBPANEL
IF RIGID NONMETALLIC CONDUIT IS INSTALLED HERE, THEN A NO. 8 SOLID INSULATED COPPER GROUNDING CONDUCTOR MUST BE INSTALLED IN CONDUIT.
A SEPARATE GROUNDING CONDUCTOR SIZED ACCORDING TO THE OVERCURRENT DEVICE FOR THE CIRCUIT SHALL GROUND TRANSFORMER ENCLOSURE, JUNCTION BOX, OR OTHER ENCLOSURE DIRECTLY, WITHOUT SPLICE TO THE SUBPANEL GROUNDING TERMINAL.
A SEPARATE INSULATED GROUNDING CONDUCTOR MUST BE RUN IN THE SAME CONDUIT AS THE FEEDER. THIS CONDUCTOR MUST BE CONNECTED BETWEEN THE GROUNDING TERMINAL IN THE MAIN SERVICE AND THE GROUNDING TERMINAL OF THE SUBPANEL. THIS GROUNDING CONDUCTOR IS SIZED AS PER TABLE 250-95, BUT NOT SMALLER THAN NO. 12.
SEAL
• WET-NICHE FIXTURE GROUNDED BY NO. 16 MINIMUM COPPER CONDUCTOR IN FLEXIBLE CORD.
• IF SEPARATE GROUNDING CONDUCTOR IS USED, IT MUST BE COPPER, MINIMUM NO. 12, AND MUST BE RUN IN CONDUIT WITH THE CIRCUIT CONDUCTORS.

Fig. 30-3 Grounding of important parts of a swimming pool installation.

tective device in the event of a fault in the circuit. Grounding is covered in other units of this text.

Grounding Conductors

Grounding conductors *must*

- be run in the same conduit with the circuit conductors, or be part of an approved flexible cord assembly, as used for the connection of wet-niche underwater fixtures.
- terminate on equipment grounding terminals provided in the junction box, transformer, ground-fault circuit interrupter, subpanel, or other specific equipment.

It is important to note that metal conduit by itself is NOT considered to be an adequate grounding means for equipment grounding in and around pools.

BONDING (*SECTION 680-22*)

Section 680-22 of the Code requires that all metal parts of a pool installation *must* be bonded together by connecting the parts to a common bonding grid, figure 30-4. This grid may be

- the steel reinforcing bars in the concrete, or
- the wall of a bolted or welded metal pool, or
- a solid copper wire no smaller than No. 8, insulated or bare.

Bonding Conductors

Bonding conductors

- need *not* be installed in conduit.
- may be connected directly to the equipment that requires bonding, by means of brass, copper, or copper-alloy clamps.

ELECTRIC HEATING OF SWIMMING POOLS (*SECTIONS 680-9* AND *680-27*)

The Code requirements for pool heating are illustrated in figure 30-5.

A radiant heater *must* be

- located at least 12 feet (3.66 m) above the deck.
- suitably guarded.
- securely fastened.
- located at least 5 feet (1.52 m) back from the edge of the pool.

- permanently wired.

A unit heater

- *must* be rigidly mounted on an existing structure.
- *shall not* be located over the pool.
- *must* be located at least 5 feet (1.52 m) back from the edge of the pool.

SPAS AND HOT TUBS

Outdoor Installation

Spas and hot tubs installed outdoors *must* conform to the installation requirements discussed previously for regular swimming pools.

Listed packaged units are permitted to have a cord not longer than 15 feet (4.56 m). Such units must be GFCI protected.

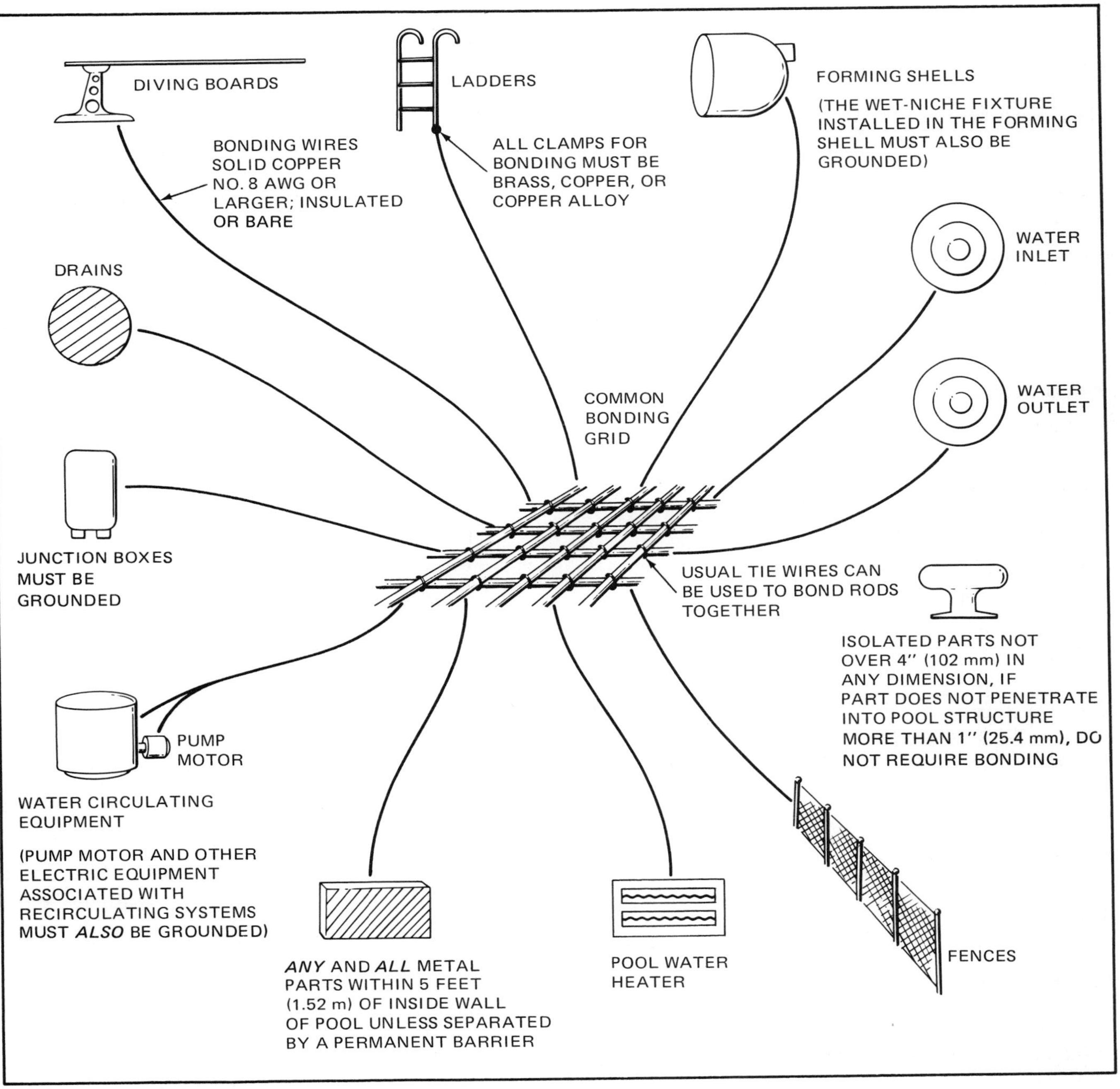

Fig. 30-4 Bonding of a swimming pool installation.

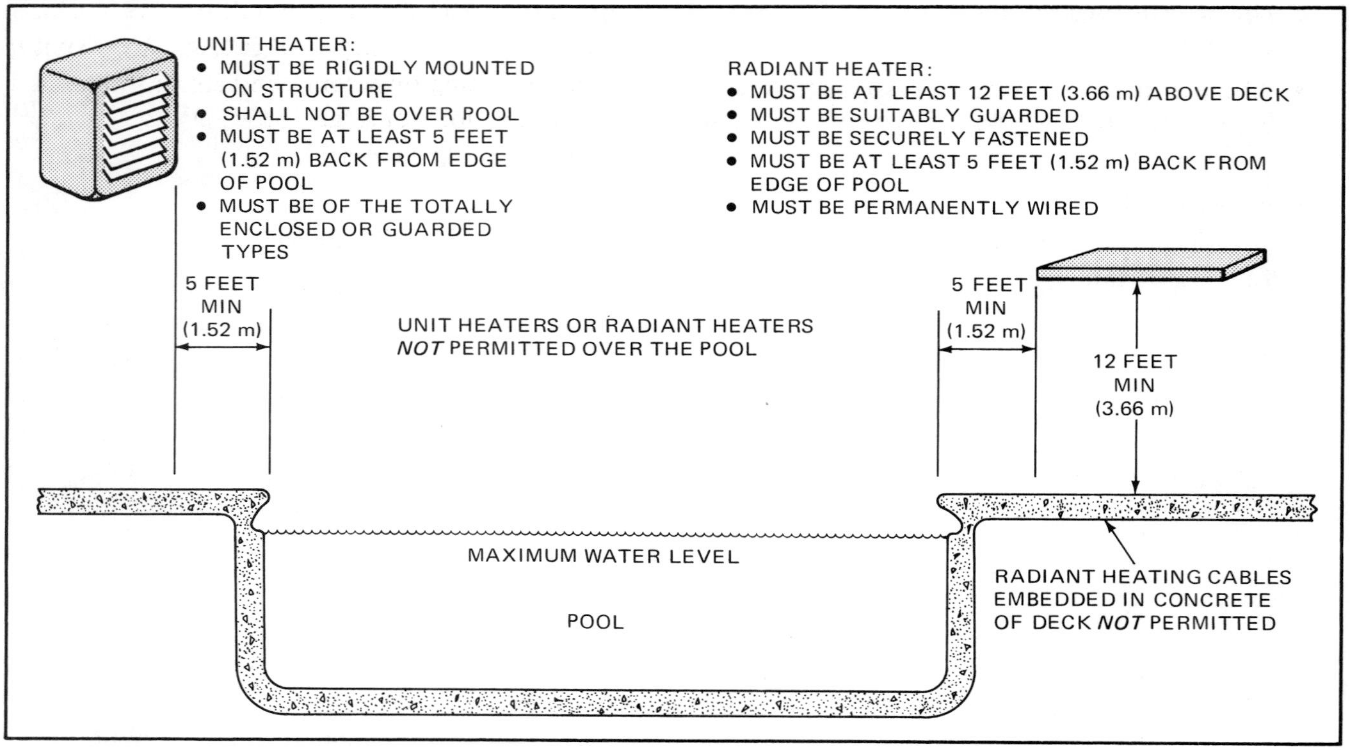

Fig. 30-5 Electric heating of a swimming pool.

Indoor Installation

Spas and hot tubs installed indoors *must* conform to the following requirements:

- receptacles are installed at least 5 feet (1.52 m) from the inside edge of the spa or hot tub.
- any receptacle located within 20 feet (6.1 m) of the inside edge of the spa or hot tub must be GFCI protected.
- any receptacle that supplies power to the spa or hot tub must be GFCI protected no matter how far it is from the spa or hot tub.
- all lighting outlets and fixtures within 5 feet (1.52 m) from the inside wall of the spa or hot tub, or located directly over the spa or hot tub, must be GFCI protected.
- if any underwater lighting is to be installed, the rules discussed previously for regular swimming pools apply.
- wall switches must be located at least 5 feet (1.52 m) from the inside walls of the spa or hot tub.

GROUNDING

The requirements for grounding spas and hot tubs are:

- ground all electrical equipment within 5 feet (1.52 m) of the inside wall of the spa or hot tub.
- ground all electrical equipment associated with the circulating system, including the pump motor.
- grounding *must* conform to all of the applicable Code rules of *Article 250.*
- grounding conductor connections *must* be made according to the applicable requirements of *Chapter 3* of the Code.
- if equipment is connected by means of a flexible cord, the grounding conductor must be part of the flexible cord, and must be fastened to a fixed metal part of the equipment.

BONDING

The bonding requirements for spas and hot tubs are similar to those for regular swimming pools. Bond together:

- all metal fittings within or attached to the spa or hot tub.
- all metal parts of electrical equipment associated with the circulating system, including pump motors.
- all metal pipes, conduits, and metal surfaces within 5 feet (1.52 m) of the inside edge of the spa or hot tub. This bonding is not required if the materials are separated from the spa or hot tub by a permanent barrier.
- all electrical devices and controls *not* associated with the spa or hot tub, unless located more than 5 feet (1.52 m) from the inside edge of the spa or hot tub.

Bonding is to be accomplished by means of threaded metal piping and fittings, by metal-to-metal mounting on a common base or frame, or by means of a No. 8 or larger, solid, insulated, covered or bare bonding jumper.

SUMMARY (FIGURE 30-6)

Figure 30-6, a plan-size diagram included in the back of the text, presents a detailed overview of the Code requirements for pool wiring.

Consult this diagram, as well as the others in this unit, for a thorough review of the many specific requirements relative to the wiring of residential pools.

REVIEW

Note: Refer to the code or the diagrams where necessary.

1. The article of the National Electrical Code that covers most of the requirements for wiring of swimming pools is *Article* ________________ .
2. Name the two ways in which a person may sustain an electrical shock when in a pool.

   ____
3. The Code in *Article 680* discusses two types of pools. Name and describe each type of pool. ____

   ____

   ____

   ____

   ____
4. According to *Section 680-24* of the Code, the following items must be grounded: true or false? (Check one.)

	True	False
a. Wet- and dry-niche lighting fixtures	________	________
b. Electrical equipment located within 5 feet (1.52 m) of inside edge of pool.	________	________
c. Electrical equipment located within 10 feet (3.05 m) of inside edge of pool.	________	________
d. Recirculating equipment and pumps.	________	________
e. Lighting fixtures installed more than 15 feet (4.56 m) from inside edge of pool.	________	________
f. Junction boxes, transformers, and GFCI enclosures.	________	________
g. Panelboards that supply the electrical equipment for the pool.	________	________
h. Panelboards 20 feet (6.1 m) from the pool that do not supply the electrical equipment for the pool.	________	________

5. Grounding conductors (must) (may) be run in the same conduit as the circuit conductors. (Underline the correct answer.)
6. Grounding conductors (may) (may not) be spliced with wire nut-types of wire connectors? (Underline the correct answer.)

7. The purpose of grounding and bonding is to ______________________________

______________________________.

8. What parts of a pool must be bonded together? ______________________________

____

____

____

____

9. May electrical wires be run above the pool? Explain. ______________________________

____

____

10. What is the closest distance that a duplex receptacle may be installed to the inside edge of a pool? ______________________________

11. Receptacles located within 15 feet (4.56 m) from the inside edge of a pool must be protected by a ______________________________

______________________________.

12. Lighting fixtures installed over a pool must be mounted at least (10 feet) (3.05 m), (12 feet) (3.66 m), (15 feet) (4.56 m), above the maximum water level. Underline the correct answer.

13. Grounding conductor terminations in wet-niche metal forming shells, as well as the conduits entering junction boxes or transformer enclosures where the conduit runs directly to the wet-niche lighting fixture, must be ______________ with a(an) ______________ to prevent corrosion to the terminal and to prevent the passage of air through the conduit which could result in corrosion.

14. Wet-niche lighting fixtures are accessible (from the inside of the pool) (from a tunnel) (on top of a pole). Underline the correct answer.

15. Wet-niche lighting fixtures operating at above 15 volts must be protected by ______________

______________________________.

16. Dry-niche lighting fixtures are accessible (from a tunnel, or passageway, or deck) (from the inside of the pool) (on top of a pole). Underline the correct answer.

17. In general, it is *not* permitted to install conduits under the pool, or within 5 feet (1.52 m) measured horizontally from the inside edge of the pool. True or false? Explain. ______________________________

____

____

____

____

____

18. Junction boxes, wet-niche lighting fixtures, and transformer and GFCI enclosures have one thing in common. They all (are made of bronze) (have threaded hubs) (must be mounted at least 8 inches, or 203 mm, above the deck). Underline the correct answer.

19. Lighting fixtures are permitted to be mounted less than 5 feet (1.52 m) measured horizontally from the edge of the pool only if they are (made of plastic) (rigidly fastened to an existing structure) (controlled by a wall switch). Underline the correct answer.

20. The Code permits radiant heating cable to be buried in the concrete deck of a pool. True or false? ______________________

21. For spas and hot tubs, the following statements are either true or false. (Check one.)

	True	False
a. Receptacles may be installed within 5 feet (1.52 m) from the edge of the spa or hot tub.	______	______
b. All receptacles within 20 feet (6.1 m) of the spa or hot tub must be GFCI protected.	______	______
c. Any receptacles that supply power to pool equipment must be GFCI protected.	______	______
d. Wall switches must be located at least 10 feet (3.05 m) from the pool.	______	______
e. Lighting fixtures above the pool, or within 5 feet (1.52 m) from the inside edge of the pool, must be GFCI protected.	______	______

22. Bonding and grounding of electrical equipment in and around spas and hot tubs (are required by the Code) (are not required by the Code) (are decided by the electrician). Underline the correct answer.

Specifications for Electrical Work— Single-Family Dwelling

1. GENERAL

The "General Clause and Conditions" shall be and are hereby made a part of this division.

2. SCOPE

This contractor shall furnish all labor and materials to complete all electrical wiring as shown on the drawings and/or specified herein. The contractor in accepting the contract agrees to have all equipment and wiring in working order at the completion of the job. This contractor shall verify with the architect or owner the location of all fixtures and outlets. If additional outlets are desired by owner, this contractor is to quote on same on a per outlet basis.

3. MATERIALS

All material used shall be new and shall be listed by Underwriters' Laboratories, Inc., as conforming to its standards in every case where such a standard has been established for the type of material under consideration. The material shall be of the size and type specified on the drawings and/or in the specifications. This contractor is to assist the owner in the selection of lighting fixtures, fans and chimes.

4. WORKMANSHIP

All electrical work shall be done in accordance with the standards of the National Electrical Code, local codes and ordinances, and the requirements of the local electrical utility. All work shall be executed in a neat and workmanlike manner. In setting outlet boxes, care shall be taken that same are securely fastened, set true and plumb, and flush with finished plaster, wall panel, or trim. Where tile back-splashes or the like are encountered, wall plates shall be entirely within the tile area or entirely out of the tile area.

5. WIRING

In general, not more than ten (10) outlets shall be connected to any one lighting branch circuit. Exceptions may be made in the case of low-current-consuming outlets. Wiring is to be nonmetallic-sheathed cable or armored cable, adequately sized and installed according to the National Electrical Code and local ordinances. Minimum size is No. 12 AWG. Throughout the entire installation, all metal boxes, fixtures, appliances, etc., shall be grounded in accordance with the methods set forth in the National Electrical Code.

Basement wiring is to be installed in electrical metallic tubing except in the recreation room, passage, and lavatory which may be wired in cable as previously mentioned.

In utility room, all wiring shall be concealed, and all wiring devices shall be flush-mounted on the Southeast, Southwest, and Northwest walls. Outside (Northeast) wall and ceiling of utility room to be exposed electrical metallic tubing.

General lighting branch circuits to be protected by 15-ampere overcurrent devices.

6. TELEPHONES

Furnish and install 4-inch square boxes, 1 1/2 inches deep with suitable raised plaster covers at each of the five telephone locations as indicated on the plans. Install a 1/2-inch conduit from the living room, kitchen, and recreation room outlet boxes and terminate the three 1/2-inch conduits in the utility room. A 1/2-inch conduit shall be run from each of the bedroom outlet boxes and terminated in the basement storage and workshop respectively. Furnish four telephone jack plates to match the other faceplates. Kitchen phone will be wall mounted. All other wiring and equipment will be furnished and installed by the telephone company. A short fish wire shall be left in each conduit for the convenience of the telephone installer.

7. FIXTURES

An allowance of $750.00 shall be made for all lighting fixtures and lavatory medicine chest. The Heat-A-Vent light is not to be included in this amount.

All lighting fixtures and the lavatory medicine chest are to be selected by the owner and installed by the electrical contractor. Ceiling outlets are to be centered in the room or area that they are intended to illuminate unless otherwise noted. Lamps are to be furnished by the electrical contractor as part of the $750.00 fixture allowance. Should the fixtures and medicine chest exceed the allowance, an additional charge will be made for the amount in excess of $750.00.

The electrical contractor shall furnish and install 4-inch porcelain keyless receptacles:

Workshop	2	
Storage	1	
Utility room	2	
Pump room	1	
Garage	3	
Garage storage	1	
Attic	1	
Kitchen closet	1	pull-chain

The electrical contractor shall furnish and install six recessed-type, 60-watt flush, solid glass lens fixtures, complete with lamps in closets in bedroom areas as indicated on the plans.

8. HEAT-A-VENT LIGHT

Furnish and install one (1) Heat-A-Vent light where indicated on the plans complete with switch assembly required to perform the heating, venting and lighting operations as recommended by the manufacturer.

9. SPECIAL-PURPOSE BRANCH CIRCUITS

Furnish and install all necessary material for the special-purpose branch circuits. Make all necessary connections to the equipment.

Note: Owner will furnish all of the appliances except the two (2) 120-volt, 1750-watt electric heaters and the bathroom Heat-A-Vent light which shall be furnished and installed by the electrical contractor. The owner shall furnish the range hood in the kitchen and the exhaust fan in the utility room but the contractor shall install this equipment.

SCHEDULE OF SPECIAL-PURPOSE OUTLETS

Symbol	Use	Voltage	Amp Rating	Poles	Wire Size	Circuit No.
A1,2	For connection of two 1750-watt heaters	120	20	1	12	A14 A16
B	Water pump (8 amperes)	240	20	2	12	B (13-15)
C	Water heater. Top element 2000 watts, bottom element 3000 watts connected for limited demand	240	20	2	12	B (18-20)
D	Dryer (4700 watts)	120/240	30	2	10	B (17-19)
E	Garage door openers (two) Each 5.8 amperes	120	20	1	12	A17
F	Refrigerator-freezer 5.8 amperes	120	20	1	12	B3
G	Counter-mounted cook top 7450 watts	120/240	40	2	8	B (14-16)
H	Wall-mounted oven 6600 watts	120/240	30	2	10	B (10-12)
I	Food waste disposer 7.2 amperes	120	20	1	12	B7
J	Dishwasher. Motor 7.2 amperes, Heater 1000 watts	120	20	1	12	B5
K	Bath Heat-A-Vent light 1475 watts	120	15	1	12	A23
L1,2,3	For connections of three air conditioners - 8 amperes each	240	15	2	12	A (9-11) A (10-12) A (13-15)
M	Attic exhaust fan 5.8 amperes	120	15	1	12	A24

10. SMALL APPLIANCE CIRCUITS

Install 20-ampere circuits according to the following:

Dining area	1 circuit
Kitchen	4 circuits (install split-circuit receptacles)
Workshop	1 circuit
Utility room	4 circuits (each receptacle on separate circuit)

11. SIGNAL CIRCUITS

Furnish and install two (2) recessed two-tone door chimes where indicated on the plans, complete with two (2) push buttons and suitable chime transformer. Allow $150.00 for above items. Chimes and buttons to be selected by owner. If owner selects chimes and buttons resulting in total sum higher than $150.00, an additional charge will be made for the amount in excess of $150.00.

12. LUMINOUS CEILING

Furnish and install diffusing plastic in valance in bathroom. Owner to select pattern of diffusing plastic for a sum not to exceed allowance of $100.00. Wood framing by carpenter contractor.

13. PLUG STRIP

Where noted, furnish and install multioutlet assemblies with outlets 18″ (457 mm) on center. In living room, switch alternate outlets as noted. In workshop, furnish grounding-type receptacles.

14. TELEVISION OUTLETS

Furnish and install 4-inch square, 1 1/2-inch deep outlet boxes with single-gang raised plaster covers at each television outlet where noted on the plans. Mount at the same height as convenience receptacles. Furnish and install 75-ohm coaxial cable to each television outlet from a point in the corner of the workshop near the main service-entrance switch. Cables may be looped from outlet to outlet. Allow eight feet (2.44 mm) of cable in workshop. Furnish and install television plug-in jacks at each location. Faceplates are to match other faceplates in home. All remaining work done by others.

15. SWITCHES, RECEPTACLES, AND FACEPLATES

All flush switches shall be of the quiet ac-rated toggle type. They shall be mounted 50 inches (1.27 m) to center above the finished floor except where otherwise noted.

Convenience receptacles shall be mounted 12 inches (305 mm) to center above the finished floor except where otherwise noted. All convenience receptacles shall be of the grounding type. Furnish and install where indicated, ground-fault circuit interrupter receptacles to provide ground-fault circuit protection as required by the National Electrical Code. All wiring devices are to be provided with ivory handles or faces and shall be trimmed with ivory faceplates except in the kitchen, bath and lavatory where satin finish, stainless steel faceplates shall be used.

Convenience receptacles in the living room, bedrooms and kitchen shall be of the split-circuit design.

16. ELECTRIC HEATING

Where indicated on the plans, furnish and install all resistance-type, baseboard electric heating units complete with receptacle outlets where required (240 volts) plus two 120-volt, 1750-watt portable electric heaters suitable for ceiling mounting in the workshop and utility room. These portable heaters to have integral thermostats.

Furnish and install wall-mounted thermostats where indicated on the plans. Dining area heating unit to be controlled with living room thermostat.

Wattage ratings of heating units to be determined in accordance with established methods of the local area, and must be of sufficient rating to maintain comfortable temperatures within the home under the most adverse condition expected in the locality. For this dwelling, the heating requirements have been determined as shown in the chart on page 267.

This contractor shall furnish and install overcurrent devices and conductors sized according to the National Electrical Code.

This contractor shall instruct the owner as to the amount of insulation and/or vapor barriers

Room	Wattage	Volts	Circuit No.	Ampere Rating	Poles	Wire Size	Thermostat
Living Room	2500	240	A5-7	20	2	No. 12	Yes
Dining Room	750	240	A5-7	20	2	No. 12	LR thermostat
Kitchen	1250	240	A10-12	15	2	No. 12	Yes
Front Hall	500	240	A9-11	15	2	No. 12	Yes
Bedroom No. 1	1500	240	A2-4	15	2	No. 12	Yes
Bedroom No. 2	1500	240	A1-3	15	2	No. 12	Yes
Bathroom	500	240	A9-11	15	2	No. 12	Yes
Recreation Room	3000	240	A6-8	20	2	No. 12	Yes
Lavatory	500	240	A13-15	15	2	No. 12	Yes
Workshop	1750	120	A14	20	1	No. 12	Built into heater
Utility Room	1750	120	A16	20	1	No. 12	Built into heater
TOTAL	15 500						

neccessary to obtain optimum comfort for the occupants. Such insulation and/or vapor barriers shall be furnished and installed by the carpenter-contractor.

This contractor shall furnish and install a humidistat for attic fan.

17. DIMMERS

Furnish and install one (1) 200-watt, flush-mounted, autotransformer-type dimmer where indicated on the plans to control the living room valance lighting. Furnish and install one (1) 600-watt, three-way electronic solid-state dimmer where indicated on the plans to control the dining room ceiling fixture.

18. SERVICE ENTRANCE

Furnish and install one (1) 200-ampere, 120/240-volt, single-phase, three-wire combination main pullout disconnect complete with 200-ampere Class T fuses in workshop where indicated. Branch-circuit protection in panel to incorporate Type S plug fuses, Type SC cartridge fuses, Fusetron® cartridge fuses, or circuit breakers. Panel to have 100 000-ampere interrupting rating. Service-entrance conduit to be 2-inch rigid conduit extending through roof not to exceed three feet (914 mm). Service-entrance conductors to be No. 2/0 RHW or equivalent. Bond and ground service equipment in accordance with the National Electrical Code.

Furnish and install all fittings necessary to complete the service entrance, including meter socket. Utility company to furnish and install meter.

Furnish and install one (1) twenty-circuit 120/240-volt, single-phase, three-wire load center in utility room. Load center to have 125-ampere mains. Feed load center with three (3) No. 3 RHW conductors or equivalent protected by a 100-ampere, two-pole overcurrent device in main panel. Install conductors under basement floor in 1 1/4-inch rigid conduit. This panel to be fusible similar to main panel.

Circuit Directory cards shall be neatly typed and attached to their respective panels.

ALTERNATE

19. LOW-VOLTAGE, REMOTE-CONTROL SYSTEM

The electrical contractor shall submit an alternate bid on the following.

Furnish and install a complete low-voltage remote-control system to accomplish the same results as would be obtained with the conventional switching arrangement as indicated on the electrical plans.

In addition, furnish and install one (1) 12-position master selector switch in bedroom No. 1 or as directed by the architect or owner. Outlets to be controlled by this switch to be selected by the owner.

In addition, furnish and install two motorized 25-circuit master controls. These motor-operated controls shall be controlled from the front hall, bedroom No. 1, rear hall and garage, or as directed by the architect or owner. Connect motor-operated master control in such a manner that each and every switch-controlled lighting outlet and switch-controlled convenience outlet may be turned *off* or *on* from the above-mentioned control stations.

All low-voltage wiring to conform to the National Electrical Code.

Appendix

USEFUL FORMULAS

TO FIND	SINGLE PHASE	THREE PHASE	DIRECT CURRENT
AMPERES when kVA is known	$\frac{\text{kVA} \times 1000}{E}$	$\frac{\text{kVA} \times 1000}{E \times 1.73}$	not applicable
AMPERES when horsepower is known	$\frac{\text{hp} \times 746}{E \times \%\text{eff.} \times \text{pf}}$	$\frac{\text{hp} \times 746}{E \times 1.73 \times \%\text{eff.} \times \text{pf}}$	$\frac{\text{hp} \times 746}{E \times \%\text{eff.}}$
AMPERES when kilowatts are known	$\frac{\text{kW} \times 1000}{E \times \text{pf}}$	$\frac{\text{kW} \times 1000}{E \times 1.73 \times \text{pf}}$	$\frac{\text{kW} \times 1000}{E}$
KILOWATTS	$\frac{I \times E \times \text{pf}}{1000}$	$\frac{I \times E \times 1.73 \times \text{pf}}{1000}$	$\frac{I \times E}{1000}$
KILOVOLT AMPERES	$\frac{I \times E}{1000}$	$\frac{I \times E \times 1.73}{1000}$	not applicable
HORSEPOWER	$\frac{I \times E \times \%\text{eff.} \times \text{pf}}{746}$	$\frac{I \times E \times 1.73 \times \%\text{eff.} \times \text{pf}}{746}$	$\frac{I \times E \times \%\text{eff.}}{746}$
WATTS	$E \times I \times \text{pf}$	$E \times I \times 1.73 \times \text{pf}$	$E \times I$

I = amperes E = volts kW = kilowatts kVA = kilovolt-amperes

hp = horsepower % eff. = percent efficiency pf = power factor

EQUATIONS BASED ON OHM'S LAW:

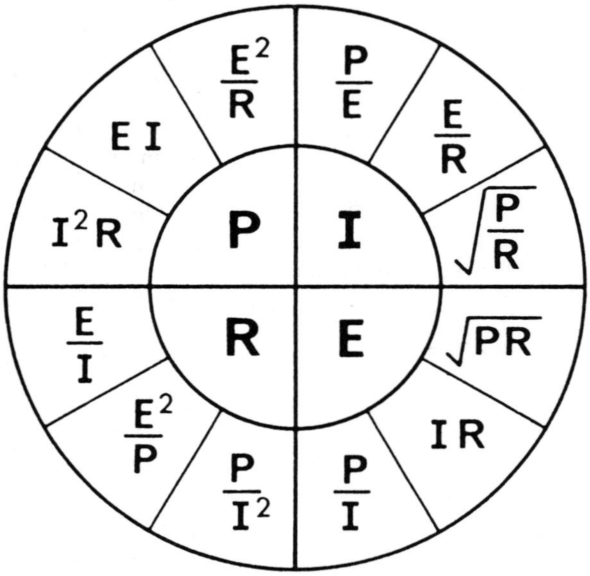

P = POWER, IN WATTS
I = CURRENT, IN AMPERES
R = RESISTANCE, IN OHMS
E = ELECTROMOTIVE FORCE, IN VOLTS

INDEX

The back of the text contains the following prints (10):

Sheet 1	Plot Plan (Front)
Sheet 1	Basement Plan (Back)
Sheet 2	Floor Plan (Front)
Sheet 2	Northeast Elevation (Back)
Sheet 3	Southeast Elevation, Northwest Elevation (Front)
Sheet 3	Southwest Elevation (Back)
Sheet 4	Kitchen Fireplace Details (Front)
Sheet 4	First Floor Electrical (Back)
Sheet 5	Basement Electrical (Front)
Sheet 5	Figure 30-6: Code requirements for pool wiring. (Back)

(Continued from front flap)

With its wealth of revised information, the Seventh Edition of *Electrical Wiring—Residential* is much more than the most current guide to household wiring available. It is the most comprehensive storehouse of working know-how on the subject.

About the Author

Ray C. Mullin is currently Vice President for a large electrical components manufacturer. He has utilized his expert knowledge of all aspects of electrical installations—gained while working as both a journeyman and supervisor—throughout an extensive teaching career. He was electrical circuit instructor for the Electrical Trades, Wisconsin Schools of Vocational, Technical, and Adult Education. Mr. Mullin has also taught both day and night electrical apprentice and journeyman trade extension courses as well as conducted engineering seminars. A former member of the International Brotherhood of Electrical Workers, he is now a member of the International Association of Electrical Inspectors, the Institute of Electrical and Electronic Engineers, Inc., and the National Fire Protection Association, Electrical Section.